'READING' GREEK CULTURE

'READING' GREEK CULTURE

Texts and Images, Rituals and Myths

Christiane Sourvinou-Inwood

CLARENDON PRESS · OXFORD
1991

Oxford University Press, Walton Street, Oxford OX2 6DP
Oxford New York Toronto
Delhi Bombay Calcutta Madras Karachi
Petaling Jaya Singapore Hong Kong Tokyo
Nairobi Dar es Salaam Cape Town
Melbourne Auckland
and associated companies in
Berlin Ibadan

Oxford is a trade mark of Oxford University Press

Published in the United States
by Oxford University Press, New York

© C. Sourvinou-Inwood 1991

All rights reserved. No part of this publication may be reproduced,
stored in a retrieval system, or transmitted, in any form or by any means,
electronic, mechanical, photocopying, recording, or otherwise, without
the prior permission of Oxford University Press

British Library Cataloguing in Publication Data
Sourvinou-Inwood, Christiane
'Reading' Greek culture: texts and images, rituals and myths.
1. Greek culture, ancient period
I. Title
938
ISBN 0-19-814750-3

Library of Congress Cataloging in Publication Data
Sourvinou-Inwood, Christiane.
'Reading' Greek culture: texts and images, rituals and myths
Includes bibliographical references and index.
1. Greece—Civilization—To 146 B.C. I. Title.
DF78.S67   1991   938—dc20   90-21986
ISBN 0-19-814750-3

Typeset by Dobbie Typesetting Limited, Tavistock, Devon
Printed in Great Britain by
Biddles Ltd., Guildford and King's Lynn

PREFACE

The chapters of this book are updated and to a greater or lesser extent restructured and/or modified versions of seven essays which were originally published as separate articles. They form three thematic units, but they also make up a coherent whole in so far as they are the product of, and illustrate, a particular methodological approach. I discuss this methodological approach in section I which was written specially for this book—though it incorporates material which had been included in the previously published version of some of the essays.

The idea of this book originated with Professor Sally Humphreys to whom I am very grateful for the suggestion as I am for her support and encouragement. Besides her, and the scholars acknowledged in the individual chapters, I would also like to thank Professor Pat Easterling, Professor Fergus Millar, Dr Robert Parker, and Professor Martin Robertson for their comments and encouragement.

For permission to reproduce the seven essays, or versions of them, I am indebted to the Council of the Society for the Promotion of Hellenic Studies, to the Director of the Institute of Classical Studies and editor of the *Bulletin of the Institute of Classical Studies* Professor J. P. Barron, to Professor C. Collard and the Classical Journals Board, to the editorial committee of *Opuscula Atheniensia*, to Professor C. Bérard and to Routledge the publishers.

For the illustrations I am grateful to the following: to Dr I. Saverkina and the Hermitage Museum, Leningrad, for the photographs on Plates 5–10, to Mr Michael Vickers and the Visitors of the Ashmolean Museum, Oxford, for the photographs on Plates 1–2, to Dr John Prag for Plate 3, to Dr Dyfri Williams and the Trustees of the British Museum for Plate 4, to Dr P. Gercke for the photographs on Plate 13, and to Dr O. Watson for Plates 11–12.

I reproduce photographs of the Orsi drawings to illustrate the Locrian pinakes because I was unable to obtain photographs of the pinakes from the Museum of Reggio Calabria.

'Reading' is in quotation marks in the title because it refers to the attempt to make sense of a whole society and its cultural artefacts. But for the desire to avoid a neologism in the title I would have indicated the book's focus by calling it 'Ancient Anagnostics'. I would define 'anagnostics' (from the Greek word *anagnosis* = reading) as that discipline which is concerned with reading and its epistemology and methodologies. This book focuses on these questions as they pertain to our reading and making sense of the cultural artefacts of a society to which we have limited access: ancient Greek texts and images, and the myths and rituals articulated in them. A selection of such texts and images, rituals and myths, are investigated through the methodologies and strategies which are, I suggest, appropriate for ancient anagnostics.

CONTENTS

List of Plates   viii

PART I. READING GREEK TEXTS AND IMAGES, EXPLORING GREEK RELIGION AND MYTHS   1

I. Reading Greek Texts and Images, Exploring Greek Religion and Myths   3

PART II. IMAGES OF ATHENIAN MAIDENS   25

II.1. Menace and Pursuit: Differentiation and the Creation of Meaning   29

II.2. A Series of Erotic Pursuits: Images and Meanings   58

II.3. Altars with Palm-trees, Palm-trees, and *Parthenoi*   99

PART III. PERSEPHONE AND APHRODITE AT LOCRI: A MODEL FOR PERSONALITY DEFINITIONS IN GREEK RELIGION   145

III. Persephone and Aphrodite at Locri: A Model for Personality Definitions in Greek Religion   147

PART IV. MYTH AND HISTORY   189

IV.1. The Myth of the First Temples at Delphi   192

IV.2. Myth as History: The Previous Owners of the Delphic Oracle   217

IV.3. 'Myth' and History: On Herodotos 3. 48 and 3. 50–53   244

Bibliography   285

Index of Vases   303

General Index   307

LIST OF PLATES

1. Stamnos Oxford A.M. 1911.619. Courtesy of the Visitors of the Ashmolean Museum, Oxford
2. Volute-krater Oxford A.M. 525. Courtesy of the Visitors of the Ashmolean Museum, Oxford
3. Pelike Manchester III.1.41
4. Stamnos London, British Museum E 446. Courtesy of the Trustees of the British Museum
5. Bell-krater Leningrad, Hermitage 777 (St. 1786), side B. Courtesy Hermitage Museum
6. Bell-krater Leningrad, Hermitage 777 (St. 1786), side A. Courtesy Hermitage Museum
7. Pelike Leningrad, Hermitage 728 (St. 1633). Courtesy Hermitage Museum
8. Lekane fragment Leningrad, Hermitage. Courtesy Hermitage Museum
9. Neck-amphora Leningrad, Hermitage 709 (St. 1461), side A. Courtesy Hermitage Museum
10. Neck-amphora Leningrad, Hermitage 709 (St. 1461), side B. Courtesy Hermitage Museum
11. Cup London, Victoria and Albert Museum, 4807.1901, side A
12. Cup London, Victoria and Albert Museum, 4807.1901, side B
13. Alabastron Kassel 551
14. Drawing of Locrian pinax: Hades and Persephone enthroned (after Orsi, *Bolletino d'arte*, 3 (1909), fig. 8)
15. Drawing of Locrian pinax: Offering girl (after Orsi, *Bolletino d'arte*, 3 (1909), fig. 6)
16. Drawing of Locrian pinax: Child in the basket (after Orsi, *Bolletino d'arte*, 3 (1909), fig. 39)
17. Drawing of Locrian pinax: Child in the basket (after Orsi, *Bolletino d'arte*, 3 (1909), fig. 41)

PART I
Reading Greek Texts and Images,
Exploring Greek Religion and Myths

I
Reading Greek Texts and Images, Exploring Greek Religion and Myths

This collection of essays represents a significant segment of my published work in the last ten years.[1] It sets out and illustrates the methodology I have followed for the study of the ancient world, as it has developed in the course of that period.[2] It explores a series of interrelated questions: how should we proceed in order, first, to read the ancient images and texts in ways that are as near as possible to the ways in which their ancient contemporaries read them; second, to reconstruct the ways in which the Greeks made sense of their myths, which often articulate, and are always articulated in, these texts and images; finally, to reconstruct ancient realities, especially ancient religious practices, beliefs, and collective representations and attitudes. The methodology set out and applied in this volume places in a wider and more systematically articulated framework certain principles which govern also my earlier work (such as the belief that each phenomenon or cultural artefact must be investigated in the context of the society which produced it); it also, and especially, provides a concentrated focus on, and a systematic discussion of, a major problem at the heart of our efforts to reconstruct and understand the Greek world, and explores the ways in which, in my view, it is appropriate to deal with this problem.

My attempts at exploring methodological questions did not arise out of contacts with the intellectual movements, such as structuralism and post-structuralism, which eventually (albeit in a modified form) came to influence my work. This came later, as a result of these explorations, which were first inspired by a deep feeling of unease about (what were still then) the dominant approaches to the study of the Greek world, namely variations of unquestioned 'common sense' empiricism.

This unease was three-fold. First, unease about the extent to which our own assumptions about the world in general and ancient Greece in particular implicitly affected our perception and interpretation of that world. Second, unease about the extent to which the perceptions and interpretations of earlier generations of scholars often set the agenda for, centred, the interpretative discourse, creating filters which directed, and thus seriously affected, our own perception, our structuring, and understanding of the ancient evidence. Third, unease about the fact that the methodologies based on unquestioning empiricism were not designed to protect the investigation from the intrusion of such preconceptions, because they were often not neutral; that is, their validity depended, implicitly and ultimately, on the validity of the researcher's (implicit or explicit) assumptions, so that if the latter were incorrect the discourse was corrupted in self-validating ways.[3] As a result of this conjunction of factors I felt, as others did, that the study of the classical world, especially of the exceptionally tricky areas of religion, mythology, and collective representations was vulnerable to serious corruption through the intrusion of alien assumptions, and prone to the creation of logical constructs which make perfect sense to modern scholars (precisely because they are shaped by our own culturally determined preconceptions), but do not in fact reflect the ancient realities. In the essays presented here, above all in Chapter IV.2, I set out some examples of such superficially coherent explanatory constructs, and tease out their hidden assumptions—which I argue to be invalid—and the hidden circularity which sustains them.

In my view, as in that of others, unless special care is taken to avoid them through the construction of appropriate methodological strategies, distortions of this type are inevitable in any modern attempt to reconstruct, and make sense of, the ancient world. First, because, as I shall discuss in detail below, all perception and interpretation are culturally determined, which clearly creates serious problems for any attempt to make sense of an alien society and its artefacts. Secondly, because our very limited access to the ancient world (especially to its conceptual universe, with which I am primarily concerned), the paucity and the fragmentary and ambiguous nature of the evidence, makes the investigation even more vulnerable to the intrusion of the researcher's own preconceptions. This, in symbiosis with the first phenomenon, can lead to the creation of the culturally determined explanatory constructs mentioned above. It is therefore necessary to devise a methodology which can guard against these dangers.

The limitations of access to the object of our study, and, more generally, the desirability of finding non-circular means of testing the validity of our, and indeed our predecessors', discourses about the ancient world, lead inevitably to the necessity of seeking help outside classical scholarship. There are several potential sources of such help, or, to put it differently, cognitive discourses with which we need to be familiar in order to achieve our purposes: general epistemological discourses, and also disciplines concerned with the study of societies and their various artefacts, and with the processes of perception, reading, making sense, and communication—such as social anthropology and social history (including the history of mentalities), the psychology of perception, semiology, cognitive studies, and literary theory.

The help which can be—and in recent years has been—derived from these sources, is three-fold. First, we can learn epistemological principles which have important methodological implications. Thus, for example, the important principle that meaning is created in context, through relationships, is not only relevant in the obvious way to the reading of texts and images, it also has important implications for the study of all phenomena. It tells us, for example, that Greek divinities do not have 'essences', and thus should not be studied in a vacuum. As we shall see in Chapter III, they must be placed in the context which determines their divine personalities; this entails the consequence that we must not conflate a deity's Panhellenic persona with her different local divine personalities; we must focus on particular local divine personalities in specific cult contexts, and study each local personality of a deity in the context of the pantheon to which she belongs, of the religious system of each particular city, and also of the more general circumstances of that city, of the worshipping group—for religious systems, including deities, are shaped in interaction with that group and its (often changing) circumstances; finally, each local divine personality must also be studied in the context of its interactions with the personalities of the same deity in the religious systems of other cities and, above all, with the deity's Panhellenic persona. Even with regard to texts, the implications of the epistemological principle in question are wider than may at first appear; for, among other things, it entails that, if we want to reconstruct the ancient readings of a text, we must first of all place that text in its full context, social, cultural, economic, political, religious, and generally conceptual, since it was in this context that it acquired its meanings in the eyes of its contemporaries.

The second way in which other disciplines can help us in our reconstruction of the ancient realities is by suggesting models which can function as eye-openers, show up the culturally determined nature of what appear to us obvious explanations, and alert us to a variety of possibilities, thus allowing us to construct methodologies that take such possibilities into account and thus guard against distortion from a variety of potential sources, methodologies which are therefore more neutral and less vulnerable to culturally determined corruption than their more 'traditional' alternatives. (The use of models is now becoming relatively widespread in classical studies and related disciplines—though this use is not always as circumspect as is necessary in order to avoid introducing other potential sources of distortion.)

Finally, lessons learnt from other disciplines may suggest the possibility that there are some general universal modalities in some of the cognitive operations that concern us—such as, for example, the complexity and interactive nature of the process through which readers make sense of a text or an image, which brings into play the readers' own prior knowledge and assumptions. Even though we may hesitate to take the universality of such modalities for granted, and to treat it as a given in our investigation, they can at least help us construct methodologies that take that possibility and its implications into account, and are thus as neutral (and therefore non-distorting) as is possible.

The study of the processes through which we perceive and make sense suggests that the following two very elementary general strategies are conducive to neutrality and the avoidance of distortion. Their detailed operations will be discussed later on in this chapter and illustrated throughout the book.

First, the evidence should be studied in the most exhaustive detail possible, without preconceived notions of what is 'important' or 'representative', since such selections depend on *a priori*, inevitably culturally determined, as well as subjective, judgements. The cut-off points must be set after, not before, what appear to be 'areas of only marginal concern' have been investigated. It is necessary to dig very deep, and not rush into quickly constructing explanatory schemata because they appear to make sense of what seems to be the important evidence. For such judgements and selections are inevitably determined by the unconscious influence of our own preconceptions, which include assumptions derived from earlier scholarship, themselves culturally determined, and which thus shape the explanatory schemata. Nor is it an adequate strategy simply to test those schemata against further data. For in any such testing it is inevitable that the schemata under consideration, as well as our unconsciously intruding assumptions, will corrupt the investigation by functioning as unconscious organizing centres, ordering the data into patterns which reflect them, by privileging certain data and ignoring or underplaying others, by making assumptions about relationships between the data, and unconsciously striving to find answers which can fit the explanatory schemata which are in theory being tested. Readings based on procedures of this kind appear convincing because they are coherent and self-validating, and because they reflect modern assumptions, but they are, in fact, invalid and circular. This criticism is relevant to all investigations involving, explicitly or implicitly, the construction of schemata and their (explicit or implicit, systematic or haphazard) testing; for they are all vulnerable in the ways indicated here, whether they are the products of instinctive empiricist readings or of conscious model-building and model-testing in the context of more theoretically based approaches.

The second general strategy conducive to neutrality of method involves the structuring of the investigation into a series of separate analyses: we must conduct entirely separate and independent analyses of the different sets of data and different grids of evidence (for example, textual and archaeological), and of the different questions that come into play in any complex process of reading and reconstruction. For this strategy will prevent unconscious adjustments of parts of the evidence to make it accord with the rest (in the ways indicated in the preceding paragraph and on the basis of some unconscious implicit notion of what is a reasonable 'fit' in the researcher's mind); it will prevent preconceptions and distortions from creeping from one part of the discourse into the other and contaminating it, creating an apparently coherent whole which is in fact sustained by hidden circularity; and it will also allow us to cross-check the results of the separate analyses and thus test their validity.

Our access to the ancient world is through its archaeological remains, including artefacts, chief among which are images, and through texts of various kinds, ranging from boundary stones and grave inscriptions through sacred laws to literary texts. Text and archaeological object, with or without images, are sometimes combined—as they are, for example, on inscribed grave stelai with figural representations. The reconstruction of ancient realities, for example, of religious beliefs and practices which are a major concern in my work, depends first on our reading of texts and images 'correctly', and second on our making sense of archaeological remains in ways that reflect the ancient realities, on our reconstructing their context and function correctly. If we leave aside extremely simple texts, the function of which is to give simple information, such as boundary stones, the reading of ancient texts and images presents serious problems, if what we are attempting is to make sense of these texts and images in ways as near as possible to those in which their contemporaries did.

The readings of texts and images as nearly as possible in the ways in which their contemporaries read them is, first, important in its own right; it is clearly of great interest to reconstruct the ways in which, for example, a Sophoclean tragedy was made sense of by the fifth-century Athenian audience.[4] Second, as I mentioned, it is important for readers whose primary aim is to 'extract information' in order to reconstruct the ancient realities. For texts and images are not documents constructed with the aim of conveying information, which can then be extracted in a straightforward manner. Thus, in order to assess whether a particular action in a particular tragedy is presented as good or bad, reasonable or perverse, we must determine in a very detailed manner the precise ways in which it is articulated and coloured in the text; this can only be determined in the context of the reconstruction of the ancient readings of the whole text; this involves reading it through perceptual filters which take into account the text's idiom, aims, and particular articulations, and which are shaped by the cultural assumptions of the ancient society as they have been reconstructed on the basis of other data.[5] Of course, there is a danger of circularity here, and for this reason it is necessary to take great care in avoiding it. I shall set out in a moment a detailed set of strategies with regard to the reading of ancient images, the application of which is illustrated especially in the three essays in Part II. The same general principles apply with regard to texts, the methodology for the reading of which will be briefly considered in this chapter and illustrated most clearly in the essays in Chapters IV.2 and IV.3.[6] I shall begin with the epistemological background and general methodologies pertaining to both texts and images.

Investigations in different fields, from the psychology of perception to art history and literary criticism, have shown that we do not read, and create meanings out of, pictures (or texts) 'neutrally', but through perceptual filters made up of sets of culturally determined assumptions and expectations.[7] Thus, if we want to reconstruct (as far as this is possible) the ways in which a text or image was read by its ancient contemporaries, a 'common sense' approach is obviously unsatisfactory. For the fact that it is empirically demonstrable that perception and understanding (including, of course, 'common sense' itself) are culturally determined entails that any 'direct' reading involving an empirical confrontation with the ancient texts or images inevitably deploys implicitly our own assumptions which are culturally determined and thus vastly different from those of the ancient Greeks, and which therefore are alien intrusions corrupting our reconstruction of the ancient meanings.

Consequently, there are two basic ways of reading a picture or a text, each entailing the possibility of many variations. The first involves treating the image or text as a 'floating' artefact and thus reading it 'directly' and 'empirically': we look at it and decide what it means; that is, we 'make sense' of it by means of our own assumptions and expectations, which are different from those of the society which produced the text or image that concerns us. This approach will produce many different readings and interpretations, by different readers in different periods; it is legitimate, but it does not lead to the recovery of the meanings which had been inscribed into the text or picture by its ancient creator and 'extracted' by his contemporaries. (There is, of course, a significant irreducible content in each text and each picture, a matrix of determination which limits the freedom of reading and provides basic directions, but this does not concern us here.) The second basic way of reading a picture or text involves anchoring it in its historical context and attempting to recover the meanings which the artist or writer had inscribed in the picture and his contemporaries read into it. This goal is implicit in classical scholarship; however, for it to be achieved it is necessary to do much more than traditional classical scholarship normally does.

It is not enough to relate ancient texts or images, read implicitly through our own filters created by our own assumptions, to ancient realities and assumptions known from other evidence. Such a procedure involves an unacceptably incomplete and haphazard deployment of the ancient assumptions, and, moreover, it has an in-built 'programme' for distortion; for by the time the ancient assumptions are brought in, the reading of the text or image has already been shaped by the wrong schemata (those dependent on our culturally determined assumptions), which functioned as unconscious organizing centres—in the ways discussed above—and which will also (unconsciously) structure the perception of the ancient realities and assumptions and their relationship to the text or image. What we must do instead is try to reconstruct all the relevant ancient assumptions and expectations and fashion perceptual filters out of them, and to read the texts or images through these filters: we must reconstruct the original process of signification of the time of the production of the text or image. Only thus may we be able to approximate a reading of, for example, fifth-century Athenian texts and images through fifth-century Athenian eyes.

I say approximate because all reading and interpretation is a cultural construct, dependent on the assumptions of the readings and interpreting culture.[8]

However, not all cultural constructs have the same value[9] and it is possible to approximate the ancient realities to a considerable extent. For this we need a methodology capable of, first, preventing (as far as possible) our own assumptions from intruding in, and distorting, the reading and interpretation of the ancient pictures; and second, of reconstructing the fifth-century Athenian assumptions and expectations and shaping fifth-century perceptual filters out of them, so as to read fifth-century Athenian images through fifth-century Athenian eyes. Without such a methodology, we will be (implicitly) wrenching the images from their historical context and reading them as floating pictures by default; for unless they are blocked, our own filters are deployed by default—inevitably distorting the original meanings, because they bring alien assumptions to bear on meaning-production.

We must, then, try to reconstruct these Athenian filters by recovering the assumptions and expectations which shaped them; but first we must determine what these assumptions are and, more generally, determine the process by which meaning is created. I will set out the parameters governing the production of meaning by describing the process of signification as I see it—from a (modified) post-structuralist perspective. This perspective seems to me to offer a valuable framework in which to inscribe the empirically observed culture-determination of perception, reading, and interpretation. The following are important properties of the sign, and characterize signification.[10]

First, no sign has a fixed meaning. Its value in any signifier (such as a text or image) is determined by a complex and dynamic movement of interaction: an interaction between on the one hand the element under consideration (for example 'spear' in an image of erotic pursuit) with its semantic field of functions, associations, connotations, and on the other, first, its syntactical relationships with the other signifying elements in the representation (the value of which is also determined by the same movement of interaction) in the context of the overall signifier (in this case image); and second, its relationships with the other (semantically related) elements which might have been selected in its place but were not; for example, the value of the spears in scenes of erotic pursuit is also determined by the fact that they are not a sword, or a spear being brandished.

Second, signs are polysemic. Third, a related point, not all the meanings produced by the signifying elements in a signifier contribute to the production of one unified coherent meaning. Some can produce different perspectives, warring discourses, which deconstruct the dominant one.[11] Thus we shall see in Chapter II.2, in some images representing erotic pursuits, the implicit connotations of violence, correlative with the violent semantic facets of the (especially sexual) relations between male and female, are both defined (as sexual) and deconstructed through signifying elements producing an effect of consensual erotic intimacy.[12]

In order to reconstruct the fifth-century perceptual filters we must reconstruct the assumptions and expectations which shaped them. At this point I shall focus on, and illustrate the specific strategies I am advocating through, the study of images. (The detailed application of these strategies will be illustrated in the essays in Part II.) Subsequently, I shall discuss the specific strategies which I consider appropriate for the reading of texts.

There are two basic types of assumptions and expectations pertaining to the reading of ancient images. The first is iconographical: the conventions, codifications, and modalities of the signifying system of Greek iconography. The second is semantic: the knowledge, ideas, assumptions, and mentality which constitute the semantic fields inscribed in, and called up by, the images under consideration during the inscription of meanings by the painters when the images were created and during the 'extraction' of meanings, the making sense of the scenes, by the viewers. The iconographical assumptions can be recovered through formal analyses of various kinds which it is not necessary for me to describe.[13] The semantic assumptions can be reconstructed through semantic analyses. Here I will only stress some methodological points that seem to me important. First, as we saw, methodological rigour demands that we conduct the two sets of analyses, the iconographical and the semantic, separately and independently, to prevent the interpenetration of assumptions and fallacies from one set of analyses to the other, and block unconscious adjustments to make the two types of evidence fit. This strategy allows controls on our analyses through cross-checking between the two sets: for example, the fact that in the investigation conducted in Chapter II.1 the patterns of iconographical and semantic differentiation, recovered independently, match confirms the validity of the analyses and the reconstructions.

I will now signal certain dangers to which, in my view, formal analyses are vulnerable. Because their apparent objectivity is to some extent illusory, formal analyses are also vulnerable to culture-determination. Unless we are dealing with identical signifiers, even the process of comparison itself is vulnerable to the (normally implicit and unquestioned) intrusion of our own culture-determined notions of, for example, what constitutes a similarity, and whether or not an apparently minor and 'insignificant' formal divergence constitutes a difference, and especially a 'significant' difference. The measure of culture-determination in the recognition and assessment of formal similarities may be gauged when we consider that the recognition of resemblance between an iconic sign and the represented object is culture-dependent.[14] Moreover, cultural traditions create certain expectations, a mental set affecting significantly the perception and understanding of pictorial representations,[15] and perceptual selections favour and stress the familiar as determined by the perceiver's cultural tradition.[16]

The danger of underestimating the importance of apparently small divergences is especially great. First, because small divergences are more immediately obvious to the members of the 'cultural community' in which the images were produced.[17] And second, because the differences in the relevant cultural conditioning between us and the fifth-century Athenians reinforce this tendency. For, on the one hand, in a conventionalized codification system based to a great extent on the use of codified signs like that of Athenian iconography small divergences are the means by which iconographical schemata are individualized and differentiated—and the viewers who shared the same assumptions as the painters were conditioned to read images in this way. On the other hand, our own cultural conditioning leads in the opposite direction. Living, as we do, in a representational universe which is far from being dominated by highly conventionalized artistic forms, we may disregard, or even not notice, small divergences. But apparently small divergences can sometimes signal two different signifying elements which, for the ancients, produced significantly different meanings. Perhaps we can approximate more the perceptual cast appropriate for fifth-century ceramic iconography if we think more in terms of letters of the alphabet, where minor formal divergences involve radically different values.

Another consideration which should be spelt out explicitly is this: since making sense of a picture is a complex process involving continuous to-ing and fro-ing between the picture and the reader's knowledge and assumptions which were called up by it,[18] the fifth-century Athenians did not read images on the basis of the signifying elements' formal characteristics alone. The interaction between image and assumptions means that an operation such as, for example, the individuation of topic,[19] which takes place during the reading of the image, generates certain expectations and directs the reading in certain ways, highlights certain elements, reduces the polysemy in certain directions. Thus, for example, during this process, what seems to us an insignificant divergence can have a crucial role and lead to the production of very different meanings from those which we may have assumed on the basis of the scene's formal similarity to other scenes.

This type of culture-determination can be minimized by making the formal analyses as exhaustive as possible.[20] But this strategy is not sufficient, for the evaluation of the resulting data would still be open to culture-determined judgements based on our own assumptions concerning logical categories, operations, and so on—which cannot be presumed, indeed are unlikely, to be the same as those of the fifth century. The intrusion of our own assumptions can be reduced significantly if (through preliminary analyses) we recover models of comparison which can help us define in fifth-century terms the concepts directly relevant to the central investigation.

Thus, in my investigation in Chapter II.1, in order to evaluate the comparative analysis of the representations of erotic pursuit and of the iconographical theme which I call 'Theseus with a sword' painted by the same artist, I will try to recover some models for comparison, recover the parameters of the concept 'thematic differentiation', by considering the parameters of, first, variability within one theme, and second, variability between different themes, within the work of one artist. The first investigation will involve the comparison of representations of erotic pursuit within the work of individual artists; the second involves the comparison of representations of erotic pursuit to representations formally related to the theme 'Theseus with a sword' and known to represent attacks by the same artists, and also the comparison of its results to the results of the first set of comparisons. These analyses cannot themselves be wholly free of culture-determination; but if they are exhaustive they, and their proposed use, make up a strategy which allows as much freedom from culture-determination as it is practicable to expect.

In my view, structural analyses of iconographical themes are also culturally determined, and thus vulnerable to the dangers which I have outlined. I will now illustrate this through an example involving two iconographical themes which will be discussed in detail in Chapters II.1 and II.2. The iconographical scheme representing 'Theseus with a sword' (Pls. 3–4, 6) is very similar to that representing 'erotic pursuit by a youth with spears' (Pls. 1–2, 5): in both Theseus with an offensive weapon[21] is pursuing a fleeing woman; this may be taken to mean that the two themes are semantically closely related. This hypothesis seems confirmed when another related series is brought into play, erotic pursuit by gods,[22] which appear (when seen from our present point of view) in two basic iconographical variants: one similar to that of Theseus' erotic pursuit, the other representing the divine pursuer in a threatening stance, virtually as an attacker. The structural analysis, then, could be taken to indicate that 'Theseus with a sword' is also an erotic pursuit. However, this hypothesis—as I will argue in Chapter II.1—is a cultural construct which distorts the ancient realities. The following are some of the nodes through which our own assumptions have seeped in, and distorted the discourse.[23]

First, especially given the 'economy' and conventionality of Attic ceramic iconography, the similarity of the basic iconographical matrix—which is, in any case, the same as that of attack—involves no more than the common semantic matrix: 'A is going after B in order to impose something (punishment, death, sexual intercourse) on him/her.' Even when this becomes 'Theseus, offensive weapon in hand, is going after a woman who is fleeing, in order to impose something on her', it still does not tell us anything very precise about the relationship between the respective meanings of the two themes.

Second, to describe both iconographical schemes as 'youth/Theseus with an offensive weapon' is to make an arbitrary decision and obscure a fundamental difference. As I will show in Chapter II.1, while the drawn sword in 'Theseus with a sword' is an offensive weapon used as such in an attack, the spears are definitely not; on the contrary, their character precisely as not being weapons which are in use or about to be used is often stressed. Third, the sign 'fleeing woman' (when read through fifth-century assumptions and expectations) belongs to the same 'spacing' in the two themes, and can be filled by the same codified sign, even if we assume that 'Theseus with a sword' is indeed an attack and not an erotic pursuit. For even if 'Theseus with a sword' represents an attack, there is a common semantic core pertaining to the two women's roles, and it is that common core that is encoded in the codified sign. The core consists in the definition of the woman in both themes as the fleeing prospective victim of the male who is going after her to impose something on her. The sign—like all signs—acquires its full meanings in context, through the interactive process of meaning-production:[24] the different contexts give 'fleeing woman' different meanings in the different iconographical themes.

A fourth assumption which I hope to prove fallacious through a detailed discussion in Chapter II.1 pertains to the relationship of 'god' and 'hero'. At this point I will only say that it is methodologically unsafe to exclude the possibility that what may appear to us as 'the same' signifier, 'erotic pursuit by a god/hero', might produce importantly different meanings, if in the sets of assumptions of the time the semantic fields 'god' and 'hero' were significantly different in ways affecting the concept 'erotic pursuit'.

This type of analysis, then, is also vulnerable to distortion through the intrusion of our culture-determined assumptions, and needs the strategies outlined above to minimize the dangers.

As we saw, the correct strategy for reading ancient texts also involves the detailed reconstruction of the cultural assumptions which the text's contemporaries had shared with the author and by means of which meaning was created, followed by the attempt to read the text as much as possible through the perceptual filters created by these assumptions, which had provided the parameters for the author's selections. I set out, and illustrate the application of, a detailed methodology for the reading[25] of ancient dramatic texts elsewhere.[26] In the present collection the methodology I propose for the reading and understanding of ancient texts in general is illustrated most clearly in Chapter IV.3.

As with the iconographical conventions in the case of images, the assumptions that need to be reconstructed include the conventions of the genre governing the creation of the text under consideration. An important textual element which is not limited to one genre but articulates, and helps create meaning in, many ancient texts, is the element I call 'schemata'. 'Schemata' are particular models of organizing experience which structure myths and collective representations—such as the 'patricide' schema, which structures all the myths involving patricide—and are themselves structured by, and thus express, the society's realities, collective attitudes, beliefs, and ideologies, its cultural assumptions. This will be illustrated below in this chapter and in the discussion of the schemata articulating 'father–son conflict' myths, of which patricide is one variant. As we shall see most clearly in Chapter IV.3, the deployment of such schemata helped to articulate the texts, functioning as matrices shaping the elements that make up the story.[27] Another schema which will be discussed in Chapter IV.3, the 'initiatory schema', which is derived from ritual practice, structures several myths and also articulates some non-mythological texts.[28]

The final question to be briefly considered here concerns the methodology appropriate for the study of myths. First, there is a general strategy which follows unambiguously from all that has been said so far. Myths are multidimensional, polysemic, and multifunctional cultural products, shaped by a variety of factors and circumstances, realities and idealities, pertaining to the society that generated them and reshaped them over time; therefore, in order to make sense of Greek myths, it is necessary to investigate a variety of questions pertaining to all aspects of the Greek world, through a variety of methodological tools and approaches. Each problem must be investigated in depth and in its own right through its appropriate methodologies, not as a quickly examined appendage to another set of data. It must be investigated through the different tools which are best suited to the nature of each particular heuristic enterprise and to the reconstruction of the different facets of ancient realities and of the different levels of meanings that came into play.

How, then, should we set about reconstructing—as far as is possible—the meanings which the myths under investigation had in the eyes of their ancient contemporaries? Some important strategies pertaining to this question have, I hope, emerged from the above discussion. To begin with, it is clear that it is necessary to read systematically (in the ways discussed above) every image and text in which the myth under consideration is articulated, and to attempt to reconstruct the meanings of each of these versions separately. Chapter II.2 illustrates most clearly this type of procedure and the reasons for its necessity. We must not conflate elements from different versions, and construct 'the main story', the outline of 'the most important aspects', of the myth; for this creates nothing but a cultural construct which has no reality, no reference in the ancient world, and which is made up of elements wrenched from their original contexts, that is, separated from the sets of relationships that had in fact (in the reality of the particular ancient articulations of the myth) helped to ascribe meanings to them. Consequently, we must read each individual articulation of the myth as it stands; we must not import narrative elements and meanings from one mythological articulation to another, and assume that one part of a myth in a text necessarily evoked for the ancient reader (and thus allows us to take account of) all the other variants. On the contrary, since, as we saw, meaning is also created through each element's relationships to the alternatives which were not chosen in its place, the exclusion of a particular narrative element from a particular articulation of the myth played a significant role in the process of meaning-creation and must often have contributed to the creation of meanings very different from those in which that narrative element was present. (See, for example, the various versions of the myth of the Previous Owners of the Delphic oracle discussed in Chapter IV.2). We must, then, study each articulation of the myth under discussion in each individual text or image on its own, and consider whether it contains any elements which may have evoked for the ancient reader other variants of the myth; if it does not, it is not legitimate to import into its readings elements from other versions. Different variants of a myth may focus on different aspects or potentialities of that myth, they may stress, or understress, or elide, particular elements; they thus create their own version of the myth, with its own specific meanings. This is illustrated, for example, in the variants of images of erotic pursuit which include strong nuptial elements discussed in Chapter II.2.

Clearly, one factor which affects the selections leading to the creation of a particular variant is the context; each variant is shaped also by the text (or, when appropriate, nexus of images) in which it is articulated and which it helps to articulate and by the function which this myth has in the text (or nexus of images).[29] It is frequently the case that a particular mythological articulation involves more than one theme and is structured by more than one schema, so that the myth is shaped through the interaction of these schemata—an interaction which is, of course, itself determined also by the overall context of the text which articulates the myth. Since myths are polysemic, the different variants of a myth may put emphasis on different facets of the myth's signification.

In these circumstances, it is unambiguously clear that the investigation must begin with the analyses of the different articulations of the myth under consideration, beginning with the reconstruction of the ancient readings of the relevant texts and images.[30]

For ancient readers the process of making sense of a text or image and that of making sense of the myth articulated in them were intimately intertwined, part of one complex interactive process. But since we do not share the ancient assumptions we cannot replicate those interactive processes; we can only attempt to reconstruct the meanings which the ancient readers created through them, by artificially reconstructing the different facets of the signification process; this entails the articulation of those complex interactive processes into distinct operations. The strategies for reconstructing the meanings of the individual mythological articulations must be based on, and reflect, the modalities through which meaning is created. Thus, we must determine the semantic, the conceptual, field of each element in the myth by considering all the meanings, functions, connotations, and associations which it had in the social, cultural, religious, and generally conceptual, world of the society that produced the myth. We must then attempt to reconstruct the complex interactive relationships between the form of each element in that particular articulation of the myth and its semantic field, its syntactical relationships with the other elements that make up the particular articulation of the myth, and its relationships to the other (semantically related) elements which might have been selected in its place but were not.

It is clear from what has been said above that these operations must be conducted through perceptual filters which do not reflect our own culturally determined preconceptions, but are (as far as possible) shaped by the relevant ancient assumptions. For example, the 'evaluation' of the syntactical relationships between the narrative elements, the attempt to reconstruct the meanings created by those relationships (and thus also the meanings of the individual elements and of the myth), must be set in the overall context of the culture which generated the myths and involve also comparisons with comparable syntactical relationships and, when these exist, at the more specific level, with comparable configurations in other myths—but not in the other versions of the myth under consideration, since this would introduce the danger of circularity. Thus, for example, we shall see in Chapter IV.3 that a certain set of narrative elements in certain sets of syntactical relationships in the text under discussion corresponds to, and can be seen to be governed by, the schema 'ephebic initiation', which carries, or rather helps to create, certain meanings. The specific value of these meanings is determined by the interaction between the schema and the particular form and content of the particular elements that make up the articulation. This procedure will bring out the schemata which structure that particular articulation of the myth.

At the end of the analyses of each individual articulation of the myth, its reconstructed meanings must be placed in the context of, and tested against, an independent reading of the overall text (or nexus of images) of which that mythological articulation is part. The application of this methodology is illustrated most clearly in Chapters IV.2 and IV.3 with regard to texts and II.1 and II.2 with regard to images.

After the analyses of the individual articulations of the myth have taken place these different articulations should be compared to each other; this will determine the parameters of variation in the meanings articulated in that myth or mythological nexus. And this brings us back to the schemata. For when such analyses are conducted, the inescapable conclusion is reached that the different versions of a myth, for example the myths of father—son hostility, are all shaped by a basic underlying schema which structures (with variations) all variants. This schema is itself structured by, and thus expresses as 'messages', perceptions which correspond to the social realities and ideologies of the society that produced them: for example, in the myth just mentioned, the perception that the son must privilege his relationship with his father above all else, otherwise catastrophe will follow (cf. Chapter IV.3).[31] For these mythological articulations in texts and images are not wholly 'individual' constructs independent of cultural constraints; they are shaped by the parameters created by the social realities, collective representations, and beliefs of the society that generated them. They are articulated by, and thus express, those realities and idealities.

Consequently, one aim of the analyses of myths must be to recover, to reconstruct, the perceptions that articulate them, both the perceptions articulating the different individual versions of a myth and those articulating the schema which is common to all versions of that myth: for example, the schema which structures all myths involving patricide. The latter will give us the ancient society's significant perceptions concerning the semantic area to which the myths under discussion pertain—in the case of patricide, the areas of father–son relationships and the related question of the young male's accession to full adult status.

It clearly follows from the methodological principles and strategies discussed above that the analyses of the myths must be conducted separately from the investigation of the social realities and collective representations of the society that produced the myths, and which come into play in those myths. If the two coincide, if the analyses of the myths has produced 'messages' which express the significant perceptions of that society pertaining to the semantic area concerned as they are known from other sources, this constitutes some validation, some confirmation of the validity of the analyses.

NOTES TO CHAPTER I

1. Chs. II.3, III, IV.2, and IV.3 have been updated, and have had additions and cross-references inserted. Chs. II.1, II.2, and IV.1 have also been further modified somewhat (cf. introductions to Parts II and IV).
2. The two early chapters, III and IV.1, represent a less mature version of that methodology (which was itself the end-result of earlier developments, since both originated in lectures delivered in 1975, themselves the result of earlier research), while the other essays articulate the fully developed version. I am not here concerned with comparing my

methodology to that of others, who have anticipated or shared some or all of my concerns and have proposed and explored similar or comparable methodological strategies. This would involve a separate enterprise of a different nature. Here I set out my own perceptions of the *Problematik* and the methodology (the general strategy for making sense of the ancient world, and the particular versions pertaining to the particular object of study, text, image, myth, ritual) which I consider best suited to the task, and which is deployed in the rest of the book.

3. I illustrate the ways in which this type of non-neutral methodology operates through hidden circularity in my critique of certain interpretations of the myth of the Previous Owners of the Delphic oracle in Ch. IV.2.
4. For an elaboration and refinement of this concept of reconstructing the ways in which the fifth-cent. Athenian audience made sense of tragedies cf. Sourvinou-Inwood 1989a.
5. Cf. ibid. for examples of such evaluations of both particular actions and general conduct.
6. Cf. also, for a combination of texts and images, Sourvinou-Inwood 1988a.
7. A few references to works discussing culture-determination in sensory perception and in the perception of pictorial representations: Gregory 1966: *passim*; cf. pp. 204–19, and esp. 220–8. Gombrich has written extensively on the subject: cf. 1977: *passim*, and cf. esp. 76–7, 170–203, 231. On perceptual controls in general cf. e.g. Douglas 1982: 1–8 (cf. esp. 1). On *Rezeptionsforschung* in art cf. also Schneider, Fehr, and Meyer 1979: 7–41; Meyer 1980: 7–51.
8. Cf. Boon 1982: 27–46. On the impossibility of eliminating completely cultural dependence and reconstructing precisely the ancient processes of meaning-creation through which the classical vase-painters created their images cf. also Schmidt 1980: 753.
9. Cf. Derrida 1967: 414.
10. On the post-structuralist notion of the sign and process of signification cf. Derrida's work. Cf. e.g. Derrida 1972: 29–46, 105–30; 1974; 1976: *passim*, esp. 11–15, 44–73; 1973: 129–60; 1967: *passim*, esp. 311–14; 1982. Cf. also Culler 1981: 41–2; Sturrock 1979.
11. For a handy discussion of the multifaceted concept of deconstruction and its operation cf. Culler 1983, with extensive bibl. Cf. also, very briefly: Culler in Sturrock 1979: 172; Culler 1981: p. ix.
12. In Sourvinou-Inwood 1989a I offer an example of a complex deconstruction of the dominant meanings in a Greek tragedy.
13. Cf. e.g. Bérard 1976: 61–73.
14. Cf. Lyons 1977: 102–5: Culler 1981: 24; cf. also Kaplan 1970: 275–6; Eco 1976: 204–5; Gombrich 1977: 73–8, 230–1.
15. Gombrich 1977: *passim*; cf. esp. 53–78.

16. Cf. also Gombrich 1982: 36–7; Douglas 1975: 51–2.
17. Cf. Gombrich 1977: 53.
18. Cf. Eco 1981: 3–43 on reading texts.
19. On this cf. Eco 1981: 24–7.
20. The desirability of exhaustive analyses which do not neglect any of the elements of the representation is also urged by Schmitt-Pantel and Thelamon 1983: 17.
21. I discuss the identity of the protagonists in detail in Ch. II.2 and argue that when there is one pursuer and the scene is mythological, in the absence of an additional sign specifying a different identity, the pursuer is Theseus. On the non-mythological versions of the theme and the relationship between the mythological and the non-mythological versions cf. Ch. II.2.
22. On erotic pursuits by gods cf. Kaempf-Dimitriadou 1979. Cf. also Schefold 1975: 87–94.
23. Schefold 1982: 233 warns of the danger that the fine distinctions between erotic pursuits by heroes and erotic pursuits by gods may be overlooked.
24. Cf. also Gombrich 1982: 86. I discuss the sign 'fleeing woman' in more detail in Ch. II.2.
25. I am using the word 'reading' partly metaphorically, to include also the notion 'making sense in performance'.
26. For tragic texts cf. Sourvinou-Inwood 1989a; 1988b; 1989b. For comedy: Sourvinou-Inwood 1988a: 136–52, cf. 23–4, and *passim* for the reconstruction of the relevant assumptions. I also discuss the whole question of cultural determination in the reading of ancient texts in a book now in preparation entitled *Charon Who? An Exercise in Reading—Focussing on Death and Epitaphs.*
27. For the role of such schemata in tragedy cf. my discussion in Sourvinou-Inwood 1989a.
28. Cf. Bowie 1987: 112–25. I should mention that, as will be illustrated in the essays presented here, the distinction between mythological and non-mythological texts and images is not as sharp and clear-cut as it is sometimes assumed.
29. Cf. Sourvinou-Inwood 1988b: 167–83 and Sourvinou-Inwood 1989b.
30. The summary presentation of some of the results of the analyses of the individual articulations of some of the myths discussed in Chs. IV.1, IV.2, and IV.3, which concentrates on the schemata, for brevity's sake and to avoid too much distraction from the already extensive foci of the investigations, must not be confused with the conflation of elements from different variants into one myth. In general, while in this chapter I set out what I consider to be the ideal methodology step by step, because of constraints pertaining to publication, my presentation does not always set out explicitly and articulate in detail every single step

of the reading and interpretative operations, though these have been conducted on the basis of the step-by-step procedure advocated here—as any attempt at reconstructing them will demonstrate.
31. Cf. also on the 'messages' structuring myths my discussion in Sourvinou-Inwood 1979: 14–15, 65–6 nn. 62–78.

PART II
Images of Athenian Maidens

INTRODUCTION

The three papers* in this section have a common double focus, methodological and thematic. They explore and illustrate some aspects of the methodologies by means of which we read fifth-century Athenian images through the systematic study of a nexus of iconographically related themes. These investigations (of the images representing these themes and their meanings, and also of the relationships between the iconographical themes) bring into play, and I hope illuminate, important aspects of the *Problematik* pertaining to the reading of ancient images and also important facets of the fifth-century Athenian realities and perceptions pertaining to *parthenoi*, marriage, and the relations between the sexes in general. Because, as we saw in Chapter I, it is necessary to investigate exhaustively and in its own right every facet of the phenomenon, text, or image under consideration, they also bring up and consider a series of other questions pertaining both to the fifth-century Athenian ceramic images and their modalities and to the realities, beliefs, and perceptions of the society that produced those images. They also bring out some important modalities through which fifth-century Attic red-figure iconography operates.

Chapter II.1 is the most complex and demanding, because it is the most dense and technical. Since all essays are self-contained it can be skipped by the casual reader. Other readers may prefer to begin with the investigation of erotic pursuits in Chapter II.2. But I have placed this chapter at the very beginning of the section both because I believe that this is the most satisfactory position for it from the point of view of the section's thematic articulation and also, and especially, because it provides the most extensive and multifaceted illustration of the methodologies articulated in Chapter I, here applied to the solution of a complex problem involving the need to deploy a variety of strategies.

NOTE

* Ch. II.1 was first published in C. Bérard, C. Bron, and A. Pomari, eds., *Images et société en Grèce ancienne: L'Iconographie comme méthode d'analyse*.

Actes du Colloque international, Lausanne 8–11 février 1984 (Lausanne, 1987), 41–58. Ch. II.2 was first published in *JHS* 107 (1987), 131–53. Ch. II.3 was first published in *BICS* 32 (1985), 125–46. The essays on which Chs. II.1 and II.2 are based have been not only updated but also somewhat modified here, primarily in order to fit the present format, in which they are published together, and are also accompanied by a separate methodological essay (Ch. I). This format in fact reflects the original version of Chs. II.1, II.2, and II.3, which began life together as parts of one very long essay, which eventually evolved into three self-contained papers that were published separately. I would like to thank Prof. Mary R. Lefkowitz who commented on the version of it which had been circulated as part of the Lausanne Colloquium Preliminary Proceedings. I am also grateful to Prof. C. Bérard for providing the inspiration and intellectual stimulation for this nexus of papers.

II.1

Menace and Pursuit: Differentiation and the Creation of Meaning

1. INTRODUCTION

The aim of this paper is to study the relationship between the iconographical theme 'erotic pursuit' in which a youth with or without spears is pursuing a girl (Pls. 1–2, 5 and 7) and the scenes showing Theseus pursuing a woman, drawn sword in hand (Pls. 3–4 and 6). The theme 'erotic pursuit' is investigated in great depth in Chapter II.2. The theme representing Theseus pursuing a woman drawn sword in hand (henceforth 'Theseus with a (drawn) sword') was a central concern in my book *Theseus as Son and Stepson* (= Sourvinou-Inwood 1979). The starting-point of the present study was a suggestion and criticism made by Claude Bérard in his review of that book[1] in which he observed that I should have considered how 'Theseus with a sword' relates to the representations of erotic pursuit involving a youth usually identified as Theseus. This criticism was methodologically justified, so I take the opportunity to explore this relationship here, and in the process explore also some important questions pertaining to this thematic area, and also, and above all, some important methodological questions pertaining to the reading of Greek images.

Bérard's suggestion was that the possibility that 'Theseus with a sword' may also represent an erotic pursuit deserves further consideration. I will now investigate this possibility.

2. ANALYSES

1. Introduction

The suggestion that the scenes showing Theseus with a drawn sword pursuing a woman may represent an erotic pursuit[2] can be articulated into two different hypotheses. First, the supposition that Theseus' holding the sword has exactly the same significance as his holding the spears in erotic pursuits. This in its turn may mean either that Theseus with a drawn sword does not represent an actual attack, but something similar to erotic pursuit; or that the spears in erotic pursuits also denote attack. We shall see that 'Theseus with a sword' is an unequivocal attack, and 'erotic pursuit with spears' is unequivocally not an attack. Thus neither alternative of the first hypothesis can stand. The second hypothesis supposes that 'Theseus with a drawn sword', acknowledged as an attack, may 'stand for' an erotic pursuit; that the latter can be represented also through the metaphor of attack. It is here that belongs the argument that the threatening stance of some erotically pursuing gods provides support for the view that a hero's erotic pursuit can also be shown as an attack. It will be shown that this second hypothesis cannot stand either.

II. First hypothesis

(a) The iconographical schemes

i. Theseus with a sword: The sword

As I have pointed out elsewhere,[3] the closest iconographical similarities to the theme 'Theseus with a drawn sword' are found in two themes representing an attack against a perfidious woman which did not end in murder (the subject which, on my hypothesis,[4] was also represented in 'Theseus with a sword'): Odysseus attacking Circe and Menelaos attacking Helen.[5] The basic elements of Theseus' stance in 'Theseus with a sword' can be paralleled in other scenes of attack; I know of no instance in which the sword is being held in this way and the denotation is not attack. Moreover, the version in which Theseus is striding, left arm outstretched, is one of the conventionalized stances of attack in Attic iconography.[6] More especially, the variant in which the outstretched left on which is thrown the chlamys is holding the scabbard[7] belongs together with, and calls up, not only other iconographical schemes of attack in which this stance is found, but also, and especially, that of Aristogeiton in the Tyrannicides group. The latter was undoubtedly the model for the codification of this particular attack-schema, and a very potent schema in the sets of assumptions through which fifth-century Athenians painted and read iconographical signs. This stance, then, carried the strong denotation 'attack'.[8]

The basic stance and the position of the drawn sword in the theme 'Theseus with a sword' resemble in a general way those seen in some scenes depicting Ajax attacking Kassandra,[9] representing sacrilegious rape.[10] It may be argued that this weakens my case, since here it is rape that is involved, and is rape not just a stronger version of 'erotic pursuit'? Kassandra and Ajax were sometimes represented through the 'erotic pursuit' scheme, and, after all, Menelaos' and Helen's relationship was erotic, and didn't Odysseus' and Circe's become just that? This objection depends on an erroneous methodological assumption. It assumes a fixed signified 'rape of Kassandra', evoked by all representations of that theme. This disregards the (relative) autonomy of the signifier[11] and the nature of the dynamic process of meaning-production. Let us consider the variant which shows Ajax and Kassandra through the scheme of erotic pursuit. The element 'erotic pursuit' is dominant in the scene on the cup Leningrad Hermitage N 658 (Waldhauer) (*ARV* 817.3; *LIMC* i, pl. 263.64). Correlative with the use of this scheme is the absence of the Palladion, that is of the element denoting the sacrilegious aspect of Ajax's action. Correlative with both is the absence of the drawn sword which denoted—on my thesis which is being tested here—a serious attack. That is, the perception of Ajax's action here is different from that expressed in the Ferrara cup (n. 10). Here it is presented as 'a bit of a lark'—a perfectly plausible male perception of rape. This perception is also seen in (and for the viewers reinforced through) the exaggerated jerky movement of the characters. As Touchefeu comments:[12] 'la vivacité d'A. est presque poussée jusqu'à la caricature'. Correlative with this perception of the act, and helping define it, is the absence of the sword and Palladion. The fact that the adoption of the scheme 'erotic pursuit' for Ajax and Kassandra involves the exclusion of the drawn sword confirms that the latter denotes serious attack. Indeed, the drawn sword is also absent from scenes in which, though the erotic pursuit aspect is still dominant, elements of attack are included.[13] It clearly belongs to the 'serious attack' pole of the spectrum of perceptions expressed in fifth-century Attic representations of Kassandra's rape, together with the motif 'Kassandra kneeling on the steps of the Palladion and being dragged away by Ajax'—though, of course, the sword is not always included in these versions.

Consequently, we may conclude that the drawn sword held in the way in which it is held in the theme 'Theseus with a sword' is represented in the role of an attacking weapon in action, and denotes serious menace and attack. It is clear from the above that the sword thus held is a highly codified sign;[14] its codification is, of course, strongly motivated.[15]

ii. Erotic pursuit: The spears

I will begin with a comparison between the two iconographical types of erotic pursuit; in the first, which I call type 1, the youth is carrying spears, in the second, type 2, he is not. The two types are closely related. First, leaving aside the spears, the two schemes are almost identical, and involve the same pursuer (cf. Chapter II.2, section 3). Second, as I will now show, the spears in type 1 do not denote attack; and the implicit connotations of violence they produced are also connoted by the iconographical scheme itself and by other individual elements. Third, Peleus' pursuit of Thetis is represented through both iconographical schemes,[16] which suggests that here also the same subject is depicted in two iconographical versions. Fourth, the different versions of the two types show the same variations in the syntactical relationships between the iconographical elements (e.g. in both types sometimes the youth grabs the girl and sometimes there is no physical contact), and also a similar appearance of other variables (e.g. the presence/absence of petasos). This suggests that the painters were operating through assumptions in which the two were closely related—a hypothesis which a detailed investigation will now confirm.

One of the strategies that will allow us to assess how the painters and their public perceived the relationship between the two types consists in investigating how the representations of the two types relate to each other in the work of individual artists. In order to prevent our own assumptions from intruding in the assessment of, for example, what constitutes a significant similarity or difference in the signification system of fifth-century Athenian ceramic iconography we must try to recover the parameters of thematic differentiation in that system. The parameters of differentiation within the work of individual artists is a firm foundation for this. I begin here with the recovery of the parameters of variability within one theme in the work of the same artists.[17]

Space prevents me from setting out the detailed comparisons of the various scenes of erotic pursuit of type 1 painted by the same artist. I will illustrate the types of relationships between such scenes through the comparison of type 1 scenes by the Painter of the Yale oinochoe, whose treatment of the themes of erotic pursuit and 'Theseus with a sword' on the same and on different vases will be considered below. The relationships between type 1 pursuits painted by this painter exemplify, and are characteristic of, all such relationships.[18] I shall consider the following type 1 pursuits by the Painter of the Yale oinochoe: (*a*) stamnos Warsaw 142353 (ex Czartoryski 51) (*ARV* 501.2; CVA pl. 28) side B; (*b*) stamnos Krefeld Inv. 1034/1515 (*ARV* 502.5; CVA Germany 49 pls. 37 and 38) side A; (*c*) bell-krater Leningrad 777 (St. 1786) (*ARV* 502.11; Peredolskaya pl. 116; here Pl. 5) side B.

We note the following. The stance of the youth and the position of the chlamys is very similar in all three, with the following differences. First, on (*a*) and (*b*) the chlamys is thrown over his extended left arm, covering it almost completely, while on (*c*) it is thrown over the same arm like a wrap. Second, on (*b*) he has a petasos thrown at the back of his neck, while on (*a*) and (*c*) he has no petasos. Third, on (*a*) and (*b*) he grabs the girl—her shoulder on (*a*), her wrist on (*b*)—while on (*c*) he does not. All these differences between type 1 pursuits by this painter are of the type generally found between scenes of type 1 painted by the same artist.

Divergences between individual renderings by the same artist of the same type of erotic pursuit pertain to the direction of the pursuit,[19] details of dress of the type of those observed in the work of the Painter of the Yale oinochoe, small variations within definite parameters in the youth's stance,[20] the presence or absence of additional characters such as the father and companions,[21] and finally, divergences in the position of the spears, though in all the variations the latter are shown as not weapons in use or about to be used in attack. Some artists consistently use motifs which make this character of the spears explicit, others use them in some scenes and not in others. The representations of erotic pursuit of type 2 by the same artist show the same variations as those pertaining to type 1.[22]

The comparison of type 1 with type 2 erotic pursuits by the same artist reveals striking similarities. If we leave aside the spears, the divergences are of the type observed between representations of pursuit of the same type painted by the same artist. I will begin with the work of the Painter of the Yale oinochoe. Side B of the stamnos Krefeld Inv. 1034/1515 (*ARV* 502.5; CVA Germany 49 pls. 37, 38) shows an erotic pursuit of type 2—its side A, we saw, depicts a type 1 pursuit. I will now compare these two representations by the same painter on the same pot. Leaving aside the fundamental difference defining the two types, the presence of the spears in 1, their absence from 2, the following differences may be noted. i. On A the youth has short hair, on B long, falling on his shoulders. ii. On A he is striding, on B he is running. iii. On A he is grabbing the woman by the wrist, on B by the shoulder. iv. On A his stance is three-quarters, on B profile. v. On B his chlamys has a black border, on A it does not.[23] All these are divergences which are also observed between representations of erotic pursuit of the same type by the same artist. As to the defining difference, the presence–absence of spears, it only produces a difference in emphasis. We shall see that the meanings produced by the spears are not denotative of attack, but connotative of hints of violence. Meanings of this type are also carried by the theme of erotic pursuit itself, the spears only increase the emphasis. Moreover, the motif of capture, signified through the iconographical motif of grabbing the woman, produced the same more emphatically 'violent' meanings as the spears. Both scenes considered here are 'captures' of this kind. Thus the fact that on A this 'violent' facet of the signified semantic field is implicitly reinforced through the spears can only produce a small difference of emphasis between the two types, not a difference of meaning. Though the grabbing of the girl does not occur in all pursuits of type 2, it does appear in a significant number of them.[24] These divergences between sides A and B of the Krefeld stamnos exemplify the differences observed between pursuits of type 1 and type 2 in the work of other artists.[25] If we leave aside the spears, the divergences are of the same type as those observed between representations of pursuit of the same type painted by the same artist.

Consequently, the comparison of types 1 and 2 in the work of artists who painted both shows that these artists were operating through assumptions in which the relationships between representations of types 1 and 2 were the same (apart from the defining difference presence–absence of spears) as those between representations of erotic pursuits of the same type.

This conclusion is reinforced by the existence of a third, 'intermediate' type, very closely related to both: the scene on the skyphos Providence 25.072 (ARV 973. 10; CVA pl. 20.1) by the Lewis Painter, in which the spears are resting on the ground, not carried by the youth; he has just put them down and is running after the girl. This, surely, is a stronger version of the type of motif depicted in type 1 pursuits, which, we shall see, make explicit the fact that the spears were not shown as weapons in action in the course of an attack. This intermediate type makes their inactive character even more explicit and emphatic; it confirms that the role of the spears was to characterize the youths, being part of their equipment, and suggests that types 1 and 2 are different segments of a continuum of signifiers signifying erotic pursuit, with variations in the emphasis put on the violent facets of that semantic field. The pursuit on the Providence skyphos is closely related to a type 1 pursuit by the Lewis Painter (skyphos Reggio 3877 (*ARV* 974. 25) and to a type 2 pursuit in his manner (skyphos Reggio 4134 (*ARV* 975. 3). The striking similarities between these three types, and, especially, the intermediate position of the Providence skyphos, provide further confirmation for the conclusion that, in the sets of assumptions which shaped the painters' manipulation of the established iconographical schemata when individual scenes were painted, the semantic relationship between types 1 and 2 was very close; the two types signified the same semantic field, with a small difference in the emphasis put on one of its facets.[26] For only such a context could have provided the matrix for the creation of a scene in which the youth has just put down his spears to run after the girl.

I will now consider further the role and significance of the spears in type 1 pursuits.

A preliminary point: the fact that the spear in attacking action is interchangeable with the sword in attacking action[27] does not invalidate my argument, which consists in showing that the position of the spear(s) in erotic pursuits is not that of a weapon in action. On the contrary, precisely because 'spear in attack' is equivalent to 'sword in attack', the spears which are simply being carried and not portrayed as an offensive weapon in action should be equivalent to a sword which is simply being carried, not being used in an attack, that is, to a sword in its scabbard. The youth in erotic pursuits does sometimes wear a sword in its scabbard, either 'instead of' the spears in type 2[28] or in addition to them in type 1.[29] The spears carried by the erotic pursuers are held in a variety of positions: horizontally (or almost horizontally),[30] diagonally,[31] vertically (or almost vertically),[32] in the right[33] or in the left hand.[34] But the forearm is never raised in an attacking position. These positions of the spears in erotic pursuits do not ever correspond to the positions in which spears are held when they are used as attacking weapons in scenes of combat.[35]

Furthermore, and most importantly, I will now show that this character of the spears as 'not weapon in use or about to be used in an attack' is sometimes stressed iconographically through motifs which emphasize it. This, of course, entails that the signification space in which the spears thus portrayed were inscribed included the element 'not attack'; and also that this character 'not attack' was made explicit to viewers. The following motifs signify the character of the spears as 'not weapons in use in the course of an attack in progress'. First, of the fifty-seven scenes of erotic pursuit of type 1 which I have considered in detail, in twenty-six, just under half, the spears were held in the left, and not the right, hand. Obviously, the fact that the vase-painters depicted the spears held in the right about as often as they did in the left indicates that they were not treating them as weapons in action. There is no red-figure example known to me where a spear functioning as a weapon in action is held in the left.[36] Sometimes the spears are held in the left because the vase-painter has reversed the more frequent left to right direction of the pursuit into right to left, and reversed the hand holding the spears accordingly. But even this shows that it was a matter of indifference to the painters in which hand the spears were held; and precisely this shows that they were not treating them as weapons in action, to be held in the hand in which they were held when in action. For the opposite is the case with the drawn sword. In the representations of 'Theseus with a sword', when the direction of the attack is reversed the sword remains in the right hand.[37] Moreover, in fifteen out of the twenty-six cases in which the spears are held on the left the pursuit is in the more frequent left to right direction.[38]

Second, in a few scenes the spears are held with the spearheads turned away from the girl, and thus also away from the direction of movement of the youth who is holding them.[39] This is a most unnatural arrangement, especially when (as on the Ferrara oinochoe, n. 39) the spears are held horizontally or almost horizontally. For while it may seem natural (and is paralleled in other scenes) to have the spearheads turned away when the spears are in certain diagonal positions, the arrangement is most peculiar when the spears are held horizontally. Given the codification of the 'normal' arrangement, and given also that the spearheads form the attacking part of the weapon, their being turned away inevitably produced the meaning 'not attack', for this is how these two facts define the value of this signifying element—through opposition.

The selection of this arrangement tells us that the artist had been thinking of, and treating, the spears, as 'not weapons in use/about to be used in attack'. And because the selection of each—even trivial—signifying element is determined (in an interactive process) by the 'spacing' created by the other elements that come into play in the complex movement of meaning-production, we conclude that the emphasis on the spears' character as 'not weapons in use in an attack' was dictated by the semantic field of the theme 'erotic pursuit' and the perceived function and meaning of the spears in it. This is clear in the case of the horizontally held spears with the spearheads turned away; but even the choice of the diagonal arrangement is not without significance. For when a comparable arrangement (spears being carried diagonally) is found in scenes where context and circumstances involve hostility and (potential) attack/hostile action, the spearheads can be shown turned downwards and in the direction of the hostile presence, as in the case of Eos and Kephalos in a representation such as that on the volute-krater Bologna PU 283,[40] where the natural direction of the spearheads would have been upwards, in the direction in which Kephalos is moving. The fact that they are turned towards Eos is correlative with the context of hostility that prevails in this theme. Kephalos' attitude towards the pursuing Eos is never friendly, and can be hostile and threatening.[41] This type of arrangement, a selection which might have been made in our scenes but was not, and which, like all the choices that were not selected, also helps give value to the signifying element which was deployed, confirms that even when the spears are held diagonally, the spearheads turned away from the girl are not so much 'natural', as correlative with the (non-hostile) context and the fact that the spears are not perceived as weapons being used or about to be used in an attack.

The arrangement in which the spears are held vertically, as, for example, on the stamnos Oxford 1911.619 (*ARV* 629.16; here Pl. 1) is another motif which makes explicit that the spears in erotic pursuits are not perceived and treated as weapons in use, or about to be used, in an attack, do not denote 'attack' as the drawn sword does. For the vertical position of the spears denotes that they are being carried and not about to be used, that they are in repose. This is clear from scenes like, for example, those on the cup Harvard 1917.149 (1642.95),[42] on the white-ground lekythos Athens NM 1818,[43] or on the stamnos Munich, Museum Antiker Kleinkunst 2415.[44] Another motif which makes explicit that the spears are not portrayed as weapons in an attack (in progress or about to take place) is that in which the spears, held horizontally in the right, are lowered have the spearheads inclined a little towards the ground.[45] Yet another motif with a similar meaning is that in which the spears are held diagonally in the right, resting against the crook of the youth's arm[46] or against his upper arm.[47]

In these circumstances, it is clear that the spears in representations of erotic pursuit are not depicted as weapons in use, or about to be used in an attack; on the contrary, the fact that they are not in use as offensive weapons in an attack (in progress, or about to begin) is emphasized. It could be argued that I am jousting at windmills, that no one would want seriously to maintain that the spears in erotic pursuits are actually used to attack the girl; only that they add an element of menace and threat, and it is this that may be comparable to 'Theseus with a drawn sword'. So the fact that the spear is not shown in the position of attack tells us nothing. And there are, in fact, representations of warriors not in a position of attack, but alert and preparing for it, where the spears' position does not seem dissimilar to that of some of the spears in erotic pursuits. One such example is the scene on the kantharos Athens N.M. 1236:[48] the Greek (Theseus) is advancing, on the alert for an attack, or the opportunity to attack, while an Amazon is represented on the other side. However, first, the position of the spear may appear to be 'the same' as in some erotic pursuits, but it is not. For on the kantharos it is held by a man whose stance and context differ significantly from those of the pursuers; compare, for example, the way in which he stoops forward as he is advancing. And since each element's value is also determined by its syntactical relationships to the other elements, the value of the spear cannot be 'the same' when the stance, and the overall context, are different.

A more important point is that, once a radical difference has been acknowledged (here that we are not talking about the spears being represented as a weapon in action in erotic pursuits—as the sword is in the theme 'Theseus with a sword'), then what is or is not comparable, and what does or does not constitute a criterion for believing that diverging iconographical schemes have 'the same' meaning, is a matter of individual, culturally determined judgement. In this case there are several other, independent, arguments supporting the thesis that the two themes are semantically different. But even in purely 'common-sense' culture-determined terms, there is a very great difference between 'attack' and 'implicit menace of violence'. Of course, the presence of the spears in the hands of the pursuer did carry implicit connotations of menace. In fact, one particular stance sometimes used for the pursuer (striding, left arm extended forward with the chlamys thrown over it, spears in the right)[49] resembles the attacking stance and thus produces definite connotations of menace. And the choice of this stance for some pursuers indicates that the painter was thinking in terms of some kind of menace. But what is involved here is connotation of menace, not denotation of attack. For the attacking stance, like all signs, did not have a fixed meaning, but acquired its value through the complex interactive process of meaning-production; therefore, it cannot mean the same thing in scenes where it involved the drawn sword and the scabbard held in the left as in those where it involved the spears defined as 'not weapons in use during an attack'. In the latter case it produces the meaning 'implicit connotations of menace/violence'. The use of motifs stressing that the spears are not depicted as weapons in use in an attack shows that the artists perceived this character of the spears ('not weapons in action') as a significant part of their value in these scenes, and diminishes the strength of the implicit connotations of menace produced by the spears.

The fact that the spears are not represented as weapons in action in type 1 pursuits confirms the semantic equivalence of types 1 and 2. And the conclusion, reached on the basis of independent arguments, that the two types are semantically equivalent, in its turn strengthens the conclusion reached through the consideration of the role and value of the spears in type 1, that any meanings pertaining to menace that may be produced by those spears are connotative and implicit—which implies that the meanings produced by the spears in type 1 pursuits differ only in emphasis from those produced by the spearless type 2 pursuit. Implicit connotations of violence are also produced by the iconographical scheme itself and the semantic field it activates, especially the notion 'capture of the girl', which is represented in many scenes, through the motif 'grabbing of the girl',[50] which therefore is another element carrying intimations of violence; at the same time this motif, which depicts the physical contact between youth and girl at the moment of the capture which will lead to the sexual act, gives sexual colouring to that implicit violence/menace. There is a great difference between denotation of attack and connotation of violence/menace; and it is especially important when attempting to read a signification-system which we only understand imperfectly not to blur distinctions and not to conflate diverse—if related—meanings into vaguely defined concepts.

The connotations of violence carried by this iconographical scheme representing erotic pursuit express the violence of the sexual aggression and possession.[51] Perhaps the spears also allude directly to a part of this semantic content: Athenian viewers, reading these scenes through assumptions and expectations telling them that the girl towards whom the horizontally held spears were pointing was captured or about to be captured and (about to be) submitted to sexual intercourse, would also read the spears, especially in some of these scenes,[52] in terms of a thrusting penis about to enter her. I do not mean that they would necessarily perceive this consciously; I am indicating a contribution to the creation of the meaning 'erotic pursuit' out of this signifier through a complex interactive process. There is, then, a correspondence between the allusive representation of 'violence' in these scenes and the subject's semantic connotations, the sexual act as an act of physical domination of the woman, inherent in the notion of pursuit and capture with the purpose of abduction/rape.

Positive parallels for the spears held by the youths in erotic pursuits do exist: their various positions find their closest parallels in representations of various youths, both Theseus and generic ephebes, wearing, like the pursuers, chlamys and petasos and carrying the spears as part of their equipment. Such youths appear in various contexts, in movement, stationary, and about to set out, and they are not using but simply carrying the spears, for example, in scenes of an ephebe's departure.[53] Ephebes (whose characteristic garment was the chlamys and who also wore the petasos) carried spears as part of their equipment, and the spears played an important part in their training and activities.[54]

Consequently, for both Athenian artists and viewers the elements 'youth', combined with 'chlamys, petasos, and spear(s)' made up the sign 'ephebe' and/or 'Theseus as ephebe'. That is, the spear(s) carried by these youths, in erotic pursuits and elsewhere, were part of the iconographical scheme which represented the signified 'ephebe';[55] and, of course, the spears are also an important element of the iconographical sign 'Theseus'. This role of the spears in the semantic field and sign 'ephebe/Theseus' does not neutralize the connotative significance of the spears. But it provides a solid foundation for, and so confirms further, my case that the violent hints carried by the spears are implicit and connotative and not strong.

iii. Conclusions

The above analyses lead us to conclude that the character and role of the spears in representations of erotic pursuit by Theseus are radically different from those of the sword in the scheme 'Theseus with a sword' and produce radically different meanings.

(b) *Comparison of representations of 'erotic pursuit' and of 'Theseus with a sword' painted by the same artist*

This analysis aims at recovering the ways in which the Athenian artists thought about, and manipulated, the two iconographical schemes, in order to determine whether or not they represented the same subject.

We have already established the parameters of variability in the treatment of the same theme (of the same type and of different types) by the same artist. Now, before comparing 'erotic pursuit' and 'Theseus with a sword' in the work of the same artist, we must consider how 'erotic pursuits' relate to representations by the same artist which are formally close to 'Theseus with a drawn sword' and known to depict an attack. I will focus this comparison very precisely by considering representations painted on the same vase. This offers the most closely defined context of production possible for the two scenes and an exact context of comparison, an identical context of production for the two sets of relationships, the control relationship and that between 'erotic pursuit' and 'Theseus with a sword'.

The calyx-krater New York 41.83 by the Persephone Painter (*ARV* 1012.3) represents on side A above Odysseus attacking Circe and on side A below an erotic pursuit of type 1. Leaving aside the fundamental difference between the drawn sword in the attack and the spears in the pursuit, we note the following differences. 1. In the Odysseus scene the movement is from left to right, in the pursuit from right to left. 2. Odysseus has the drawn sword in his right, the youth is carrying the spears in his left. 3. Odysseus' non-functional left arm is covered by the chlamys and slightly bent at the elbow, the youth's non-functional right arm is extended towards the girl and uncovered. 4. Odysseus is striding, the youth is running. Differences 1, 3 and 4 are of the type which also differentiates representations of the same theme and type painted by the same artist. The two scenes, then, were differentiated from each other in the following modality. First, through the fundamental divergence 'drawn sword' as opposed to 'spears being carried'; this fundamental divergence is intensified through the motif of holding the spears in the left in erotic pursuit, which stresses the character of the spears as not weapons in use in attack. And second, through a clustering of small divergences, of the type also differentiating representations of the same theme by the same painter, which underpin the central difference. The same type of divergences differentiates also the representation of erotic pursuit from that of attack on the volute-krater Bologna 269 by the Niobid Painter (ARV 599.8), side A of which shows an erotic pursuit on the neck and an Ilioupersis, including a scene of Menelaos attacking Helen, on the body.

Turning to the comparison of 'Theseus with a drawn sword' and 'erotic pursuit', I will consider the work of the Painter of the Yale oinochoe, and summarize the results of my detailed comparisons of scenes painted by eight others (Hermonax, the Agrigento Painter, the Geneva Painter, the Sabouroff Painter, the Penthesilea Painter, the Lewis Painter, the Painter of the Louvre Centauromachy, and the Hasselmann Painter).

A comparison between the two sides of Leningrad 777 (ARV 502.11; *Para.* 513; *Add.* 123; here Pls. 5–6), of which A depicts 'Theseus with a sword' and B 'erotic pursuit', reveals the following differences.[56] 1. The fundamental divergence 'drawn sword' versus 'spear'. 2. This fundamental divergence is reinforced a little through the motif of the spearhead being slightly inclined towards the ground, which belongs to the category of motifs making explicit the character of the spear as not weapon in use in attack. 3. On A Theseus has long hair with a fillet, on B the youth has short hair. 4. On A he is wearing a petasos slung at the back of his neck, on B he has no petasos. 5. On A he has a chlamys thrown over his outstretched left which it covers. On B the chlamys is worn like a wrap, thrown over the left shoulder, hanging at the back, and looped over the forearm. 6. On A the chlamys has a black border, on B it does not. 7. While on both A and B the youth's left hand and the woman's right hand are extended towards each other, palm upwards, and almost meet, there is a difference in the way they relate to each other. On A the youth's hand is above the woman's. On B they are at exactly the same level and their fingertips almost meet. This produces an effect of consensual erotic contact about to take place, which both defines the erotic content of the violence implicit in the theme as erotic, and also conflicts with, and thus deconstructs, the predominant meaning 'force about to be used on the girl who is being pursued and will soon be caught'. This effect distances and differentiates the erotic pursuit on B from the attack on A. With regard to the other differences, 3, 4, and 6 are of the type also found between different representations of erotic pursuit by the same artist; the same is true of 5, with the additional element that the arrangement of the cloak on A recalls the similar arrangement in the similar codified attack-stance of Aristogeiton in the Tyrannicides group, and thus it reinforced visually the denotation 'attack' of A. This had the effect of differentiating A from B more sharply by stressing that A belongs to the pole of serious attack; and this distancing is further intensified by the fact that on B the erotic pursuit's own inherent nature (and consequent radical thematic divergence from A) was stressed through the element of consensual eroticism produced by the arrangement of the hands, which stresses the fact that B pertains to the erotic sphere. The visual distancing between the two scenes is increased by the differentiation of the two women through details of dress, hair arrangement, and gestures.[57]

Thus the modality in which 'Theseus with a sword' and 'erotic pursuit' are differentiated on the Leningrad pot is the same as that in which our control erotic pursuit–attack scenes were differentiated, only stronger: the central and radical divergence 'sword in attack' versus 'spears being carried' (role explicated through the lowered spearheads) is doubly reinforced through elements that intensify it and distance the two scenes by stressing the attack character of the scene on A and the erotic character of the one on B. It is also underpinned by a clustering of other divergences of the type also found between different representations of the same theme by the same artist.

When we compare these two scenes on the Leningrad vase with the Painter of the Yale oinochoe's treatment of 'erotic pursuit' and 'Theseus with a sword' on other vases we note the following. First, the arrangement of the hands which produces the effect of consensual eroticism is not found in any of the other representations of erotic pursuit by the Painter of the Yale oinochoe known to me. Instead, on the Warsaw scene and the two sides of the Krefeld stamnos[58] the youth is grabbing the girl, a motif denoting capture and connoting the violent facets of 'erotic pursuit/abduction' and also the erotic/sexual character of that violence. But signifying elements acquire meaning in context; thus the grabbing of the woman is not confined to erotic pursuits, it is also found in 'Theseus with a sword' scenes, where it has a different meaning, determined by the context which guided the reading of the individual elements. In some versions of 'Theseus with a sword'[59] the grabbing of the woman is explicitly of the 'grievous bodily harm nature', for it takes the extreme form of 'grabbing by the hair', a motif which (at least until the fourth century) denotes serious attack (attack with intent to kill or sacrilegious rape.[60] But even in 'normal' versions of grabbing in 'Theseus with a sword', the meanings produced, because determined by the context, are radically different from those in erotic pursuits.[61] The fact that the Painter of the Yale oinochoe did not include the 'grabbing the woman' motif in the erotic pursuit juxtaposed to 'Theseus with a sword' may be correlative with the differentiating parameter which guided his manipulation of the two schemata here: that motif, by producing connotations of implicit violence, would have pulled B a little towards A, while the present choice, which produces connotations of consensual eroticism, pulls them further apart. A similar selection determined by the operation of the differentiating parameter can be seen in another element in which the erotic pursuit on B differs from the other pursuits by the Painter of the Yale oinochoe, the arrangement of the chlamys: in his other erotic pursuits, the Painter of the Yale oinochoe has the chlamys thrown over the youth's arm in a scheme similar to that of Theseus on Leningrad side A. On Leningrad B, however, it is thrown like a wrap, thus differentiating the pursuer of B from Theseus on A whose stance and arrangement of chlamys recall Aristogeiton's, and thus reinforce the denotation 'attack'.

The Painter of the Yale oinochoe produced two more representations of the theme 'Theseus with a sword': on the stamnos London E 446 (*ARV* 502.4; Sourvinou-Inwood 1979, no. 8; here Pl. 4) and the neck-amphora New York 41.162.155 (*ARV* 502.14). All three are closely related, with a small number of minor divergences of the type which, we have seen, characterize different representations of the same theme by the same artist: presence of petasos (Leningrad pot), absence of petasos (the other two); direction of movement (left to right on the Leningrad and New York scenes, right to left on the London one); two additional characters on the stamnos, none on the other two pots. There are two further differences, two motifs which intensify the denotation 'attack'. On the neck-amphora Theseus is holding the scabbard on his outstretched left which has the chlamys thrown over it. The arrangement of the arm and chlamys is similar in the other two scenes, and especially on Leningrad side A; but the additional motif of the scabbard being held in the left hand brings this stance even closer to the Aristogeiton codified attack stance and thus strengthens the denotation 'attack'. On the stamnos the motif 'grabbing by the hair' denotes the attack's seriousness. Thus, this painter, who regularly shows the girl being grabbed in erotic pursuits (except when desiring to pull the theme away from the connotations of violence on the Leningrad pot), when the woman is grabbed in 'Theseus with a sword' he adopts the most extreme, violent version of this motif, the grabbing by the hair—never deployed in 'erotic pursuit'.

The comparative analyses of representations of 'Theseus with a sword' and of 'erotic pursuit' in the work of the other eight painters I have considered produced results—which I will now summarize— similar to those pertaining to the Painter of the Yale oinochoe. The two themes are differentiated from one another in a different modality from that in which the different representations of the same type of the same theme, or of different types of the same theme, differ from each other. 'Theseus with a sword' and 'erotic pursuit' diverge in a way comparable to, but sharper and stronger than, that in which other attacks diverge from erotic pursuits painted by the same artist. Though the different painters vary in the strength and intensity of their differentiation of the two themes, they share a common, composite differentiating mode, which is as follows. The fundamental divergence 'drawn sword'—'spear not in use in attack', which distinguishes the two themes and produces radically different meanings, is underpinned by a (larger or smaller, depending on the artist) clustering of other divergences, of the type also found between different representations of erotic pursuit by the same artist, and this clustering increases the visual distancing between the two themes. The fundamental divergence itself is sometimes intensified through motifs—e.g. holding the spears in the left—which stress the spears' character as 'not weapons in use in attack'. In addition to these there is a third type of diversification: at least one element is deployed which differentiates the two themes more sharply than those that make up the clustering, because it underlines and strengthens the inherent nature of one or the other theme, either attack (e.g. through the adoption of the Aristogeiton stance for the attacking Theseus) or eroticism (as in the Leningrad pursuit).

The fact that this consistent mode of differentiation is stronger than that differentiating other attacks from erotic pursuits must be due to the fact that the two schemes with which we are concerned have the same (mythological) male protagonist; and this, which pulls them nearer each other in one way, creates the (semiological) need to push them further and more strongly apart in others. This consistently emphatic mode of differentiation reveals that a differentiating parameter determined the manipulation of the two closely related iconographical schemes which shared a common schema-matrix; and this shows that, in the sets of assumptions through which the fifth-century Athenian painters operated, the two themes signified through the iconographical schemes 'Theseus with a sword' and 'erotic pursuit' were very different; and this semantic divergence gave rise to a strong differentiating parameter which entered the sets of assumptions shaping the creation of images and determined the relationships between the two iconographical themes in the various representations. There is one exception to the differentiating mode I have just described. The Hasselmann Painter differentiates only through the fundamental divergence (in its intensified form) and the clustering of smaller divergences. This shows a relative decrease in differentiation, probably due to a deintensification of the differentiating parameter resulting from the routinization of the two iconographical schemes. But even in this case the differentiation is not smaller, but the same as that between the pursuits and the other attacks in our control scenes. Type 2 erotic pursuits are even more strongly differentiated from 'Theseus with a sword'. It is the central divergence that is very much stronger here: 'drawn sword'—'absence of weapons', except sometimes for the sword in its scabbard which is both inactive and low profile.

In these circumstances, we conclude that the second, independent, part of the investigation of the first hypothesis confirms the conclusions of the (independently conducted) analysis of the first part. And this shows conclusively that this first alternative articulation of the suggestion that 'Theseus with a sword' may represent an erotic pursuit is untenable.

III. Second hypothesis

The second hypothesis acknowledges the fundamental difference between the iconographical themes 'Theseus with a sword' and 'erotic pursuit', recognizes that the former represents an attack while the latter does not, but maintains that 'Theseus with a sword' can represent erotic pursuit. It is here that belongs the suggestion that the attacking demeanour of some erotically pursuing gods[62] provides support for the view that erotic pursuits by heroes (or mortals) can also take the iconographical form of an attack. This case depends on fallacious methodological premises. For the existence of the iconographical scheme 'divine pursuer as attacker' can only provide an argument for postulating an equivalent scheme 'heroic/mortal pursuer as attacker' if we assume that there is a fixed meaning 'erotic pursuit' which remains unchanged even though the identity and nature of the pursuer change; or, alternatively, if we assume that in fifth-century Athenian mentality the semantic field 'god' had the same value as that of 'hero'. The notion that the meaning of one element can remain unchanged while the composition of the others has changed is—as is clear from Chapter I—untenable on semiological grounds. If it is correct—and we shall see that it is—that the value 'god' is different from that of 'hero', the semantic field 'erotic pursuit by a god' is necessarily also different from the semantic field 'erotic pursuit by a hero'. It is not 'the same' semantic field, with 'only' the identity of the pursuer changed. This is a theoretical deduction, deriving from the properties of the sign and the nature of the process of signification. Its validity can also be demonstrated through an independent argument.

Since iconographical schemata are read through culturally determined assumptions and expectations, and the representations of erotic pursuits by deities activated the whole semantic field 'erotic pursuit by a deity', we must not assume that the representations of divine pursuits resembling erotic pursuits by a hero produced the same meanings as the latter. The (relative) autonomy of the signifier (cf. n. 11) entails that the meanings emphasized by this scheme were different from those produced by the 'attacking erotically pursuing gods' variant. But the latter, which were part of the wider semantic field 'erotic pursuit by gods', and dominant in the 'attacking' variant, could not be totally excluded from the reading of the other variant.[63] In any case, even if we leave aside these particular polarized meanings, the fact that the sign 'deity' had a different value from the sign 'hero'[64] entailed that different meanings were produced in the two cases.

This demonstrates the limitations of purely formal analyses: they can only tell us that erotic pursuits by deities appear in two iconographical variants, 'not-attacking' and 'attacking', while erotic pursuits by heroes are only known to us in the former; but the interpretation of this pattern is vulnerable to culture-determination. Of course, semiology should guard us against the fallacy that the meaning of erotic pursuit can be 'the same' when the nature of the pursuer changes, and thus against assuming that the existence of the two variants in one case allows us to postulate a second variant also in the other. But only the consideration of the relevant semantic fields and assumptions can tell us whether or not 'god' and 'hero' differ in ways that affect significantly the notion of erotic pursuit; and thus whether or not the second iconographical variant of divine pursuit produces (and expresses) meanings which are contained also in the semantic field 'hero's pursuit', and can thus be postulated for that pursuit also.

Heroes stand between men and gods, and it depends on the context towards which of the two poles they will drift.[65] While in cultic contexts, for example, the heroes drift towards the pole 'gods' as cult-recipients, in opposition to men who are cult-givers, in the context of mythological themes such as that of erotic pursuit they drift towards the pole 'men'. For the erotic pursuits involved took place in the heroes' lifetime, not after their death and heroization, and they did not involve contact with 'the other world' or its powers. I argue in Chapter II.2 that the theme 'erotic pursuit by Theseus' functioned also as a mythological paradigm with special reference to ephebes and youths, and carried meanings pertaining to the relationship between the sexes and marriage. The implicit intimations of violence carried by this theme correspond to the 'violent' aspects of these relationships.

As for the gods, the myths represent erotic contact between gods and mortals as extremely dangerous for the latter. This notion is sometimes expressed explicitly in the myths through the motif of the sexual encounter with the god leading to the girl's death, Semele being the most explicit example.⁶⁶ Another motif expressing the same notion that erotic contact with deities is dangerous for mortals is that in which the result of the erotic contact is that the woman loses her humanity and drifts to the other pole of animality, like Io and Kallisto.⁶⁷ In other myths erotic contact or even simple pursuit, with the girl escaping actual erotic contact with the god, leads to her being turned into a plant.⁶⁸

In the heroic times in which the myths are situated, 'normal' contact with the divine included social contacts with the gods. The representation of erotic contact between gods and mortals expresses a more intimate contact, the most intimate contact possible, and is thus, among many other things, a metaphor for excessive contact between gods and humanity. In Greek religious mentality (even leaving aside the dangers inherent in all excess) excess in the contact with the divine involves a transgression of an important principle—the limits of humanity—and is therefore negative, as is excessively low contact. Thus the mythological representation of erotic contact between gods and mortals leading to loss of the human condition through death or loss of humanity is correlative with, and articulates in narrative terms, the belief that excess of contact with the divine transgresses the proper relationship between men and gods, and transgresses humanity's circumscribed limits,⁶⁹ which this representation helps express and define. It is, then, a religious representation pertaining to the relationship between men and gods, and of humanity's circumscribed limits. Of course, this is only one of the semantic facets of this polysemic mythological theme; there are others, and they also function as narrative articulators, organizing centres shaping the articulation of the mythological narratives in the various mythological versions. As a result, not all myths involving erotic contact between gods and mortals give narrative expression to the notion of danger and the mentality which underlies it. In some, which put the emphasis elsewhere (for example on the subsequent fate of the deity's son)⁷⁰ the danger is not articulated in narrative terms, though it is implicitly there since it expresses an important aspect of Greek religious mentality.

This dangerous and threatening aspect of the erotic contact between deities and humans corresponds precisely to the iconographical scheme of erotically pursuing gods represented as attackers. Consequently, the latter scheme must represent this particular perception of divine erotic pursuit as dangerous and threatening. That these representations of the attacking erotically pursuing gods signify death is not a new (or *ad hoc*) hypothesis; it has already been suggested by Hoffmann.[71] I would stress that this 'divine erotic pursuer as attacker' scheme is correlative with aspects of the semantic field 'erotic pursuit by gods' which express important aspects of the Greek religious mentality; the representations showing the divine erotic pursuits according to the same scheme as the heroic ones put the emphasis on different facets of that semantic field, in the same way as many literary narratives do.

Consequently, the iconographical theme 'erotically pursuing attacking gods' corresponds exactly with, and must be seen as signifying, a facet of the semantic field 'erotic pursuit by gods' which was not, and could not be part of the semantic field 'erotic pursuit by heroes'—or mortals. Thus the corresponding iconographical scheme of the divine erotic pursuit as an attack could not be used to represent an erotic pursuit by a hero—or a mortal. On the contrary, precisely because 'erotic pursuit by a hero' is also determined and defined through its differentiation from other, related but different themes (in this case 'erotic pursuit by a god'), we should expect a differentiating tendency in the representations, a denotation 'not attack' differentiating erotic pursuits by heroes from those by gods. And this is indeed what we do find. For it is in this context that we must understand the use of motifs stressing that the spears are not represented here as weapons in use in an attack. This confirms our conclusions and demonstrates unambiguously that the fact that some erotically pursuing gods are shown as attackers not only does not provide support for the view that erotic pursuits by heroes can be represented through the metaphor of an attack, but actually helps invalidate that notion. The same differentiating parameter distancing divine and heroic pursuits operated at the level of reading. Thus (even without the explicating motifs stressing the not-attack character of the spears) the theme of erotic pursuit was understood as 'not attack'. For in the fifth-century assumptions, first, danger (expressed as attack) characterized divine pursuits which it differentiated from heroic ones; and second, the spears were part of the iconographical scheme 'ephebe' and 'Theseus as ephebe'.

Consequently, the hypothesis that heroic erotic pursuits could be represented as attacks in fifth-century Attic iconography is untenable. Therefore 'Theseus with a sword', which is an unequivocal attack, could not possibly represent an erotic pursuit.

3. CONCLUSIONS

In these circumstances, and given that the results of our independently conducted sets of analyses coincide totally, thus confirming each other's validity, we can confidently conclude that the theme 'Theseus with a sword' does not represent an erotic pursuit.

There are some correlations between the two themes. In each Theseus (the Athenian hero and paradigm for Athenian male) is 'confronting' and pursuing a woman who seeks to escape and whom, on one way or another, he will 'defeat'—a favourite motif of Athenian men who (at least to some extent and some level) feared and distrusted women. Another possible semantic connection between the two themes is this. In 'erotic pursuit' Theseus is confronting a future sexual partner and potential wife. If the woman attacked by Theseus is Medea as I suggested,[72] in the scenes showing 'Theseus with a sword' Theseus (the paradigm for Athenian ephebes) is confronting a censored version of the mother, whom (in the fifth-century Athenian collective representations) every ephebe had to overcome and 'defeat', in order to become fully male and join the world of men.[73]

It is, I believe, because of this semantic facet that Medea is not shown in Oriental dress in this episode,[74] even after that dress became a regular part of her iconography: it would have produced 'noise', interference in communication, and hindered the production of this—not necessarily consciously articulated—important nexus of meanings. If I am right, Medea who is attacked and defeated by Theseus in these scenes is at the same time the mother, the (dangerous) female, and the symbol for the Persians, in a rich and polysemic process of meaning-production involving fundamental Athenian perceptions and ideas which were not necessarily all part of a coherent whole, but operated at different levels of signification. To give her Oriental dress would have entailed blocking her signification of 'woman' and 'mother' in general, and reducing her—and the scene—to a monosemic propaganda symbol for the Persian wars; to put it differently, it is because in the artists' perceptions she stood also, in this episode, for 'woman' and 'mother', that the selection 'Oriental dress' was blocked for them.[75]

NOTES TO CHAPTER II.1

1. Bérard 1980: 616–20.
2. Bérard 1980.
3. Sourvinou-Inwood 1979: 41 and 70 nn. 147–50, with references.
4. Cf. Sourvinou-Inwood 1979; cf. also below sect. 3.
5. For references to relevant representations cf. Sourvinou-Inwood 1979 nn. 147, 149.
6. Cf. e.g. Theseus and the Minotaur: skyphos New York x. 22.25 (GR 585): *ARV* 559.150; the death of Argos: column-krater Oxford 527 side A (CVA Oxford 1, pl. 23.3).
7. Cf. e.g. the neck-amphora New York 41.162.155 (*ARV* 502.14).
8. On the Tyrannicides' statues cf. Brunnsåker 1971. On the Tyrannicides' imagery and its ideological functions cf. also Bérard 1983: 27–31.
9. On this iconographical theme cf. O. Touchefeu, *LIMC* i, s.v. Aias ii; Moret 1975: 11–26.
10. Cf. e.g. the cup Ferrara T. 264 (*ARV* 1280.64; *Add.* 178) side A.
11. On the relative autonomy of the signifier: Kristeva 1981: 267; for this concept in the work of structuralist and post-structuralist thinkers: Sturrock 1979: 15; Bowie 1979: 127–8; Culler 1983: 189–90.
12. *LIMC* i. 344, no. 64.
13. Cf. e.g. *LIMC* i, pl. 263, no. 65.
14. On codification cf. Guiraud 1975: 24–5.
15. On motivation cf. Guiraud 1975: 25–7.
16. Cf. e.g. for type 1, pedestal of lebes gamikos in the Robinson Collection (CVA Robinson Collection 2, pl. 51a–c); for type 2, stamnos Villa Giulia 5241 (*ARV* 484.9).
17. I concentrate on the youth, since, we saw, the similarities in the female figure are far less significant, given her passive role and the character of the sign 'fleeing woman' which is involved. I discuss the sign 'fleeing woman' further in Ch. II.2.
18. The differences between the Painter of the Yale oinochoe's representations of type 1 pursuit are fewer than those we find in some other painters' work; but such differences are never many.
19. Cf. e.g. in the work of the Phiale Painter the left to right direction of the pursuit in the Nolan amphora Bonn 77 *ARV* 1015.12 with the right to left direction in the Nolan amphora Yale 134 (*ARV* 1015.13).
20. Cf. e.g. in the two scenes by the Phiale Painter (cf. n. 19) the striding youth of Bonn 77 to the running pursuer of Yale 134.
21. Cf. e.g. in the work of the Shuvalov Painter the oinochoe Bologna 346 (*ARV* 1207.21; *Add.* 169) with a fleeing companion to the oinochoe Ferrara sequestro Venezia 2505 (*ARV* 1206.3; *Para.* 463; *Add.* 169) which shows the father instead, and to the neck-amphora Mykonos 1424 (*ARV* 1209.54) without additional characters.

22. Cf. e.g. in the work of the Phiale Painter, the Nolan amphora Syracuse 20537 (*ARV* 1015.16) to the lekythos Laon 37.951 (ex Lambros) (*ARV* 1021.106).
23. The visual differentiation is increased by the divergences pertaining to the two women: the one on B is wearing a diadem and her hair is falling on her shoulders.
24. Cf. e.g. the pelike Leningrad 728 (St. 1633) (*ARV* 843.131; *Para*. 516; here Pl. 7).
25. Cf. e.g. type 1, the cups Carlsruhe 59.72 (*ARV* 883.60; 1673), and type 2, Philadelphia, American Philosophical Society (*ARV* 880.3), by the Penthesilea Painter. Or, in the work of the Sabouroff Painter, the type 1 pursuit on the Nolan amphora Warsaw 142334 (ex Czartoryski 52) (*ARV* 842.117) and the type 2 pursuit on the pelike Leningrad 728 (St. 1633) (*ARV* 843.131; *Para*. 516; here Pl. 7).
26. There are also further variations in emphasis (e.g. a consensual element brought in, or left out) within each type, and the same variations cut across types; i.e. the same range of variations was applied to and served for both—which confirms further their close semantic similarity.
27. Cf. e.g. the cup Louvre G 22 (*ARV* 151.52, sides A and B).
28. e.g. Louvre G 423 (*ARV* 1064.6).
29. Cf. e.g. Leningrad 709 (*ARV* 487.61; *Para*. 512).
30. Cf. e.g. the bell-krater Leningrad 777 (St. 1786) (*ARV* 502.11; here Pl. 5).
31. Cf. e.g. the calyx-krater Geneva MF 238 (*ARV* 615.1; *Para*. 397).
32. Cf. e.g. the stamnos Oxford 1911.619; *ARV* 629.16; here Pl. 1.
33. Cf. e.g. the stamnos Krefeld Inv. 1034/1515 (*ARV* 502.5; *CVA* Germany 49, pls. 37.1, 38.1.).
34. Cf. e.g. Oxford 1911.619; here Pl. 1.
35. A few random examples of spears used as an offensive weapon in the course of fighting: neck-amphora London E 285 (*ARV* 530.25); oinochoe Louvre C 10729 (*ARV* 1160.2; *Add*. 166); calyx-krater New York 08.258.21 side A, below (*ARV* 1086.1); volute-krater Palermo Museo nazionale G 1283 (*ARV* 599.2. 1661); cup Louvre G 115 (*ARV* 434.74); volute-krater London E 468 (*ARV* 206.132; *Add*. 96). And cf. also a paradigmatic stance of attack with spear: that of the statue of a striding warrior being made in the foundry on the cup Berlin 2294 (*ARV* 400.1; 1651; 1706; *Add*. 114). The similarity between some marginal and inactive figures in a few mythological collective hunt scenes and some of our pursuers does not invalidate this. For example, on the volute-krater Louvre G 343 (*ARV* 600.17; Durand and Schnapp 1984: 64–5, fig. 98) the figures in question (one on the extreme left, the other on the extreme right—on the arrangement of the figures see Durand and Schnapp 1984: 65–6) are

not taking part in the attack, nor will they do so, since the animal has been transfixed by spears, which signifies death; thus they are not shown attacking the animal, but approaching the action. The same is true of the second youth on the left, whose non-participation in the action is shown by his holding two spears and a club in his left, all parallel to the ground. The stance of the two youths attacking the animal and the position of the spears have some similarities with that of the pursuer in some scenes—due primarily to the fact that the hunters have been represented in the heroic codified stance of attack recalling Aristogeiton, in order to stress their heroic status (cf. Durand and Schnapp 1984: 65–6); but their meanings, determined also, and mainly, through their relationship to the animal, are radically different.

36. Prof. A. M. Snodgrass kindly tells me that in some very early representations of combat (cf. Lorrimer 1947: 95, cf. 97, fig. 8; I owe this reference to Professor Snodgrass) the spear can be inferred as being held on the left, but this is motivated by compositional considerations or sheer confusion.
37. Cf. e.g. the stamnos London E 446 (*ARV* 502.4; here Pl. 4).
38. Cf. e.g. neck-amphora Berne 12214 (*ARV* 1148).
39. Cf. Chicago University hydria fragment by the Agrigento Painter (*ARV* 579.86); pelike once Coghill near the Phiale Painter (*ARV* 1024.1); oinochoe Ferrara sequestro Venezia 2505 from Spina by the Schuvalov Painter (*ARV* 1206.3; *Para.* 463; *Add.* 169); Lekane fragment in Leningrad (*Compte rendu de la commission impériale archéologique* 1877, pl. 5b—thanks to the generosity of Dr I. Saverkina I am able to reproduce a photograph of this fragment as Pl. 8); Cup Louvre C 10932 (*ARV* 837.6).
40. *ARV* 260.8; Boardman 1975: fig. 203.
41. Cf. e.g. cup formerly in the Basle market Kaempf-Dimitriadou 1979: no. 93, pl. 7.3, p. 84; calyx-krater London E 466 Kaempf-Dimitriadou 1979: no. 117, pl. 9.1–2, p. 86. Cf. also Kaempf-Dimitriadou 1979: 18. The hostility, fear, and resistance of the youths/boys pursued/abducted by Eos are correlative with the fact that this abduction signifies—among other things—death (cf. Roberts 1978: 179–80).
42. *ARV* 402.16; *Add.* 114.
43. *ARV* 998.161; *Para.* 438; *Add.* 152.
44. *ARV* 1143.2; *Para.* 455; *Add.* 164.
45. Cf. e.g. side B of the volute-krater Oxford 525 (*ARV* 1562.4; here Pl. 2); stamnos Krefeld Inv. 1034/1515 (*ARV* 502.5; CVA Germany 49, pls. 37.1; 38.1); calyx-krater in the Lucerne market in the manner of the Niobid Painter (*ARV* 1661); neck-amphora London 1928.1–17.58 (*ARV* 1010.5; CVA, pl. 59.3).
46. Cf. the volute-krater Bologna 275 (*ARV* 1029.18).

47. Cf. cup Ferrara from Spina T.264 (*ARV* 1280.64; 1689; *Add.* 178).
48. *ARV* 1213; *Add.* 172.
49. Cf. e.g. the hydria Florence 4014 (*ARV* 1060.144).
50. Cf. e.g. the pelike Leningrad 728 (St. 1633) (*ARV* 843.131; *Para.* 516; here Pl. 7).
51. For the semantic analysis of the theme 'erotic pursuit' cf. Ch. II.2.
52. Cf. a few examples: Nolan amphora Warsaw 142334 (ex Czartoryski 52) (*ARV* 842.117); calyx-krater New York 41.83 (*ARV* 1012.3); pelike Naples RC 155 (*ARV* 1079.3); Nolan amphora Munich 2334 (J. 257) (*ARV* 1081.9); oinochoe fragment Louvre 11029 (*ARV* 1167.113; *Add.* 166).
53. Cf. e.g. for Theseus: the fragment of a hydria Malibu S80.AE.185 (Robertson 1983: 65, fig. 17); for ephebes: calyx-krater Oxford 1924.929 (*ARV* 1056.88); oinochoe Marseilles Mus. no. 2093 (inv. Roberty 3593) (*REA* 42 (1940), pls. ID, II); bell-krater Louvre A 488 (*ARV* 1067.2); the lekythoi Oxford 1938.909 and 1920.104 (*ARV* 993.93–4).
54. Cf. Pélékidis 1962: 231–2.
55. Dr M. Schmidt kindly mentioned that she also thinks that the pursuers are holding spears in erotic pursuits simply because they are part of their equipment and characterize them.
56. I am, again, concentrating on the youth.
57. The woman on A is far more majestic, which would fit well with her being Medea (cf. sect. 3), a (somewhat older) queen, differentiated from the nubile girl on B. Perhaps the difference between the two women and, especially, divergences 2 and 3 between the youths pertain to the identity of the latter; while on A the petasos, sword, and long hair are part of the characterization of the youth as Theseus, the absence of these elements from the scene on B may be correlative with a conception of the pursuer as 'primarily' (cf. below, Ch. II.2) a generic youth. If this tentative suggestion is correct, the erotic pursuit here represents not the mythological paradigm, but the genre version.
58. Cf. above 2 II (*a*) ii.
59. Sourvinou-Inwood 1979: nos. 8 (on which cf. below) and 33; here Pl. 4.
60. On the motif of grabbing by the hair in archaic and classical Greek iconography cf. Moret 1975: 191–225.
61. Cf. e.g. the grabbing on Pl. 3 (pelike Manchester iii.I.41; *ARV* 486.42) and on Pl. 7 (pelike Leningrad 728; *ARV* 843.131; *Para.* 516).
62. Cf. e.g. the Nolan amphora London E 313 (*ARV* 202.87; *Add.* 96 (Zeus)); column-krater New York 96.19.1 (*ARV* 536.5; *Add.* 125 (Zeus)); lekythos New York 17.230.35 (*ARV* 1020.100) (Poseidon)); hydria in the Peiraeus Archaeological Museum, *AR* for 1981–2, 11, fig. 14 (Zeus). On the iconographical scheme of erotically

pursuing gods in an attacking pose cf. also Kaempf-Dimitriadou 1979: 22.
63. Cf. for a related approach to the reading of such images: Schefold 1975: 94, on the hydria London E 170 (*ARV* 1042.2; *Add.* 156) which represents Apollo pursuing Daphne.
64. Different deities had different values; but this does not concern us here.
65. On this concept of drift cf. King 1983: 110–11, 124–5, 125 n. 2.
66. Cf. e.g. Koronis, Ariadne in one of the versions of her myth (e.g. Hom., *Od.* 11.321–5), Leukothoe. On mortal men's sexual encounters with goddesses leading to their death cf. *Od.* 5. 118–28; to other harm: Hom. *H. Aphr.* 189–90. Possibly already in the 5th cent. some representations of divine abductions may have been perceived as signifying also death: on the abduction of youths and boys by Eos signifying death: Roberts 1978: 179–80; on Boreas' abduction of Oreithyia signifying death as well as marriage: Roberts, p. 179, with bibliography, to which add Simon 1967: 117 (on this myth cf. also Ch. II.2 n. 21). On the association between erotic contact with gods and death cf. also Kaempf-Dimitriadou 1979: 25, 175 n. 65.
67. These myths are complex, and it is beyond my scope to discuss them here; I only want to mention that, of the many themes which interact and articulate narratively the myth in its various variants, the theme of erotic contact with a god is here correlative with abnormality, animality, and death. On Kallisto cf. the discussions in: Bodson 1978: 136–9; Borgeaud 1979: 48–55, 60–1; Henrichs 1987: 254–67.
68. Cf. e.g. Daphne, Pitys, Syrinx (on the last two cf. Borgeaud 1979: 123–5).
69. On this type of transgression of humanity's limits cf. Buxton 1980: 34–5.
70. Cf. e.g. the story of Iamos' birth in Pind., *Ol.* vi. 35 ff.; or indeed the myth of Heracles. Another narrative articulator may e.g. be the image of the rape of a woman by a god as an image of divine possession (see Padel 1983: 12–14, 16). This motif is not unrelated to the semantic facet of this theme discussed here in which the rape of the woman is also seen as a metaphor for the invasion of the human by the divine.
71. Hoffmann 1980: 748–9.
72. Sourvinou-Inwood 1979: *passim.*
73. On this see Zeitlin 1978: 149–84.
74. The criticism that if she had been Medea, and if—as I have argued—'Theseus with a sword = Theseus attacking Medea' symbolized the Athenian victories over the Persians, she would have been shown in Oriental dress, was made against my hypothesis by Bérard (1980: 620) and Brommer (1982: 134 n. 21).

75. I offer some further arguments in favour of the thesis that the theme 'Theseus with a sword' represents Theseus attacking Medea in a paper entitled 'Myths in Images: Theseus and Medea as a Case Study', in: L. Edmunds, ed., *Approaches to Greek Myth* (Baltimore and London, 1990), 395–445, where I also discuss the mythological representations 'stepmother' and 'bad mother'.

II.2

A Series of Erotic Pursuits: Images and Meanings

1. INTRODUCTION

The focus of this chapter is a series of representations depicting a youth, with or without spears, pursuing a girl who is fleeing before him (Pls. 1–2, 5, 7–10). I call the pursuits in which the youth is carrying spears 'type 1' and those in which he is not 'type 2'. I also discuss various matters pertaining to girls, marriage, and the relations between the sexes, and to myths such as Peleus' capture of Thetis.

I have discussed aspects of the theme 'erotic pursuit' in Chapter II.1, and argued, first, that the spears carried by the pursuer in representations of type 1 (as part of the iconographical scheme characterizing ephebes in general and Theseus in particular) carry implicit, muted, connotations of violence and menace (but do not denote attack); and secondly, that types 1 and 2 are closely related, and produce meanings which only differ in emphasis. The intimations of (unstressed) violence produced by the spears in type 1 are also carried by the theme of erotic pursuit itself; the spears only increase the emphasis. Moreover, the motif of the capture of the pursued girl, signified in our theme through the iconographical motif 'grabbing the girl' (found in both types), produces less muted 'violent' meanings, similar to those produced by the spears.[1]

2. THE SUBJECT

The interpretation of the representations of a youth pursuing a girl as erotic pursuits is certain. It is based on the similarity between this iconographical scheme and the scheme observed in other scenes of pursuit which, because their story is known, can be identified with certainty as erotic. In my view, generic similarities do not securely establish such an identification[2] but there is a truly close parallel to our pursuits which allows us to identify them as erotic: the representation of Peleus' pursuit of Thetis. I discuss this theme's close similarity to ours below. There are other reasons too for identifying our pursuits as erotic. The iconographical scheme of type 2 is similar to a variety of erotic pursuits involving different beings[3] and the same as the scheme representing Eros pursuing a woman,[4] an undoubtedly erotic type of pursuit. This demonstrates that our scenes represent erotic pursuits. For what the Eros scenes show is surely not a mythological incident, either an incident from Eros' own love life or one in which he is acting on behalf of another deity.[5] As Durand has convincingly argued, the representations which show Eros performing activities elsewhere performed by youths, through the same iconographical schemes, articulate the notion that those youths' actions were performed under Eros' power.[6]

Thus the identification of our scenes as erotic pursuits is certain. The analyses that follow will show that the meanings produced by these scenes correspond exactly to the semantic field 'erotic pursuit' in the semantic universe of fifth-century Athens.

3. IDENTIFICATION OF THE PROTAGONISTS

First a basic question: does it matter who the protagonists of these pursuits are? Is the search for names and identities not clinging to outdated modes of research and should we not rather concentrate on the images and situations and try to recover their meanings? In my view, the identities are important. For meanings are inscribed and read into the images through signs; and each mythological figure was a sign, carrying specific connotations which contributed to the creation of the image's complex meanings.

Here I am only concerned with the identity of the lone pursuer when he is a youth wearing a chlamys—sometimes also a chiton, rarely a chiton without chlamys—with or without petasos, with or without spears. The other variants, involving more than one youth, or a young warrior, or other types of pursuer, are considered briefly below. The identity of the pursuer in the scenes at the centre of our investigation is considered controversial by some scholars,[7] but it can be demonstrated that when there is one pursuer, and the scene is mythological, in the absence of an additional sign specifying a different identity, that youth is always Theseus. Three inscriptions in three different representations identify the pursuer as Theseus. First, the lekane fragment in Leningrad (*Compte rendu de la commission impériale archéologique* 1877, pl. 5b; here Pl. 8) in which the youth is inscribed 'Theseus' and the pursued girl 'Thetis'; second, the hydria Worcester (Mass.) 1903.38 (*ARV* 1060.143); and third, the bell-krater Louvre G 423 (*ARV* 1064.6).[8] The pursuits on the Leningrad fragment and on the Worcester hydria[9] belong to type 1, that on the Louvre krater to type 2. In both types the pursuer inscribed 'Theseus' is represented according to the same iconographical scheme as in the rest of the series. The Leningrad pursuer inscribed 'Theseus' is represented according to one of the commonest iconographical schemes used for the pursuer in our pursuits. He wears a chlamys, endromides, and a petasos thrown at the back of his neck, and he is carrying two spears.[10] The motif of the spearheads being turned away from the girl, though not common, is also found elsewhere.[11] The Louvre pursuer is also represented through a scheme common in the series. His stance and grabbing of the girl are common in pursuits of types 1 and 2.[12] His dress, hair, and gear are widely paralleled in pursuers of both types. The chlamys worn as the only garment is the overwhelmingly common dress for our pursuer; the arrangement here, chlamys thrown over the outstretched left, is common in both types.[13] The short hair characterizes several pursuers as does the absence of petasos.[14] The wreath worn by the Louvre Theseus is also worn by other pursuers.[15] Finally, the Louvre Theseus has a sword in its scabbard hanging at his side, as do some other pursuers in both types of our series.[16] The sword is closely connected with Theseus through the stories of the *gnorismata* and the recognition by Aigeus[17] (thus it is especially linked with Theseus' ephebic persona), and this connection is reflected in the hero's iconography. Theseus on the Worcester hydria is also represented through a scheme common in the series: he is wearing a chlamys (which covers his outstretched left) and a petasos thrown at the back of his neck; he is carrying two spears in his right. We conclude that all the elements making up the sign 'Theseus' in the inscribed scenes are closely paralleled in the rest of the series.

The pursuer can also be identified as Theseus with some certainty on some uninscribed scenes, through contextual associations. For example, the cup Frankfort, Museum V.F., X 14628 (*ARV* 796.117) is decorated with the representation of Theseus and Skiron on the tondo and of a youth with a spear pursuing a woman on B. Theseus in the tondo is identical with the pursuer on B. One indication of this is that both are wearing the petasos on their head, a much rarer arrangement than having it thrown at the back of the neck. Both youths (wearing chiton and chlamys) are holding one spear; the way in which they hold it is different, but this depends on the context. In the tondo Theseus is holding the spear vertically in his left and very lightly leaning on it; the pursuer on B is holding his spear in the right and horizontally, one of the regular positions in which spears are carried when the figure is moving. This pursuer is undoubtedly Theseus: unless the painter had been thinking of the two youths as the same person he would not have used the same sign for both on the same cup; and a viewer looking at the pursuer who was identical to the juxtaposed Theseus, and shown in a role, story, and iconographical scheme in which Theseus at least sometimes appeared,[18] would inescapably identify that pursuer as Theseus. Leaving aside other cases in which the identity of the youth as Theseus is established through various types of contextual associations, we note that in all our erotic pursuits the pursuer is represented through the scheme which in Athenian iconography characterizes Theseus—though, we shall see, not only Theseus. In the vast majority of scenes the pursuer is wearing a chlamys, usually on its own, sometimes over a chiton. Chlamys, or chiton and chlamys, characterize Theseus in fifth-century Attic iconography,[19] with the sword and the spears and a hat (either a petasos or a pilos)[20] completing the schema. When a hat is worn in our pursuits it is normally the petasos, more rarely the pilos—as on the Chicago University hydria fragment (*ARV* 579.86). The fact that both petasos and pilos were appropriate for our pursuer confirms further that the mythological hero of the pursuit is Theseus. Readers produce meanings out of pictures with the help of their assumptions and expectations, and the mythological hero of these pursuits would certainly be identified as Theseus—*unless additional evidence was given to the contrary*. The youths named 'Theseus' in the inscribed scenes cannot be iconographically differentiated from the pursuers in the uninscribed scenes, and the iconographical schemes of the inscribed scenes cannot be differentiated from those in the uninscribed; it follows that without such additional information the viewers understood the pursuing youth to be Theseus because *that is* who he was in the established scheme, and thus also in the sets of assumptions through which they read these images.

The conclusion that the mythological male protagonist of these pursuits is Theseus is confirmed rather than impugned by the existence of scenes representing Peleus pursuing Thetis according to the same scheme of erotic pursuit as that of our series, and with Peleus shown according to the same scheme as our pursuer. For there *is* such an additional element in these latter scenes: the sign 'dolphin' which identifies them as representations of the abduction and capture of Thetis by Peleus. For example, on the pedestal of a lebes gamikos in the Robinson Collection (CVA Robinson 2, pl. 51a–c) a dolphin is added at the end of the picture, identifying the pursuit as that of Thetis. And on the stamnos Villa Giulia 5241 (*ARV* 484.9) the representation is identified as the pursuit of Thetis through the addition of two dolphins, one of them held by a fleeing companion of the abducted girl, thus identified as a Nereid. That is, through the addition of this sign the iconographical scheme which usually depicts Theseus' pursuit is transformed, with great economy, into the erotic pursuit by Peleus.[21] The pursuit of Thetis by Peleus is represented both through erotic pursuits of type 1 in which Peleus is carrying spears, and of type 2 in which he has no spears.[22] Apart from the dolphin, our pursuits and the pursuit of Thetis by Peleus are identical. The fact that the pursuit/abduction of Thetis by Peleus could be thought of, and so shown, also in the same terms as the pursuit of a girl by Theseus suggests the possibility of a semantic similarity between the two. We shall see below that there was indeed such a similarity, despite the fact that Thetis is a goddess, and Theseus pursues erotically on his own behalf only mortal girls (he only took part in the attempt to abduct Persephone to help his friend Peirithous). As we shall see, in the iconographical versions which resemble Theseus' pursuit, Thetis' divine identity is 'neutralized'; other meanings and aspects of this myth are emphasized, so that the goddess' capture by a mortal is semantically close to a mortal's capture by a mortal. The close similarity between the pursuits by Theseus and Peleus is presupposed by, and explains, the inscriptions on the Leningrad lekane fragment (Pl. 8), where the pursuer is inscribed 'Theseus' and the pursued 'Thetis'. Beazley[23] took this to be a slip, and this is, indeed the most likely explanation. But for this to have been possible, the two themes had to be iconographically and

A Series of Erotic Pursuits   63

semantically closely related, so that they became momentarily scrambled, and the names of the two pursued girls were switched. Alternatively, it may be a deliberate conflation, some kind of joke or play. But this would also presuppose a close relationship between the two themes.

We may thus conclude that the mythological male protagonist of our pursuits is Theseus. But this theme did not, in my view, only signify the mythological narrative in which Theseus pursued and abducted a girl. All signs are polysemic, and moreover in fifth-century Athens myths and mythological scenes functioned also as paradigms. Theseus was, among other things, the Athenian ephebe *par excellence*. The chlamys, petasos, and spears which characterize the pursuer in our pursuits also characterized Athenian ephebes in general.[24] The chlamys was the characteristic garment of Athenian ephebes who also wore the petasos and in whose training and activities the spear played an important part.[25] Consequently, for both Athenian artists and viewers, the signifying elements 'youth' combined with 'chlamys', 'petasos', and 'spear(s)' made up the sign 'ephebe' and/or 'Theseus as ephebe'. The sign 'pursuer' in 'erotic pursuit' produced for fifth-century Athenians the meaning 'ephebe' as well as 'Theseus', especially since the context, which contributed to the definition of the sign 'pursuer', was of direct relevance to all ephebes. I shall discuss the meanings of our scenes below; but it is immediately clear that the erotic pursuit of a girl is of great interest—at least at the level of the imagination—to real-life ephebes; in fact the connections were deeper. So the scenes depicting the mythological narrative of Theseus' pursuit of a girl were at the same time read as 'Theseus as ephebe'/'ephebe' pursuing a girl, with all the meanings which such representations carried. In my view, it is likely that the 'generic' reading of the theme was eventually consciously articulated and iconographically established. For not all representations of erotic pursuit are mythological in the first instance. In some there are reasons for thinking that the protagonist is a generic youth;[26] this would make the representation in the first instance 'generic'—though when read through fifth-century Athenian eyes it would also refer to, and acquire its value through, the mythological paradigm of Theseus. By 'generic' I do not mean 'everyday life', but 'emblematic', expressing certain perceptions about women and male–female relationships. The differences between fully mythological scenes and this type of 'generic' scene are only a matter of emphasis, towards the mythological paradigm or towards the generic youth. All representations of the theme carried both components, and many were probably not positively identified as one or the other, but belonged to an indeterminate part of the semantic spectrum 'Theseus as ephebe–'generic youth'.[27] Ambivalence and ambiguity are characteristic properties of signs.

The identity of the pursued girl is less important than that of the pursuer, for since the latter is Theseus he is the dominant signifying element. But the girl's identity is not unimportant, for the particular connotations and meanings with which she was associated contributed to the production of the meanings inscribed and read into these images. There is no surviving inscription to identify her, except for the Thetis 'slip'. She may be someone unknown to us—probably not Helen.[28] But we can at least define her a little, determine the kind of person that she is likely to be. We shall see that the iconographical analyses of this theme suggest that it had three major semantic facets: the ephebic one; one pertaining to male–female relationships; and, a version of the second, a semantic facet pertaining to wedding and marriage. Thus the pursued girl is likely to be one whom Theseus married, and whose acquisition was associated with Theseus' ephebic persona. Not Antiope, whose iconography is distinctive. The best candidate is probably Eriboia/Periboia/Phereboia.[29] She fits both requirements, marriage to Theseus and connection with his ephebic persona (through the ephebic exploit of the Minotaur expedition). If Barron is right[30] that Pherekydes had suggested, in the context of Kimonian propaganda, that the Philaids' claimed ancestor Ajax was the son of Theseus and Eriboia, we can even identify a context conducive to the creation and promotion of an iconographical theme representing Theseus' erotic union with Eriboia. Of course, the representation was polysemic, not a monosemic propaganda poster, and thus the relevance of its non-political meanings to Athenians, especially to Athenian youth, would have ensured its popularity even after the political content had lost its importance.[31] Whether or not this theme originated in this context, the identity of the girl may have become 'evacuated', submerged under the generic persona 'girl pursued, abducted, and possibly married by Theseus', as a result of the dominant importance of Theseus and his erotic association with a large number of women, many of whom he abducted.[32] This created a signification space in Greek myth, 'woman abducted seduced (and married) by Theseus', in which the woman's identity was not of great importance.

4. Erotic Pursuit, Capture, and Marriage

The girl is represented through the iconographical schema which may be called 'fleeing woman': a young woman running away, head turned back towards the pursuer, making gestures of supplication and/or alarm. This is a common schema, a codified sign deployed in many different themes.[33] In each case it acquires its specific meanings through the particular context, through sets of complex interactive relationships—as, we saw in Chapter I, do all signs, all elements that make up a text or image. The schema 'fleeing woman' denotes 'girl/woman fleeing in panic'. This basic semantic core, and so also the iconographical sign 'fleeing woman', is applicable to several different situations involving women (cf. n. 33). The contexts give it particular meanings in each case. In our theme, we saw, 'fleeing woman' represents a girl attempting to escape the erotic attentions of an ephebe/Theseus. Its further connotations will emerge below.

In some erotic pursuits the fleeing girl holds a flower,[34] in others one of the companions does.[35] This flower refers to the motif of the girl's flower-gathering just before the abduction, and is thus connected with the theme of abduction and through it also with marriage.[36] Flowers are held by girls also in other abductions,[37] including pursuits of Thetis.[38] The motif 'grabbing the girl' (Pl. 7) denotes capture, and carries connotations of violence, defined by the context as sexual. This erotic colouring is produced first through the activation of the frame of reference 'real-life sexual grabbings', called up by the representation of physical contact between the youth and the girl at the moment of the capture which (the identification of the topic tells the viewers) will lead to the sexual act; and secondly, through the activation during the reading of the image of established iconographical schemes of sexual grabbing.[39]

Running is associated with male and female initiations[40] and the 'initiatory' dimension is an important semantic facet of 'erotic pursuit': we have seen this with regard to ephebes, and we shall return to it. Possibly, the notion of an ephebic test and victory was part of this ephebic facet of 'erotic pursuit': the capture of the girl was perceived as a test in the context of the ephebic experience, comparable to the capture of an animal.[41] For the representation of running in pursuit to capture the girl may call up another frame of comparison, the capture of animals, which therefore also helps ascribe meaning to our theme—through both differentiation and similarity. There are strong connections between hunting and Greek initiations (including the Athenian ephebeia);[42] moreover, we shall see, unmarried girls were thought to be partly 'wild' and partaking of animality. Consquently, an expedition involving the capture of a girl by an ephebe after a pursuit was correlative, in the Greek collective representations, with the notion 'capture of animals';[43] thus, when the Athenians read the images, the representation of the former also called up suggestions of the latter—at whatever level of consciousness. This capture is not, of course, a hunt. The iconographical relationship between animal pursuits such as that on the kyathos Brussels, Musée du cinq A2333[44] and hunting scenes may be compared to that between erotic pursuits and attacks.[45] Hunting an animal :: attacking a woman (= Theseus with a sword (cf. n. 45)). Capturing an animal :: capturing a woman (= erotic pursuit). Indeed, in Athenian mentality the capture of wild animals was generally associated with the erotic sphere.[46] This is connected with their perception that women, especially *parthenoi*, were partly wild, which was expressed metaphorically through the notion that women, especially *parthenoi*, had some animal traits.[47]

The notion of the girl as a wild thing to be captured and tamed through marriage (which is, in my view, one of the perceptions expressed in our erotic pursuits) is articulated more emphatically in the paradigm of Thetis. Thetis' metamorphoses at the moment of her capture, which included animal forms, express symbolically, among other things, the wild and partly animal nature of the unmarried girl.[48] Thetis' possession of special powers allows this animality to be articulated in narrative terms. This perception of the goddess as a wild thing tamed by Peleus, a paradigm for mortal brides, is correlative with the fact that the union of Peleus and Thetis is a paradigm for marriage and is not governed by the mentality governing other god—mortal unions, which includes negative and dangerous connotations.[49] Indeed, Peleus' pursuit of, and union with, Thetis, differs radically from all other pursuits and unions involving deities and mortals.[50] First, uniquely, the mortal is the pursuer; secondly, Peleus captured Thetis to marry her; thirdly, he captured and married her with the consent and encouragement of the gods, whose will he was executing, which makes him almost, in this context, temporarily, an 'honorary god'.[51] Thus, Thetis' capture and wedding were governed by the same mentality as unions among equals. Since the difference in status between pursuer and pursued is thus neutralized here, Peleus' pursuit of Thetis is semantically very close to Theseus' pursuit of the unknown girl; for in both the pursuer was an ephebic hero, and the iconographical scheme which characterizes Theseus is also appropriate for Peleus, who is an ephebe and a hunter.[52] Given these close similarities, the nuptial associations of the pursuit of Thetis[53] suggest the possibility that Theseus' erotic pursuit also contained a semantic facet pertaining to wedding/marriage. As we shall see, there are strong additional arguments for this hypothesis; there are iconographical elements in 'erotic pursuit' which suggest that this theme connotes also marriage/wedding; and erotic pursuits are sometimes combined with scenes pertaining to the nuptial sphere. A final argument for the hypothesis that 'erotic pursuit' also alludes to wedding and marriage is that in Athenian collective representations the semantic field 'erotic pursuit/abduction' was closely related to, and expressed also, the semantic field 'marriage'. But before arguing this by further analyses I want to consider what perceptions were expressed in and through the metaphorical relationship between marriage and pursuit/abduction.

We have seen that the meanings produced by the iconographical theme 'erotic pursuit' included connotations of sexually coloured violence: sexual intercourse is about to be forced on the girl; she will be submitted to sexual violence, defloration, and sex as an act of aggression and domination of women.[54] But there are also some elements in abduction scenes which indicate the girl's consent to her abduction.[55] As we shall see, similar elements are also found in representations of erotic pursuit. These elements, which conflict with, and deconstruct, the dominant facet which presents the act as imposed by force, depend on the same mentality which regarded abduction as a paradigm for erotic union and marriage in Greek myth and ritual. On this paradigmatic relationship depend a variety of phenomena. In myth Thetis' was not the only abduction which served as a paradigm for wedding/marriage. Persephone's abduction was also a nuptial paradigm,[56] and because Hades was the Lord of the Dead, this abduction also signified death. Within the matrimonial facet of the theme's semantic field that death was symbolic, the death of the girl, to give way to the wife and mother.[57] There are also ritual phenomena which correspond to the association between pursuit/abduction and marriage in myth. Mock-abduction was part of the wedding ritual in some places;[58] in others certain ritual acts—some wedding gestures like the lifting of the bride into the chariot by the bridegroom,[59] and the *cheir' epi karpo*[60]—whether or not residues of a mock-abduction,[61] manifest the same mentality.[62] In iconography schemata related to those of abduction sometimes represent wedding scenes.[63]

There are several related Greek perceptions which can be considered to have shaped, and to be expressed in, the metaphorical relationship of abduction and marriage. These perceptions pertain, first, to the (male mental) representations of women as subordinate, 'alien', and also, metaphorically, as animals to be tamed. Secondly, to the representations of male–female relations which include that of marriage as the final stage in the 'taming' of women[64] and the notion of the subduing of the female by the male—an attitude important to the Greeks in expressing the actual and ideal state of relations between the sexes, and also a polysemic signifier articulating other values.[65] Thirdly, to certain perceptions of marriage seen from the female viewpoint, pertaining especially to the violent wrenching of the girl away from the familiar world of her father's home by a stranger who will take her to an unfamiliar place and role, and including the representation of the wedding as frightening for the woman.[66] Finally, despite the (greater or lesser) absence of free choice of partners, in the official image of marriage erotic love was deemed to play an important role.[67] In my view, the erotic pursuit metaphor also reflected this representation: it presented legitimate marriage in terms of a wild erotic union.

5. VIOLENCE AND CONSENT

The dominant perspective of the iconographical theme 'erotic pursuit', which presents the erotic union as an act imposed on the woman by force, is both defined further (the 'violence' connoted is defined as sexual) and deconstructed through signifying elements producing an effect of consensual erotic intimacy. For example, the position of the hands of the pursuer and the girl on side B of the bell-krater Leningrad 777 (Pl. 5), with their fingers almost touching, their fingertips almost meeting, produces an effect of consensual erotic contact about to take place, which both defines as erotic the mutedly intimated violence, and deconstructs the dominant meaning 'violence about to be used on the girl who is about to be captured'. Another example of an effect of consensual erotic intimacy occurs in the type 2 pursuit on the cup in Philadelphia, American Philosophical Society (from Vulci: *ARV* 880.3), where the following elements help create consensual connotations: the grabbing, here shown in a version recalling iconographical arrangements in which the erotic relationship is consensual;[68] the arrangement which makes the glances of the youth and girl appear to meet; the youth's holding up his chlamys with his right, a gesture precisely reflecting that of the girl who is holding up her himation with her right and therefore creating an effect of intimacy by binding the two figures very closely; moreover, the gesture helps characterize the youth in terms of grace and intimacy rather than violence and force. Some of the pursued girls wear the mantle over their head, or partly over their head.[69] This element in itself is polysemic, and cannot define the scene in any particular way. However, in the context of this theme it may contribute to the nuptial allusion, by calling up the figure of the bride with the himation over her head.[70] But the comparative rarity of such consensual elements in erotic pursuits is correlative with the playing down of the semantic facet 'consent', for it is the wild rather than the cultural/institutional side of erotic relations and of marriage that is stressed in this theme. Conversely, because 'erotic pursuit' is, when compared to abduction, a gentler version of this wild marriage, in which the actual physical manipulation of the girl is not shown but at most hinted at through the grabbing motif, the consensual elements were less necessary to counterbalance the 'violent' facet. Still, it is important that the signification space covered by this theme included the semantic facet 'consent'.

In Greek, and especially Athenian, mentality about marriage and erotic love (as reconstructed from other evidence) both consent and force belonged to both abduction/pursuit and to marriage as a cultural institution.[71] Of course, force predominated in the former and consent in the latter; but at the same time, each was an image, representing a facet, of the other. Thus the duality and ambivalence of the iconographical theme 'erotic pursuit', which includes elements connoting force and others connoting consent (both directly and indirectly, through allusions to marriage), correspond to the semantic field 'erotic pursuit/abduction', which includes the relationship with 'wedding/marriage'. For in marriage also there is ambivalence and duality, and this is why pursuit/abduction is an appropriate metaphor for it. As Redfield[72] noted, marriage involves both men gaining control of women, and women giving themselves to men. It has a sexual facet which belongs to nature and deconstructs its character as a social and cultural institution.[73] The theme 'erotic pursuit' represents and crystallizes a perception of marriage which is drifting towards the former poles, nature and men gaining control of women.

6. SPACE, THE GIRL, AND ARTEMIS

In most erotic pursuits known to me there are no spatial indicators. The scenes are undoubtedly perceived as happening outdoors—either in a specific outdoor space known from the story and read into the scene by the viewers (whose assumptions included such knowledge and who, the story once identified, supplemented the missing elements from their knowledge); or in an unspecified outdoor space, unspecified either because the story did not specify it, or because it is 'emblematic', correlative with the emblematic and metaphorical facet of signification of the iconographical theme 'erotic pursuit'. The scenes which include spatial indicators will help us decide among these alternatives. The following indicators[74] are represented.

1. Column on its own: calyx-krater Geneva MF 238 by the Geneva Painter (*ARV* 615.1; *Para.* 397); column on the extreme right of the scene; the girl and a companion are fleeing towards it.

2. Door (with architrave): hydria London E 198 by the Niobid Painter (*ARV* 606.79); the door is on the left.

3. Column combined with door: volute-krater Bologna 269 by the Niobid Painter (*ARV* 599.8); the column is on the left of the scene, the door on the right; the girl and two companions are fleeing towards the latter.

4. Column combined with door and altar on the left, column combined with altar on the right: volute-krater Naples 2421 by the Niobid Painter (*ARV* 600.13, *Para.* 395); on the extreme left a door, then an altar partly hidden by it, to the right of the altar and between it and the column a bearded man with sceptre; on the right a column, on its left, between it and the altar, a bearded man with sceptre; the pursuit takes place between the two groups of indicators.

A Series of Erotic Pursuits 71

5. Column combined with chair and altar with palm-tree: volute-krater Boston 33.56 (*ARV* 600.12) by the Niobid Painter; the chair is on the extreme left, the column on the right; the pursuit is moving away from a (partly restored) flaming altar with a palm-tree behind it (situated between the chair and the column).

6. Altar: cup in Philadelphia, American Philosophical Society (*ARV* 880.3): the pursuit is moving away from the altar. Skyphos Reggio 3877 (*ARV* 974.25): on B a girl is fleeing towards an altar (from the pursuer on A). Altar + column: bell-krater in a private collection (Brommer 1982, 95, 94 n. 2).

7. Palm-tree: stamnos Brooklyn 09.3 (*ARV* 1084.15; 1682).[75]

Indicators 1–5 are confined to the Niobid Painter and his group, who appear to like a particular version of 'erotic pursuit', involving a particular category of spatial indicators, and thus—we shall see—representing certain perceptions of this theme. The altar on its own denotes a sanctuary. On Reggio 3877 the girl is running towards the altar; running to an altar for protection when under threat is a stock iconographical motif, corresponding to a stock semantic motif—one which fits the theme 'erotic pursuit'.[76] But this does not mean that the altar's inclusion here is without significance; it entails either that the spatial location of the pursuit was unspecified, and there was nothing to block the artist's selection 'inclusion of altar'; or that the spatial location was determined, and the altar fitted it.

Of the other spatial indicators[77] the column usually denotes 'house', but can also signify a temple or other building. It acquires its particular value in a scene through its relationships with the other elements, especially the other spatial indicators. The spatial indicators denote space through the *pars pro toto* trope. Thus, the way in which they are combined in the image cannot be assumed to reflect their real-life spatial relationship. Indeed, in some cases it is unambiguously clear that the mode in which they are combined is non-naturalistic (cf. e.g. pyxis Louvre CA 1857; *Cité* 98, fig. 141). When, as on 3, the spatial indicators frame the scene, that scene is represented as taking place inside the space denoted by the indicators. The combination 'column and door' on 3 denotes 'house'; in so far as it represents a particular part of the house we should expect it to be the courtyard, which had a colonnade (or pillars or posts). The fact that on 3 the pursuit is contained between the door and the column means that it is located within the space enclosed by the two indicators; that is, 'column + door' here do not signify a generic 'house', but locate the pursuit in the house courtyard.

In scene 5 the pursuit is contained in a space between a chair on the left (and an altar with a palm-tree further on) and a column on the right. The combination 'chair and column' denotes the house courtyard; thus, if we leave aside the 'altar with palm-tree', the pursuit in 5 is located in the house courtyard, as in 3. An altar did stand in the house courtyard, but it was not of the type shown here; in Attic ceramic iconography the combination 'altar with a palm-tree' makes up a particular established sign which I discuss below. Here I note only that the spatial indicators in 5 locate the pursuit in the house courtyard, but include an element which belongs to a different space. In 4 there is a group of indicators on each side of, and framing, the space in which the pursuit takes place. The combination 'altar and column' can, depending on the context, denote either 'sanctuary' or 'house'; in so far as it denotes a particular part of the house it should be the house courtyard, for the altar is almost certainly the altar of Zeus Herkeios, situated in the courtyard and symbolizing the centre of the *oikos*. Thus, since 'column and door' denote 'house' in general and 'house courtyard' in particular, the combination 'altar + column + door' here must denote 'house/house courtyard'. So the spatial indicators on the left denote 'house', through indicators referring to the house courtyard. In theory, the column and altar on the right could denote either 'house' or 'sanctuary'; but the close correspondence in the arrangement of the two sides, with the bearded, sceptred, wreathed/diademed man standing between column and altar on each, suggests that almost certainly they indicate 'house'. Thus in 4 the scene takes place between two houses.

We may now conclude that the column on its own on 1 must also denote 'house', rather than a temple or other building. For in the work of the Niobid Painter, in whose group the Geneva Painter belonged, 'column + door', 'column + door + altar', and 'column + chair + altar', alternative indicators in this scene and fuller versions of the polysemic 'column', denote 'house' in general and/or 'house courtyard' in particular. The door on its own in 2 also denotes 'house'. A door locates the scene either inside the house or in the courtyard, or (if it stands for the street door) in the street. On 2 the fifth-century viewers would have taken it to indicate an outdoor space, for all other versions of this pursuit are situated outdoors. The fact that elsewhere in the work of the Niobid Painter the door is combined with a column (3) to locate a scene in the house courtyard, or with a column and altar (4) to denote 'house' and locate the scene just outside the house, suggests that in 2 the door indicates either the courtyard or the space just outside the house in the street, or, ambivalently, both, creating meanings we shall consider below.

Thus scenes 1–5 are located in the following categories of space: in a house courtyard (once with an additional element referring to a different space); in the street between two houses; in a space which is either the courtyard or the street outside the house or ambivalently both. Unlike abduction from a sanctuary or a meadow, abduction from a house courtyard is not an established motif in Greek myth. The courtyard must be that of the abducted girl's house, that is, her father's house. For in fifth-century Athens her father's house is the space and world to which the *parthenos* belongs; and her removal from her father's house is the essence of abduction. Moreover, if the house is her father's, these scenes produce meanings similar to those of the more common variant of erotic pursuit which includes the girl's father. For in that case these scenes would show the girl being taken 'from inside the father's house', a crystallization of the notion 'girl's abduction', but also a representation of a most important aspect of marriage and of the female experience of it, the girl's removal from her father's home. Women circulate in marriage[78] from one male-owned domestic space to another; our scenes show the uncivilized form of that transaction—forcible removal, 'theft', of the girl from the authority and domestic space of the father by the pursuer/husband. Interestingly, in 4 'house' and the figure of the father are associated. The choice of the courtyard as the location of the pursuits was doubtless inspired by two factors: the courtyard's accessibility, pertaining to the action and narrative; and the fact that the courtyard location contributed to the creation of the nuptial meanings, for it was also the location of certain nuptial rites relating to the passage of the bride from one house to the other. As we shall see, on 3 and 5 the scene on side B of the neck (the position most closely related to that of the pursuit) depicts a nuptial rite also taking place in the house courtyard, thus counterpointing the pursuit and confirming our interpretation. This is not a naturalistic depiction of space; it does not re-present the environment in which the erotic pursuit took place in the myth—or, at least, not only that. If it does reflect a mythological motif (itself shaped by the perceptions discussed here), it also expresses some important perceptions of marriage which had become attracted to the representation of marriage through the paradigm pursuit/abduction.

The figure of the father produces similar meanings, and expresses similar perceptions, as the spatial indicators locating the pursuit in the house courtyard. Our erotic pursuits, like others, often include a mature or old man, usually shown as a king with a sceptre, who represents the girl's father.[79] This figure also helps express the notion 'girl's removal (with connotations of a violent wrenching) from the familiar world of her father's home'. But there is a difference in emphasis between this and the 'house courtyard' variant. Scenes which include the father stress the girl's ties with him, rather than with the paternal house and the unmarried girl's world—though the latter is inevitably also associated with the father. They express the notion that the ties with the father (as well as with her life as a *parthenos* in her father's house) will now be severed,[80] and the pursuer/husband will take over; the notion 'wrenched away from her father' is most clearly articulated when the direction of the pursuit is away from the father.[81] The versions in which the pursuit is in the direction of the father[82] do not dwell on the severance of the ties, but stress instead that the girl runs to her father's protection—from which she is being removed. The fact that marriage does not involve the same severing of ties with the mother[83] contributes, I believe, to the rarity of the mother's presence in scenes of pursuit;[84] this rarity reflects primarily the father's dominant and socially significant presence in the unmarried girl's life: it is from his house and authority that she is transferred to those of another man.[85] The companions, stock figures in mythological abductions, also symbolize the familiar world of the *parthenos* with its companionships and activities. This is stressed in representations in which they are running to the father (the abducted girl's, who is sometimes also their own—if she is their sister—and who, in any case, fills the role 'father (of unmarried girls)' in these images). They run to him for protection, for they, unlike the pursued girl, still belong to his sphere and authority. This motif, then, produces meanings similar to that showing the companions running into the house. Both reflect, and express, the notion that the companions still belong to the world of the *parthenos* from which the pursued girl is being wrenched; this iconographical expression of her separation from her friends and companions throws into further relief the trauma, the psychological wrenching, of the experience of marriage here articulated through the metaphor of forcible removal.

I now consider the significance of the combination of the house courtyard location with an altar with a palm-tree on 5. There was, we saw, an important and relevant altar in the courtyard, but not one which could have been depicted in combination with a palm-tree; for 'altar + palm-tree' is an established sign with certain values in fifth-century ceramic iconography, and these do not include 'altar of Zeus Herkeios in the courtyard'.[86] The courtyard altar may have triggered off the Niobid Painter's choice to combine the house courtyard location with what is denoted by the 'altar + palm-tree'. But, though he had a predilection for this 'altar + palm-tree',[87] his selections were determined by assumptions which included knowledge of the altar's meanings, and so he could not use the sign in inappropriate contexts which would evacuate it of its particular meaning and turn it into a simple 'altar'. Even if he did, the viewers could only make sense of the sign in terms of its established value. I argue below, in Chapter II.3, that there is a close connection between the sign 'altar + palm-tree' and the Attic cult of Artemis in her persona as protector of *parthenoi* and of their preparation for marriage and transition to womanhood (especially focused on her sanctuaries at Brauron and Mounichia and the *arkteia*); and that the 'altar + palm-tree' depicted in many erotic pursuits involving different protagonists represents an altar of Artemis and connotes her connection with the theme of erotic pursuit, also manifested in the mythological motif 'girls abducted from sanctuaries of, or from choruses of girls dedicated to, Artemis'.[88] An Attic version of this motif locates the abduction of girls in the sanctuary of Artemis Brauronia: girls and women taking part in the ceremonies are said to have been abducted from the sanctuary.[89] Given the Brauronian cult's concern with *parthenoi* and their preparation for marriage, the association with abduction/pursuit may reflect a connection between the Brauronian ritual and the notion 'erotic pursuit/abduction of girls'.[90] Another connection between pursuit/abduction and the Brauronian cult is that both relate to the 'wild', metaphorically partly animal, unmarried girl, and both do so in connection with marriage. The *arkteia* was a pre-nuptial rite, preparing girls for marriage. I now believe[91] that it related to the notion of the *parthenos*' animality, and that an important aspect of its initiatory function pertains to the 'domestication' of the partly wild girl, purging her of animality and thus taming her for marriage.[92] On my analysis, 'erotic pursuit' also relates to that mentality. It represents the pursuit/capture of the girl as a wild version of marriage, in which the wild, partly 'animal' girl is captured by the 'wild' ephebe—a wild, animal metaphor (appropriate to her wild, 'animal' nature) for the cultural institution which will integrate the *parthenos* into society. Thus, 'erotic pursuit' reflects the same perception of the *parthenos* as that articulated in the *arkteia*, which stresses the girl's animality in order to purge it, as the pursuits represent metaphorically the institution which will complete that purge. Through her 'stay with Artemis' the wild girl was partly domesticated and ready for the marriage which would complete her 'taming'—for which, in the circumstances, the 'wild marriage', the pursuit and capture, is an appropriate metaphor.[93]

The motif 'girl abducted from a sanctuary of Artemis' is a narrative articulation of the notion 'girl being taken away from a place, a realm, belonging to Artemis where *parthenoi* belong'. Representations of pursuits/abductions which include an 'altar + palm-tree' express the same notion pictorially. Thus the abduction metaphor for marriage can be articulated through the image of a girl being forcibly removed from Artemis' realm. This placing of the pursuit and capture in Artemis' realm reflects and connotes certain Athenian perceptions of girls and marriage: the girl's animality, the goddess' involvement in the transition, and also, through the image of the girl being taken away from the very altar, wrenched from the protection of the goddess, it produces meanings of trauma, the trauma of the removal from a familiar and protective world.

In our scene 5 the 'altar + palm-tree' is not shown on its own (which would have located the scene in a sanctuary/realm of Artemis); it is combined with the column and chair which place the scene in the house courtyard. It will now be clear that 5 represents the girl being taken from her father's house and from Artemis' realm, combining the two important perceptions of marriage discussed above by constructing a conceptual space in which the iconographical elements expressing the notion 'pursuit and abduction out of her father's house' and 'pursuit and abduction out of Artemis' realm' are juxtaposed. Another perception reflected in 5 is the duality of the nature of the *parthenos*, who belongs both inside civilized society and outside it—inside but not yet 'of' the civilized world.[94] For in 5 her wild nature is alluded to through the theme itself and the notion 'taken from Artemis' sanctuary' with its connotations of animality (but also of its purging, so that here also we have duality and ambivalence); and the 'civilized' nature is alluded to through the notion 'taken from her father's house' which locates the girl in her place in the civilized space of the polis and the *oikos*.

Given Artemis' relationship to *parthenoi*, and the motif 'abduction from an Artemis sanctuary/chorus', the altar on its own (which also occurs in other pursuits/abductions)[95] was probably thought of by the painter and his contemporaries as an altar of Artemis—unless Theseus' myth specified otherwise. If so, the scenes in 6 located the abduction in Artemis' realm. Perhaps the altar also helped produce implicit connotations of violence, by representing (when the girl is fleeing towards the altar), or evoking, the motif 'taking refuge at an altar'.[96] The palm in our pursuit—as in others—also denotes Artemis' realm and reflects and connotes the same perceptions of the *parthenos* and of marriage as the altar + palm-tree.[97]

Let us now consider the pursuit shown in the space between two houses on 4. (If, as is most unlikely, the building on the right is a temple, the space is of the same type as 5, represented through a different scheme.) At one level, the second house may have functioned, and been read as, the companions' house, and the fact that the pursued girl and one of the companions are moving towards one house, and the other two towards the other, may support this reading. But, viewed though fifth-century assumptions, the second house surely acquired a further significance. A girl is being pursued, taken away from her house and father; she is in the street, forced by her pursuer to move towards another house, in which another father-like figure stands by another house altar. This calls up another series of scenes and the situation they represent. A girl taken from her father's house by a man (soon to possess her sexually) who leads her to another house is the essence of marriage, crystallized in the ritual act of the movement between the two houses, the wedding procession, which—and the representations of which—include gestures pertaining to abduction and coercion (cf. nn. 59–62) and are located between the bride's and the bridegroom's houses (cf. e.g. pyxis London E 1920.12–21.1: *ARV* 1277.23; *Add.* 178). I suggest that the pursuit located between the two houses in 4 called up the notion, and the iconographical frame of reference, 'wedding procession', and this helped relate this pursuit to marriage, stressing that it was a metaphor for marriage. The door in 2 may denote in abbreviated form the same space and produce the same meanings: outside the girl's house, between the two houses; or it may indicate the courtyard and represent the notion 'taken from her father's house'; or, ambivalently, call up both sets of meanings.

We must not assume that Athenian viewers when looking at pursuits without spatial indicators necessarily filled in a particular location from their prior knowledge and assumptions. For scenes without spatial indicators represented one particular version of 'erotic pursuit', one which, for example, did not emphasize the nuptial connotations as 4 did. The painters were creating their own versions of the myths, and those without spatial indicators were both polysemic and emblematic, not stressing any particular aspect of the theme, and emphasizing more the paradigmatic nature of the representation, depicted in an emblematic, unspecified space. It could be argued that, if the myth of Theseus' pursuit depicted in our scenes had located the abduction in one specific place, we should expect the representations to depict that space, and assume that the scenes which do not fit this scheme represents a different subject. But this is excluded by the conceptual representation of space seen unambiguously in 5. The fact that the space on 5 is inconsistent with real space shows that it is not the representation of a particular environment specified in a myth. We cannot know whether the myths located the Theseus pursuit in a specific space; and, if they did, whether there was more than one version of this localization. If the locality was specified, it is likely to have been a sanctuary of Artemis rather than the father's house courtyard; both locations express important perceptions pertaining to marriage, but only the former had crystallized into a mythological motif. In any case, the painters were creating their own versions, which were not necessarily contradicting any narrative version: if the localization had been firmly specified as an Artemis sanctuary, contemporary viewers may have read that location into the mythological representations without spatial indicators; but the emphasis would be different, depending on whether the artist had actually depicted the space 'sanctuary'. As for the scenes located in the courtyard or outside the house, they would not have been perceived to be contradicting a sanctuary location either: they expressed the relationship between 'erotic pursuit' and marriage, and the perceptions which that metaphor reflected and articulated.

Interestingly, the Niobid Painter, who created the versions stressing the nuptial facet of the pursuit, associates with them on 3 and 5 scenes which, in my view, depict nuptial rites in the courtyard of the bridegroom's house; that is, he represents on the other side (side B of the neck) the socialized version of the same rite, the wedding as a solemn cultural and religious institution. The scene on B on 5 is incomplete, but the arrangement of the figures and comparison with side B of 3 suggest that there was an altar in the missing part.[98] Thus the representations on side B of 5 and 3 show ritual activities which take place at the altar in the house courtyard (the spatial indicators are column, chair, and altar), involving the holding of branches, libations, and women with sceptres and diadems (on 3 one on 5 apparently two—the second is fragmentary). This combination of elements and overall scheme can only be paralleled in representations of wedding scenes; B on 3 and 5 therefore seems to represent some part of the nuptial rites taking place in the house courtyard. (Compare, for example, the pyxis London D 11 (*ARV* 899.146).) The bride and groom are absent, as though sides A and B were to be taken together, and the pair is 'represented' in the wild version of their union on A. This juxtaposition of a version of 'erotic pursuit' stressing the nuptial facet and of a scene depicting nuptial rites, which occurs twice in the work of the Niobid Painter, provides some support for my interpretations, as does the fact that, on my analysis, the overall decoration of 5 forms a system in which the different scenes complement, counterpoint, and help define each other.[99] On the body there is a representation of warriors, one of them an ephebe, leaving home. Vernant[100] has pointed out that marriage was to girls what war was to boys, and girls' initiations prepared them for marriage and transformation to the condition of (tamed) gyne, as boys' initiations prepared them for integration into the status of full warrior, hoplite. Thus, this vase presents a series of oppositions and counterpoints: ephebe : unmarried girl : erotic pursuit—hoplite : married woman (the mother[s], presiding over the transition of the [absent] girl to this category): marriage; the two forms of love and war: on the one hand the wild ephebic/parthenic, and on the other the domesticated, civilized form of the hoplite/gyne, fully participating in the normal order of the city.

7. OTHER ICONOGRAPHICAL ELEMENTS

The iconographical theme 'erotic pursuit' includes some further elements which also produced nuptial connotations. One such element is, we saw, the arrangement of the himation over the girl's head. Another is the wreath, which the pursuer wears in several scenes.[101] Wreaths are also worn in other pursuits/abductions, by the pursuer and others.[102] Bridegrooms wore a wreath in the wedding ceremony.[103] Of course, the wreath is not monosemic; it was worn on many festive occasions, and Theseus, for example, wears it in a variety of circumstances.[104] But its combination here with other elements relating to the nuptial sphere, the fact that it can also be worn by the girl's father (n. 113), and that it occurs in other pursuits/abductions (including ones with a known nuptial facet) suggest that here the wreath alluded to the marriage ceremony and contributed to nuptial suggestions. It is rare for the girl to wear a wreath in our pursuits. What she very frequently wears is an (ornate) diadem,[105] similar to that worn by brides in the wedding ceremony.[106] In this context this diadem, like the wreath, contributed to the nuptial allusions. A taenia is sometimes worn by the pursued girl[107] or the pursuer,[108] who can also be shown with a 'radiated' diadem.[109] On the cup Carlsruhe 59.72 (*ARV* 883.60; 1673; *Add.* 148) a taenia is shown in the field, as in some wedding scenes such as that on the loutrophoros Copenhagen 9080 (see n. 106). The taenia is associated with—among other contexts —Aphrodite,[110] weddings,[111] and the Braunonian rites.[112] Thus its meaning here is likely to be polysemic, with the nuptial (and perhaps the 'initiatory') connotations prominent. On some of our pursuits the father also wears a wreath,[113] which, like those worn by the protagonists, alludes to the marriage ceremony. The same is true of the branch held by the father in some erotic pursuits.[114] On the cup Ferrara 44886 (*ARV* 880.11; *Add.* 147) a flying heron denotes the outdoors, but at the same time, given the heron's association with scenes in the women's quarters,[115] it connotes domesticity and the space to which the gyne is fixed through marriage; it thus may also allude to the nuptial dimension of 'erotic pursuit'.

8. COMBINATIONS WITH OTHER SUBJECTS

I now consider briefly the type of scenes with which our erotic pursuits are combined, and attempt to determine whether these combinations support our conclusions—though they cannot invalidate them, since there is not always a thematic connection between the scenes on a vase.[116] For the same reason, an investigation of these relationships is vulnerable to culture-determination, to the creation of constructs of associations between different themes on a vase which may or may not be an accurate reflection of ancient realities. A systematic investigation of combinations would be lengthy and involve separate semantic analyses of the different subjects. I only mention some combinations which seem unequivocal and significant because they relate to semantic areas which are important facets of our theme. Representations of this type fall into the following categories. First, ephebic scenes, relating to the ephebic associations of 'erotic pursuit'. One such combination is found, for example, on the cup in Philadelphia, American Philosophical Society (from Vulci: *ARV* 880.3). To the same ephebic facet probably belong also the scenes showing athletes,[117] and the many scenes with youths which are combined with erotic pursuits[118]—the fact that they are banal, routine scenes does not mean that they are meaningless, or that their combinations are necessarily unmotivated by thematic connections. A second semantic axis along which connections are made between our pursuit and other subjects is Theseus. The following are examples: cup Frankfort Museum V.F. x.14628 (*ARV* 796.117), with an erotic pursuit on B and Theseus and Skiron on the tondo;[119] calyx-krater Geneva MF 238 (*ARV* 615.1; *Para.* 397; *Add.* 131), with a Thesean Amazonomachy in the upper row and a scene (boy with lyre and youths) pertaining to the ephebic realm on B below. The Amazonomachy relates to the erotic pursuit through Theseus and also along the semantic axis male–female relationships, certain perceptions of which are articulated in such scenes. Other pursuits/abductions are also sometimes combined with our theme[120] and stress, or help define, through similarities and differences, some of its aspects. Satyrs and Maenads are sometimes combined with our pursuits, representing the wildest form of male–female sexual relationship, the one furthest away from acculturated sexual activity in marriage in the context of the polis leading to the creation of children as social beings. This theme lies at the opposite pole from marriage—thus helping define our pursuit as 'in the middle', not the acculturated institution, but not the world of unbridled sexual beings either: a half-wild half-acculturated erotic union, appropriate for the not-fully civilized and integrated ephebes and *parthenoi*; sufficiently 'inside' to function as a paradigm for marriage, articulating certain perceptions of it.

Some scenes combined with erotic pursuits relate to their nuptial facet, confirming further its importance. Apart from those already mentioned, another subject with nuptial significance is depicted on the bell-krater frr. in Athens, in the manner of the Dinos Painter (*ARV* 1155.1), which shows our type of pursuit on B and a wedding scene on A. Some other scenes combined with pursuits may also pertain to the nuptial facet: libation scenes involving a bearded man with sceptre (the type who in our pursuits represents the father) and women[121] may represent emblematically the acculturated version of the erotic union, marriage and its religious rites, and thus relate the erotic pursuit to the nuptial dimension. Another type of scene combined with erotic pursuits is departure for war, which we discussed above. Perhaps themes from the life of women which are combined with erotic pursuits, such as the scene combined with erotic pursuit on the hydria Syracuse 36330 (*ARV* 1062.2) depicting women, one with a lyre, in a domestic context (chest, heron) and Eros, are a female equivalent of (hoplitic) war, pertaining to the life and pursuits of married, that is acculturated, integrated women.[122] Above we see the integrated women inside the house and Eros in his acculturated form (he is holding a lyre) in marriage and below the wild form of Eros, the pursuit and imminent capture of the wild girl who, through the marriage of which the capture is a wild version, will become fully tamed and fulfil her proper role, represented through the domestic scene. The two themes help define each other.

9. VARIANTS OF 'EROTIC PURSUIT BY THESEUS/AN EPHEBE'

The scenes here classified as variants have an iconographical scheme of the same type as our pursuits, but involve either more than one pursuer, or a different pursuer or girl. When the scene is mythological this entails that a different story is depicted, when it is not mythological it is a different type of 'genre' scene from that of our main series, involving a different type of person. In a major type of variant the iconographical scheme of the pursuit is the same, but the pursuer (who is heroic/human)[123] is shown through a different iconographical scheme. Another type of variant involves more than one pursuer; such scenes vary, in a way similar to scenes with one pursuer: some show youths similar to Theseus in our pursuits, others show other types of male. In another variant the pursuer is similar to that of our main series, but the identity of the pursued girl has changed.

Since these scenes also represent a human/heroic pursuit, their meanings should be compatible with those of the scenes at the centre of our investigation. For example, if it had been the case—which it is not—that the weapons carried by the other human pursuers in these variants were represented as offensive weapons being used as such during the pursuit, doubt would have been thrown on our conclusions about the role of the spears. But, in fact, the variants of 'erotic pursuit' are all compatible with our readings.[124]

In one variant, which, like our pursuits, occurs in two types, with and without spears,[125] the pursuer is a bearded man. The meanings of this theme, obviously, do not pertain to the ephebic sphere, as our pursuits do, but to the sexual (and perhaps also to the matrimonial) facet of 'erotic pursuit'. If the type without spears was partly modelled on Boreas[126] then perhaps the notion of the matrimonial dimension is reinforced. In a variant which shares many elements with our type I pursuits the pursuer is a young warrior. This variant shows that the element 'not attack' defines erotic pursuit (by heroes/mortals) across variants. In pursuits by young warriors the weapons are never held in a position of attack or menacingly; on the contrary, the use of the motif of holding the spear in the left makes explicit that this weapon is not represented in action in the course of an attack.[127] This confirms my views about the significance of the spears in our type I pursuits.

Another variant[128] is depicted on the skyphos in the Vatican by the Lewis Painter (*ARV* 974.28). On A a Greek youth, wearing a wreath and a chlamys worn like a wrap, is running, holding two spears in his right and extending his outstretched left towards an Amazon who is fleeing and turning back; there are Amazons running on B in the role of the fleeing companions. The pursued Amazon on A is extending her right towards the youth, in the exact stance of the pursued girls of the other scenes, with the difference that she is holding an axe in her bent left, in exactly the same position as that in which the pursued girl on B of the skyphos Providence 25072 (also by the Lewis Painter) holds a stylized flower—for this pursued female is a warrior, and must be thus characterized. The pursuer's stance resembles in a general way that of the pursuer on the Providence skyphos, and the skyphos Reggio 4134 in the manner of the Lewis Painter (*ARV* 975.3), and more closely that on another skyphos by the Lewis Painter, Reggio 3877 (*ARV* 974.25) of type 1.

It is clear that this is an erotic pursuit, and the fact that the scene does not denote an attack is made explicit both by the overall syntax of the scene, and by a detail in the arrangement of the spears which, like similar motifs in our type 1 pursuits, denotes that the signification space in which this theme was inscribed, and which determined its production, included the element 'not attack': the spearheads are well beyond the Amazon's body, reaching almost to the palmette decoration under the handles. But the combination of the youth's spears and the Amazon's axe do create the effect of implicit potential violence between them, appropriate to the hostility obtaining between Greeks and Amazons which is thus signalled. By presenting the Amazon about to be caught and raped by a Greek, this theme, like the equivalent mythological stories of Amazons defeated and Antiope erotically abducted, raped or seduced, and impregnated, reduces even the masculine dominant Amazons[129] to subordinate defeated females[130] and the objects of male sexual aggression.

This, then, is one iconographical expression of the Greek perceptions of the relations between the sexes, focusing on the threatening aspect of the woman, condensed in the figure of the abnormal male-like female who, even so, is subdued by the male—as she is in the myths, especially that of Theseus and Antiope. Since the pursuer lacks the elements which unequivocally build up the sign 'Theseus' the scene on the Vatican skyphos may represent a generic Greek and a generic Amazon, rather than Antiope's abduction; but if so that abduction was certainly the theme's mythological paradigm.[131]

Yet another variant of 'erotic pursuit' involves more than one pursuer, like the scene on the Basle amphora recently discussed by Schefold.[132] Whether mythological or 'genre', these multiple pursuits reflect, I believe, the notion 'companionship of ephebes/young males' which is also correlative with the mythological pairs of friends such as Theseus and Peirithous and Orestes and Pylades. A rare type of scene, which is not a variant of erotic pursuit but may superficially appear to be one, is that in which the youth is carrying both spears and drawn sword (e.g. on the hydria at Taranto from Ceglie (*ARV* 606.74) and the hydria Louvre G 427 (*ARV* 615.2)). In the complex process of interactive meaning-production through which images are made sense of, the drawn sword makes the theme an attack.[133] The spears are perfectly appropriate as signifying elements characterizing Theseus-as-ephebe. So this type of scene is in fact a variant of the theme 'Theseus with a drawn sword' (discussed in Chapter II.1 above).

All these variants produce meanings which are compatible with, and confirm, the meanings here suggested for our pursuits. In particular, the conclusion that the theme 'erotic pursuit by Theseus' articulates also certain Greek perceptions of male–female relationships is confirmed by the fact that the other variants of this theme seem to give concrete expression to this notion by replacing Theseus/the ephebe with other male types in the role of sexual pursuer.

10. Conclusions: The Erotic Pursuit

To sum up: the iconographical theme 'erotic pursuit' reflected, and expressed, certain perceptions about ephebes and *parthenoi*, about male–female relationships, and about marriage. These three semantic facets of the theme are, of course, intertwined and interdependent. Erotic pursuit/abduction as a metaphor for wedding/marriage, the representation of the 'wild marriage', capture (instead of the acculturated form with *engye* and proper ritual) involving the not fully integrated, 'wild-warrior' (anti-hoplite) ephebe and the wild girl—all express fundamental aspects of Greek realities and mentality pertaining to marriage. The following are some of the most important: the notion of the girl as a wild thing to be pursued and captured and tamed through marriage; the violence of the wrenching of the girl from her familiar world and transfer to an unfamiliar one and to the jurisdiction of a strange man; the subduing of the female by the male (an important Greek notion which is itself a polysemic signifier articulating other values); and defloration and the sexual relationship as acts of physical domination of the woman. Thus the metaphorical relationship between erotic pursuit and marriage also expresses the notion that the acculturated form of marriage which belongs to civilized society also contains within it the wild marriage to which on the surface it appears to be contrasted.

These conclusions gain some confirmation from the consideration of the myth of Atalanta, which proves to be a reversal of the nexus 'erotic pursuit' as I have reconstructed it. It includes all the elements which pertain to this nexus: pursuit and capture connected with marriage, the wild girl with elements of animality (Atalanta had been exposed and suckled by a bear) and connected with Artemis,[134] the wild ephebe,[135] and the marriage between the two. But in it the relationships between the elements are, when compared to 'erotic pursuit' and to what was considered 'correct' in Greek mentality, reversed. First, the wild girl (a huntress who roams outside her proper place) refuses her proper destination of marriage, refuses the transition out of the state of wildness and to the status of married woman. Second, the pursuit involving Atalanta (presented as a race, which relates the story to the initiatory races) is a test set by her; it was she who pursued the ephebe, and who ran armed: Atalanta's suitors had to run ahead of her; she killed those whom she caught up with but she would marry whoever succeeded in not being caught. Another related reversal is the fact that here it is non-capture that entails success and marriage, while capture brings death. Many failed and died until Melanion (in another version Hippomenes) slowed Atalanta down through a trick and she married him.[136] In the nexus 'erotic pursuit' the ephebe captures the girl and this stands for marriage leading to the woman's final taming and integration. In Atalanta's myth it is the girl's non-capture of the male that leads to marriage. Correlative with this reversal, which meant that Atalanta was not captured (and thus symbolically 'tamed'), is the abnormal nature of this marriage, which did not purge her of wildness and animality and did not integrate her into her proper place: instead of staying inside the house she continued to hunt in the wilderness in the company of her husband, who also continued to be a wild ephebe.[137] Thus, this was a wild marriage of a wild, not properly tamed girl, and did not bring about, as proper marriage does, the tamed gyne's integration into society in her proper place. This is reflected in Atalanta's and Melanion's behaviour, which was that of non-acculturated people, similar to that of animals: they copulated in a sanctuary, thus transgressing against the Greek religious observance which forbade intercourse in a sanctuary, and for this reason they were changed into animals;[138] that is, instead of losing their (and especially her) animality through marriage, they lost their humanity and became fully animal, because, not having become acculturated and integrated, they practised a wild form of erotic union which transgressed against a religious, cultural observance. This, then, is—among other things—the paradigm of a wild marriage in which the proper taming

A Series of Erotic Pursuits

and transition have not taken place; it helps define the ideal, the canonical marriage, metaphorically expressed through the theme 'pursuit, capture and taming of the wild girl', through an articulation of its opposite.[139]

The fact that the myth of Atalanta is the mirror-image, a consistent reversal, of the nexus 'erotic pursuit' as I have reconstructed it here provides some non-circular confirmation for my results and shows that they are not simply a modern, culturally determined construct.

NOTES TO CHAPTER II.2

1. The combination of spears and 'grabbing' (see e.g. the column-krater Göteborg 171–62 (ARV 284.5)) puts even more emphasis on the 'violent intimations'. There is a significant spectrum of differences in emphasis in the different scenes, from 'almost consensual' to 'implicit connotations of violence/menace'. The closeness between types 1 and 2 is confirmed by the existence of an 'intermediate' type: on the skyphos Providence 25.072 (ARV 973.10) the youth has put down the spears (shown resting on the ground) and is running after the girl.
2. I discuss the need not to overlook apparently small divergences in Ch. I (see also Ch. I n. 19).
3. See e.g. the satyr pursuing a maenad on the cup Oxford 1927.71 (CVA Oxford 2, pl. 52.3). (On satyrs pursuing maenads see Hoffmann 1977: 3–4).
4. See e.g. the hydria the Hague 634, Byvanck 1912, pl. xxx.634 (ARV 1209.58) and compare it with e.g. Leningrad 728 (here Pl. 7). I am not including the theme 'Eros pursuing a boy': I exclude non-exact equivalents to avoid overlooking important differences by making the culture-dependent judgement that they are not significant.
5. As Boardman in Boardman and La Rocca 1978: 20 implies.
6. In Bérard and Durand 1984: 32–3.
7. On this theme and the problem of the identification of the protagonists see Jahn 1847: 34–41; Beazley in CB ii. 81; Kahil 1955: 311; Lezzi-Hafter 1976: 73–5; Schefold 1975: 93; 1982: 233; Boardman 1958/9: 171; Alfieri, Arias, and Hirmer 1958: 32; see also Brommer 1982: 95; 1979: 509.
8. See Beazley in CB ii. 81.
9. Thanks to the kindness of Dr A. Lezzi-Hafter I obtained photocopies of photographs of this vase.
10. See e.g. hydria Syracuse 36330 (ARV 1062.2; CVA, pl. 25).

11. See e.g. oinochoe Ferrara sequestro Venezia 2505 (*ARV* 1206.3; *Para.* 463; *Add.* 169; Lezzi-Hafter 1976: pl. 102).
12. See e.g. the stance of the youth on the stamnos Krefeld Inv. 1034/1515 (*ARV* 502.5; CVA Germany 49, pls. 37.1, 38.1), side B.
13. See e.g. the Krefeld vase (n. 12), sides A and B; column-krater Louvre G 362 (*ARV* 1115.17; CVA, pl. 27.4.5).
14. See e.g. for both features the bell-krater Leningrad 777 (St. 1786) (*ARV* 502.11; *Para.* 513; *Add.* 123), side B (here Pl. 5).
15. See e.g. the stamnos Oxford 1911.619 (*ARV* 629.16; here Pl. 1).
16. See e.g. type 2, Nolan amphora Syracuse 20537 (*ARV* 1015.16); type 1, neck-amphora London 1928.1–17.58 (*ARV* 1010.5; CVA, pl. 59.3).
17. On the *gnorismata* cf. Sourvinou-Inwood 1971*b*: 94–109. On the recognition: Sourvinou-Inwood 1979: 18–58.
18. Moreover, there is no evidence that anyone else did, except Peleus, whose scenes, we shall see, were firmly signalled.
19. On Theseus' chlamys and its significance see Barron 1980: esp. 1, 3–4.
20. On Theseus' hats see Barron 1980: 1, 5 n. 4.
21. In my view, the use of dolphins in Attic iconography is not loose, as Beazley CB ii.81 thought. The dolphin was not a monosemic sign that only meant 'Peleus and Thetis'. All signs are polysemic and every sign acquires its value in context (see Ch. I above). Thus a dolphin in an abduction had, for a 5th-cent. Athenian, a different meaning when it 'qualified' an erotic pursuit involving our type of pursuer, from that which it had in the representation of Boreas abducting Oreithyia on the hydria Bowdoin 08.3 (*ARV* 606.68; *Para.* 395; *Add.* 130; Buitron 1972: no. 64) in which one of Oreithyia's companions is holding a dolphin, a scene mentioned by Beazley as an example of such looseness (with the alternative explanation that it may have been a slip). This transfer of a sign belonging to Peleus' abduction to a different abduction may be a play on the fact that there was a Nereid called Oreithyia (Hom., *Il.* 18. 48) and, through the Nereid allusion or a direct sea allusion, may hint at the naval help which Boreas gave to the Athenians in the Persian Wars (Hdt., vii. 189)— an Athenian propaganda theme which is at least one reason for the great popularity of Boreas and Oreithyia scenes in Attic iconography after the Persian Wars. (On Boreas and Oreithyia cf. Simon 1967: 101– 26; Neuser 1982: 30–87. The view that Boreas' help is the main motivation behind this popularity has been challenged (Agard 1966: 241–6; Schauenburg 1961: 78); but it cannot be doubted, since that belief was part of the assumptions through which the Athenians thought about Boreas, that it is one of the main reasons behind that popularity.) Such take-over was feasible because there was no possibility of mistaking Boreas and Oreithyia for Peleus and Thetis.

22. See e.g. for type 1, the scene on the pedestal of a lebes gamikos in the Robinson Collection (CVA Robinson 2, pl. 51a–c); for type 2, the stamnos Villa Giulia 5241 (*ARV* 484.9). On the iconography of Peleus and Thetis: Krieger 1975. As we shall see, Thetis' companions and father (on whom see Krieger 1975: 88–113) are sometimes shown, as are those of our girl.
23. *CB* ii. 81.
24. See e.g. the cup Oxford 1913.311 (CVA, pls. 4.3 and 13.1–2) and the lekythoi Oxford 1938.909 and 1920.104 (*ARV* 993.93–4).
25. See Arist., *Ath. Pol.* 42. 3–5; Pollux, *Onom.* 10. 16. On the chlamys see also Pélékidis 1962: 115–16; on the spears: Pélékidis 1962: 231–2. On ephebes and ephebeia in general see esp. Vidal-Naquet 1983: 151–75, 191–7; 1986: 126–44. See also Pélékidis 1962; Siewert 1977: 102–11.
26. Boardman 1958/9, also suggests that some of these scenes are not truly mythological (though he sees it in terms of loss of 'any specific mythological explanation').
27. On this type of indeterminacy see also Schefold 1975: 27; Krauskopf 1977: 28. An element such as the presence of Athena (e.g. on the krater Corinth C 33.129 and 138 (with new fr. added in 1979): *ARV* 592.29; Boulter and Bentz 1980: 300–1, pls. 82–3) pushes the scene more towards the mythological pole, though it is also compatible with the generic version.
28. Theseus–Helen connotations are not ideal for a paradigm concerning matrimony. (On Helen's abduction by Theseus see Kahil 1955: 305–13; Calame 1977: 281–5; Lloyd-Jones 1983: 95; see also Sourvinou-Inwood 1988a: 53, 93–4 n. 252, and below, n. 88.) I should, however, note that a recently published scene of the late 5th cent. in which the figures are inscribed shows a solemn wedding of Theseus and Helen: painted on a krater excavated at Serra di Vaglio, now in the depositi della Soprintendenza alle antichità della Basilicata (see Greco 1985/6: 5–35). (I owe this reference to Prof. Lilly Kahil.) Clearly, then, the character of Theseus as ephebe *par excellence* and of Helen as marriageable *parthenos par excellence* (on which see Sourvinou-Inwood 1988a: 53, 93–4 n. 252), in interaction with the nuptial facet of the mythological and iconographical theme erotic pursuit/abduction discussed here, has led to the creation of (at least one iconographical articulation of) a theme 'wedding of Theseus and Helen'. The late 5th cent. is the *terminus ante quem* for this development. It cannot be excluded that it may have taken place earlier and that the pursued girl in those of our scenes which are above all mythological may have been Helen. However, given the variously negative and problematic connotations associated with Theseus' abduction of Helen it is difficult to see how it could have functioned as a nuptial paradigm

in the way that our scenes do (though it cannot be excluded that this is a culturally determined judgement). If this is correct, our erotic pursuits do not articulate or reflect a nuptial representation of Theseus' abduction of Helen, but may have contributed to its creation by articulating in many images, and thus helping stress and strengthen, the metaphorical relationship between erotic pursuit/abduction and marriage, the extension of which, I submit, underlies the representation of the wedding of Theseus and Helen.

29. On Eriboia/Periboia/Phereboia and Theseus: Barron 1980: 2–3; see also Kron 1988: 301–2. My argument is not circular here; because of each figure's connotations the protagonists' identity contributes to the theme's meanings, but since we have no independent evidence for the girl's identity, I am simply considering whether we know of a girl who fits the theme's connotations as reconstructed on the basis of its other elements.
30. Barron 1980: 2–3 and n. 30.
31. Whether or not Periboia had replaced an earlier girl in our theme is not important. The theme becomes very popular starting with the Niobid Painter's generation.
32. See Athen. 13. 557a–b (= Pherekydes, *F Gr. H* 3 F 153).
33. E.g. in scenes in which she is the victim of an attack: e.g. the neck-amphora Vienna 741 (*ARV* 203.101; *Add.* 96); and the amphora London 1948.10–15.2 (Moret 1975: pl. 17.1). 'Fleeing woman' in a wedding context: see e.g. the pyxis Munich 2720 (*ARV* 1223.4; *Add.* 173; Roberts 1978: pls. 99.3, 100.1–2 (commentary on p. 182, see also p. 184) in which the woman is represented fleeing, through the 'fleeing woman' schema, away from a door which represents emblematically the notion 'wedding' (Roberts 1978: 182). One possible interpretation of this figure is that she emblematically signifies that semantic facet of wedding/marriage which allows it to be represented through the model of abduction; another is that she represents a part of the wedding ceremony, the bride's resistance in the course of a mock-abduction rite. An example of a different type of context is afforded by the Nolan amphora Leningrad 697 (St. 1628) (*ARV* 202.76, *Para.* 510): on A Athena running, on B a woman running, shown according to the 'fleeing woman' schema. On codification: Guiraud 1975: 24–5.
34. See e.g. skyphos Providence 25.072 (*ARV* 973.10).
35. See e.g. the volute krater Izmir Inv. 3361 (*ARV* 599.7).
36. See also Foley 1982: 161. On flower-picking associated with Persephone's myth and cult, especially in connection with her bridal aspect: Blech 1982: 349 (with pp. 349–51 on flower-gathering in Artemis' cult); see also below, Ch. III, sect. 2 III (*h*).
37. On this see Blech 1982: 352 n. 95.

A Series of Erotic Pursuits

38. See e.g. the lebes gamikos in the Robinson collection (cf. n. 22).
39. See e.g. the representation of a man and a hetaira which involves the same general iconographical scheme: Boardman and La Rocca 1978: 90.
40. Vidal-Naquet 1983: 166–7; Calame 1977: 67, 211–14; Detienne 1979: 31; Lloyd-Jones 1983: 94: At Brauron: Kahil 1977: pls. 18–19; see also Sourvinou-Inwood 1988a: 66.
41. On the capture of animals as part of the ephebic training see Schnapp 1984: 67–8 (and *passim*).
42. Vidal-Naquet 1983: 169–74; Vidal-Naquet in Vernant and Vidal-Naquet 1972: 161–2; Brelich 1969: 175, 199; Schnapp 1984: 67–82; Schnapp 1979: 40; Lloyd-Jones 1983: 98; Detienne 1979: 23–6. On the iconography of ephebic hunts see Schmitt and Schnapp 1982: 57–74, esp. 65–8.
43. The most famous mythological pursuit and capture of an animal, of the Kerynian hind, is closely associated in some versions (Pind., *Ol.* 3. 28–30; Schol. Pind., *Ol.* 3. 53) with the erotic pursuit of a girl. I should mention that, in my view, there is a metaphorical relationship between girls leaving childhood and deer (cf. also Sourvinou-Inwood 1988a: 102 n. 298, 104–5 n. 315).
44. *ARV* 333.2; *Cité*, fig. 120 (right).
45. On the differentiation between the theme of erotic pursuit and the attack scene 'Theseus with a sword' see Ch. II.1 above.
46. The capture of wild animals is closely associated with the erotic sphere: Schnapp 1984: 71–82; Schmitt-Pantel and Thelamon 1983: 17. On hunting and sexuality see Schnapp 1984: 67–82; Burkert 1983: 58–72; Detienne 1979: 25–52m *passim*; Borgeaud 1979: 55. Hunting as metaphor for homosexual pursuit: Dover 1978: 87–8. In this context (capture, through pursuit, of a girl) the endromides, worn by many of the pursuers, and the petasos, worn by hunters as well as ephebes (and travellers and others) would perhaps also call up the characterization 'hunter' and contribute to the allusion.
47. See Loraux 1978: 43–87, esp. 59–69; King 1983: 109–27. The unmarried girl is 'tamed', a process ending with marriage: King 1983: 111, 122–3; Calame 1977: 411–20. See also below.
48. Thetis' transformations are usually interpreted only in terms of her nature as a sea-deity. But mythological motifs are polysemic and acquire meaning in context; here the relevant aspects of this context are 1. Thetis' associations with 'erotic pursuit as a paradigm for marriage', and 2. the Greek mentality about the girl's animality and its association with the notion of the girl's capture (the iconography of which is closely related to that of Thetis'). Thus, Thetis' metamorphoses are correlative with, and so were inevitably seen as articulating, her (paradigmatic parthenic) animality, which is tamed (albeit temporarily)

through her capture by/marriage to Peleus. Her ability to metamorphose herself allows the animality to be expressed in narrative terms. In Eur., *IA* 703, the union between Peleus and Thetis is presented as resulting from an *engye*.

49. See Ch. II.1, sect. 2 III.
50. On Thetis and Peleus see Lesky 1966: 401–9; id., *RE* xix. 275 ff.
51. On the special status conferred to Peleus: see e.g. Pind., *Nem.* 4. 65–8; see also Thetis' promise to Peleus that he will be a god: Eur., *Androm.* 1253–8.
52. See Lesky 1966: 406.
53. See e.g. Roberts 1978: 178–9, and the lebes gamikos CVA Robinson 2, pls. 50–51c, where the erotic pursuit of Thetis on the pedestal is juxtaposed to the Epaulia on the body of the vase.
54. For 5th-cent. Athenian perceptions on rape and sexual violence against women see Walcot 1978: 137–47. On eros seen by the Athenians in terms of aggression and domination see Keuls 1983: 214.
55. See e.g. the abducted girl caressing Theseus' hair on the amphora Munich 2309 (*ARV* 27.4; 1620; *Para.* 323; *Add.* 75).
56. See Sourvinou-Inwood 1973: 12–20 and below, Ch. III, sect. 2 II–IV; Foley 1982: 169; L. Kahn and N. Loraux, *Dictionnaire des mythologies et des religions des sociétés traditionelles et du monde antique* (Paris, 1981), s.v. 'Mort. Les mythes grecs', 8.
57. Similar perceptions of this myth are expressed in: Kahn and Loraux (n. 56); Jenkins 1983: 142.
58. On this: Sourvinou-Inwood 1973: 17.
59. See Sourvinou-Inwood 1973; Jenkins 1983: 137–8.
60. On which see Jenkins 1983: 139–41; see also Sourvinou-Inwood 1973: 21 n. 54.
61. See Sourvinou-Inwood 1973: 17–18.
62. For other gestures and behaviour pertaining to abduction in weddings see Sourvinou-Inwood 1973: 16–17; Redfield 1982: 191.
63. See Sourvinou-Inwood 1973: *passim*; Jenkins 1983: 140 and *passim*.
64. See King 1983.
65. See Sourvinou-Inwood 1979: 10, 53–5.
66. Jenkins 1983: 141–2; Foley 1982: 169–70. Another perception that may also be reflected in this metaphor is that suggested by Redfield 1982: 191 for the mock-abduction of the marriage ceremony, which he relates to the great value of virginity and the desire for the contradiction in terms which is the chaste wife.
67. As has been noted by Calame 1983*a*: p. xxii. The importance of Eros in the ideology of marriage is illustrated, for example, in the many nuptial representations in which Eros is depicted (see e.g. the loutrophoros Boston 10.223 (*ARV* 1017.44; Oakley 1982: 115, fig. 2) which shows Eros flying and holding a taenia). This emphasis on Eros in the

ideology of marriage represents one end of a wide spectrum of ideologies and realities pertaining to marriage in classical Athens; the other end involves maximum coercion, represented by the *epikleros* who is forced to marry the close relative who has won her through *epidikasia*. The role of erotic love in the choice of marriage partners in classical Athens is being reconsidered by P. G. McC. Brown in a forthcoming paper.

68. See e.g. the cup Würzburg 479 (*ARV* 372.32; 1649; *Para.* 366; 367; *Add.* 111–12).
69. See e.g. the small neck-amphora Leningrad 709 (*ARV* 487.61; Pls. 9–10); Nolan amphora Bonn 77 (*ARV* 1015.12).
70. For such brides see e.g. loutrophos-hydria Copenhagen 9080 (*ARV* 841.75; *Para.* 423); pyxis Athens Acr. 569 (*ARV* 890.172; *Add.* 148); loutrophoros once in Berlin (ex Sabouroff) (Furtwängler 1883–7: pls. 58–9). See also the cup Louvre G 265 (*ARV* 416.1; *Cité*, fig. 39) which, in my view, probably represents Theseus and his bride, since the outside shows deeds of Theseus. This scene would be the aftermath (in narrative terms) of, and/or the other frame of reference for, our erotic pursuits. If this is right, and Theseus' wedding had been an established iconographical theme, it would reinforce the view that the pursuits had a nuptial semantic facet. Similar use of the 'himation over the head' element to that postulated here: e.g. on A of the skyphos Boston 13.186, Paris leading Helen away (*ARV* 458.1; *Para.* 377; *Add.* 119). On this motif and usage see now also Kron 1988: 299–300 who takes a view similar to the one expressed here.
71. For the notion of consent in marriage see Redfield 1982: 192; Foley 1982: 169. As for the force, the transfer of legal guardianship of the woman from one man to another, and her removal away from her familiar world to an unfamiliar one are elements of force, pertaining to the woman, which were among the perceptions animating the abduction metaphor for marriage.
72. Redfield 1982: 186; see p. 188 on a related duality.
73. See also Redfield 1982: 192.
74. On spatial indicators see Bérard and Durand 1984: 27–31; Bérard 1983: 14; Keuls 1983: 216.
75. I argue in Ch. II.3 below that the palm is a significant iconographical element in the pursuit, despite the fact that it is part of the handle decoration.
76. See also Sourvinou-Inwood 1979: 42. On 'altar + column' see below.
77. On the meanings of the spatial indicators see n. 74. On Greek houses see e.g. Laurence 1962: 240–9; Walker 1983: 81–91, 86, fig. 6.1, 87, fig. 6.2*a* and *b*, 88, fig. 6.3*a* and *b*. On the role of the house altar of the bridegroom's house in wedding rites and representations: Hölscher 1980: 176. On Zeus Herkeios: see Hdt. 6. 68; S., *Ant.* 487; Nilsson 1967: 402–3; Burkert 1985: 130, 248, 255–6.

78. See Vernant 1965: i. 132. On the trauma of leaving the father's home at marriage see Lefkowitz 1981: 20.
79. See e.g. the column-krater Göteborg 171–62 (*ARV* 284.5); hydria Florence 4014 (*ARV* 1060.144). That he is the girl's father is confirmed by his correspondence to Nereus in Thetis' pursuit. (On Nereus in such scenes: Krieger 1975: 88–113.)
80. On the breaking of the father–daughter bond brought about by the daughter's marriage see Redfield 1982: 186–8.
81. See e.g. Florence 4014 (n. 79).
82. See e.g. the volute-krater Bologna 275 (*ARV* 1029.18).
83. See Redfield 1982: 187–8.
84. I interpret as mother the woman with sceptre on, e.g. side A of the stamnos Krefeld Inv. 1034/1515 (*ARV* 502.5; CVA Germany 49, pls. 37.1–4, 38.1–4). The mother has the most important role in the ritual part of the ceremony (see Redfield 1982: 188), the father in the legal part, as in the facet of ideality represented in our scenes.
85. The fact that the father is also present in some representations of Eos' pursuit of Kephalos (see Kaempf-Dimitriadou 1979: 18–19) does not invalidate these interpretations: the schema 'erotic pursuit of a girl by a hero' was adapted (see e.g. the inclusion of both Kephalos' companions and his sisters, like the pursued girl's companions running to their father (e.g. neck-amphora Madrid 11097, *ARV* 1043.2, Kaempf-Dimitriadou 1979: no. 108)) to show a youth's abduction by a goddess, with all the connotations of danger and helplessness carried by this theme (see above, Ch. II.1, sect. 2 III).
86. See below, Ch. II.3, sect. 1.
87. See below, Ch. II.3, sect. 1 and n. 7.
88. See Calame 1977: 176–7, 189–90. Plut., *Thes.* 31. 2, tells us that Theseus abducted Helen from the sanctuary of Artemis Orthia.
89. See Schol. BT Hom., *Il.* 1. 594; Hdt. 4. 145, 6. 138; Suda s.v. Brauron.
90. Osborne 1985: 161–2, 168 also concluded, independently, and on the basis of different considerations, that the stories connecting the sanctuary of Brauron with rape had a close connection with the ritual.
91. See Sourvinou-Inwood 1988a: 111–12, 128–30, 133–4.
92. Again, a similar conclusion was reached independently by Osborne 1985: 165–9. A related perception would also seem to underlie Detienne's formulation (1979: 31) that the Athenian girls 'do the bear before marriage in honor of Artemis of Mounichia or Brauron to purify themselves, in the words of an ancient exegete, of any trace of savagery' (with refs. in p. 99 n. 43).
93. The element 'running' also connects the Brauron rites with erotic pursuits: the running of the girl, her companions, and the pursuer has a cultic counterpart in initiatory rites (see n. 40). The palm brings the two closer: the Brauron girls are shown dancing and running among,

and towards, palms, altars, and altar + palm complexes. There is thus a correspondence between the Brauron girls' ritual race and the girls (e.g. the Nereids) running during an abduction.

94. Even after she was tamed and brought into male society, the danger remained that her domestication could be reversed: see King 1983: esp. 110, 124; Lefkowitz 1981: 16–18.
95. See below, Ch. II.3, sect. 2.
96. See below, Ch. II.3, sect. 2.
97. See below, Ch. II.3, sect. 2.
98. See *CB* ii. 81.
99. I am discussing it as an illustration; a similar system of relationships, pertaining to slightly different aspects of the 'erotic pursuit', is observed on the Bologna krater.
100. Vernant 1974: 37–8.
101. Examples of a pursuer with a wreath: stamnos Oxford 1911.619 (here Pl. 1); bell-krater Louvre G 423 (*ARV* 1064.6).
102. See e.g. Thetis' abduction by Peleus (a marriage paradigm) on the cup London, Victoria and Albert Museum 4807.1901 (*ARV* 89.14) where, on side A (here Pl. 11) Peleus, Thetis, and the Nereids are all wearing wreaths.
103. See e.g. loutrophoros Boston 10.223 (n. 67); loutrophoros once in Berlin (ex Sabouroff) (n. 70). On wreaths at weddings: Blech 1982: 75–81.
104. See Blech 1982: 430–1, 450; see also pp. 124, 264–5.
105. See e.g. oinochoe Ferrara sequestro Venezia 2505 (*ARV* 1206.3).
106. See e.g. the Boston loutrophoros (n. 67) and the Sabouroff loutrophoros (n. 70); loutrophoros Copenhagen inv. 9080 (*ARV* 841.75; *Para*. 423). On the diadem worn by the bride see Blech 1982: 76–81.
107. See e.g. side B of the bell-krater Leningrad 777 (here Pl. 5); neck-amphora at Mykonos by the Oionokles Painter (*ARV* 648.24).
108. See e.g. the neck-amphora Leningrad 709 (*ARV* 487.61; *Para*. 512; here Pl. 9); stamnos Warsaw 142353 (ex Czartoryski 51) (*ARV* 501.2).
109. Calyx-krater Geneva MF 238 (*ARV* 615.1).
110. See Redfield 1982: 195.
111. See e.g. Redfield 1982: 188. Epaulia gifts including a taenia: Roberts 1978: 184, see pls. 56–7, 58.2, 59.1. That the taenia in the field on the erotic pursuit cup Carlsruhe 59.72 (see text) and in wedding scenes (e.g. the Copenhagen loutrophoros, n. 106) does not denote an indoor space is shown by scenes where the spatial indicators locate the scene in the courtyard, and a taenia is hanging in the field (e.g. the pyxis Athens Acr. 569 (*ARV* 890.172; with Roberts 1978: 84, pls. 56–7, 58.2) by the Penthesilea Painter, like the Carlsruhe cup).
112. Kahil 1965: 21, and see e.g. pl. 7.

113. See e.g. oinochoe Ferrara sequestro Venezia 2505 (*ARV* 1206.3); bell-krater Istanbul 2914 (*ARV* 603.41; Reisner, Fisher, and Lyon 1924: pls. 69, 70).
114. See e.g. the calyx-krater at Aachen, Ludwig (*ARV* 1661.7 bis; *Para.* 396; *Add.* 130). On the hydria London E 198 (see text section VI no. 2) the father wears a wreath and holds a branch. The Niobid Painter and his group who stressed the nuptial dimension of 'erotic pursuit' through spatial elements also liked to include wreaths and branches, which again allude to weddings. Dr M. Schmidt, who has kindly let me know that she agrees with my argument relating erotic pursuits to marriage, also thinks that the fathers' branches in the pursuits allude to sacral rites. For Krieger 1975: 80 the torches held by Chiron in some scenes of Thetis' pursuit/capture allude to marriage.
115. See below, Ch. II.3 n. 100.
116. See Bron and Lissarague 1984: 17.
117. See e.g. the upper frieze of the calyx-krater New York 06.1021.173 (*ARV* 1092.75).
118. See e.g. side B of the bell-krater Louvre G 423 (*ARV* 1064.6).
119. Boreas pursuing Oreithyia on A relates to the erotic pursuit along another semantic axis, that of pursuit.
120. E.g. Frankfort cup side A; calyx-krater New York 06.1021.173 (n. 117): Eos pursuing a youth on A, our pursuit on B. Our pursuit with Peleus': volute-krater Naples 2421 (*ARV* 600.13; *Para.* 395).
121. See the stamnos Oxford 1911.619 (*ARV* 629.16; here Pl. 1).
122. Similar scenes have been interpreted as showing professional musicians and/or educated hetairai by Williams 1983: 99–102 because of the presumed lack of education of respectable Athenian women. In my view, the latter is problematic; it is dependent on certain types of male ideality which may not necessarily reflect reality. Bérard 1984: 86–8 argues that a group of scenes of this type indicates that rich Athenian women could be cultured. In my view, the fact that our women alternate with the Muses and Sappho who are presented according to this type of iconographical scheme (cf. Williams 1983: 100) suggests that they are Athenian citizens.
123. As we may conclude when he is not characterized as some particular divine being, or the pursuit is not otherwise defined as representing some divine pursuit. I discuss divine pursuits and their relationship to heroic/human ones in Ch. II. 1, sect. 2 III.
124. As for the selection of a corpus (on which see Bérard 1983: 27; Schmitt-Pantel and Thelamon 1983: 17), mine is, I hope, conceptually open—though space forced me to confine my analyses to scenes involving Theseus or one youth depicted through the Theseus scheme. I left out scenes such as that on the stamnos Athens

NM 18063 (*ARV* 1028.13; 1678; *Add.* 155) which includes a chariot (an element which belongs to the iconography of some abductions, but not to our pursuit's) though it may be semantically related to our theme, because this is the methodologically neutral strategy, for if (as is probable, given the important divergence) it represents a different subject, its inclusion would distort our analyses. A comparison of such scenes to our pursuit must only take place after both have been thoroughly studied independently.

125. With spears: see e.g. column-krater Ferrara T.375 (*ARV* 957.60; *Para.* 433). Without spears: see e.g. column-krater Chiusi, Museo archeologico nazionale, già Coll. Civica n. 1822 (CVA, pls. 6–7).
126. See CVA (n. 125) where it is maintained that it is Boreas. On Boreas and Oreithyia see n. 21.
127. See e.g. Nolan amphora London E 310 (*ARV* 202.84); neck-amphora Oxford 1914.733 (*ARV* 1058.120); stamnos Brussels R 311 (*ARV* 502.6); column-krater in Leningrad from Kerch (*ARV* 532.46); oinochoe Florence 21B 308 frr. (*ARV* 1167.14).
128. Not recognized as an erotic pursuit by Bothmer (1957: 184, no. 73, and pp. 187–8).
129. On the significance of the Amazons and the rule of women myth see Pembroke 1967: 1–35; 1965: 217–47; Lefkowitz 1986: 17–20, 22–3, 26–7.
130. On the defeat of dominant women who represent chaos and misrule see Bamberger 1974: 263–80; Zeitlin 1978: 149–84.
131. Theseus and Antiope: see e.g. the amphora Louvre G 197 (*ARV* 238.1; *Para.* 349; *Add.* 100) where Antiope also carries a battle axe.
132. Schefold 1982: 233–6.
133. On the drawn sword and its meanings see Ch. II.1.
134. Exposed and suckled by a bear: Apollod. 3. 9. 2. On Atalanta: Immerwahr 1885; Fontenrose 1981: 175–81 and 202–4; Lefkowitz 1986: 35, 43–4; 58; Arrigoni 1977: 9–47; Vernant 1985: 19–21. (The Hesiodic *Catalogue* describes a straightforward race between Atalanta and her suitor—on Atalanta in Hes., *Catal.*: frs. 72–6 Merkelbach-West). There are two Atalantas in myth, one Arcadian and one Boeotian. Fontenrose 1981: 176, following Immerwahr 1885: 23–6, argues that there is only one Atalanta, who is primarily a fast runner in the Boeotian legend and a bowmaid in the Arcadian one. I prefer to say that different, if related, myths, pertaining to a wild *parthenos* who refuses marriage, were crystallized in the figure of Atalanta.
135. Melanion or Hippomenes: see Apollod. 3. 9. 2. and Frazer 1921: *ad loc.* On Melanion as a wild ephebe and his relationship to Atalanta: Vidal-Naquet 1983: 171–2.

136. Cf. Apollod. 3. 9. 2. The fact that this erotic pursuit is presented as a test supports the suggestion that the theme 'pursuit and capture of a girl' included a perception of an ephebic test.
137. A properly acculturated Greek man does not go hunting with his wife. The two paradigms of refusal to marry and become acculturated have been made to marry in myth, to express the notion 'wild marriage of not properly acculturated *parthenos* and/or youth'.
138. See e.g. Apollod. 3. 9. 2; and Frazer's commentary (Frazer 1921, *ad loc.*) with references; Vidal-Naquet 1983: 173; Fontenrose 1981: 179–80. On the prohibition of copulation in a sanctuary see Parker 1983: 76.
139. My interpretation fits the scene on the lekythos Cleveland 66.114 (*Para.* 376.266bis; *Add.* 118; Boardman 1983: 3–19) which shows Atalanta as a bride, fleeing, pursued by three Erotes holding flowers and a wreath (connoting marriage) and a whip, usually wielded by Eros in pursuit of boys: the inverted pursuit as a paradigm of wild marriage corresponds to the image of the fleeing wild girl as a bride pursued (by Erotes connoting marriage and especially) by an Eros with a whip, characteristic of schemata involving young males. (On the iconography of Atalanta in general see most recently: J. Boardman in *LIMC* ii (1984), s.v. Atalanta; Bérard 1986: 201–2; 1988: 280–4.)

II.3

Altars with Palm-Trees, Palm-Trees, and Parthenoi*

1. INTRODUCTION; ALTAR WITH PALM-TREE AND *ARKTEIA*

The purpose of this chapter is to consider some aspects of the iconographical element 'altar and palm-tree' (a palm-tree behind, or near, an altar, with the two sometimes combined into one element) in Attic ceramic iconography, especially of the fifth century, and to re-examine one particular aspect of the palm-tree itself, with special reference to iconographical schemata in which it is interchangeable with the sign 'altar and palm-tree'.[1] I will be arguing that there is an important nexus of meanings pertaining to the signs 'altar and palm-tree' and 'palm tree', the significance of which has not hitherto been noticed; and that many of the apparently disparate contexts in which these signs occur can be shown to pertain to this important iconographical and semantic category: altar/sanctuary/realm of Artemis in her persona as overseer of unmarried girls and of their preparation for marriage and transition to womanhood through marriage.[2]

The altar + palm-tree is an established sign in Attic iconography; it is a variant of a category of signs—itself a variant of 'altar'—in which the altar is combined with a tree which denotes the altar's (and scene's) association with one or more divinities. Other variants are 'altar + laurel-tree' (altar of Apollo)[3] and 'altar + olive-tree' (altar of Athena).[4] The 'altar + palm-tree' is associated with more than one deity. There is nothing problematic in that. Signs, even more complex signs like 'altar + palm-tree', are polysemic, and it is the context that determines their particular values in each representation.[5] But polysemy does not imply looseness of usage, and, in my view, the value of the iconographical motif 'altar + palm-tree' has not been properly assessed. It is not—as is sometimes suggested or implied[6]—a random variation of the motif 'altar' which could be used to denote any altar, and by extension (as a spatial indicator) any sanctuary; it is an established iconographical sign with a specialized set of values. Particular painters may have had a predilection for this 'altar + palm-tree' motif (the Niobid Painter certainly did),[7] but their selections were determined by assumptions which included knowledge of the motif's meanings, and so they could not use it in inappropriate contexts (contexts which would evacuate it of its particular meanings and turn it into a simple 'altar'). And even if they did, the viewers could only make sense of the sign in terms of its established values.[8]

In my view, the notion that the 'altar + palm-tree' is equivalent to 'altar', and may denote any altar or sanctuary, is based on a misreading of the place and role of this sign in certain scenes. This does not affect my argument,[9] which is concerned with the identification of one particular facet of the 'altar + palm-tree' and its semantic spectrum. But I want to point out that there are, to my knowledge, no representations of the 'altar + palm-tree' which can be convincingly shown[10] to be associated with a deity other than Apollo, Artemis, Leto, and (less frequently, and almost certainly through the connection with Apollo and Delphi) Dionysos.[11] Representations which may at first glance appear to invalidate this view are only apparent exceptions. The case of Thetis will be discussed below, as will the presence of an 'altar + palm-tree' in a representation which allegedly associates it with Aphrodite. Another apparent exception involves Zeus Herkeios: according to the literary evidence,[12] Priam was killed at the altar of Zeus Herkeios; in the representation of his death on the hydria Naples 2422[13] a palm is shown near the altar on which he is sitting; but this does not mean that Zeus Herkeios' altar is here associated with a palm-tree. For in this scene three separate elements are juxtaposed and function also as indicators of the sacred: from left to right: the Palladion from which Kassandra is being dragged (belonging to Athena whom the sacrilege offends); the palm-tree, which belongs with the Kassandra scene and is shown to be a separate unit from the altar of Zeus Herkeios to its right through the arrangement of its branches, which are turned away from the altar (contrast the arrangement of the branches when the two belong together, e.g. on Douris' Palermo lekythos[14]) and curving to the left and downwards towards the Palladion and a grieving woman sitting behind the tree; the latter figure is facing another woman who is squatting behind the Palladion, and this relationship between, and antithetical arrangement of, the two women[15] helps confirm that the palm-tree is part of the Kassandra scene and is a reference to Apollo and his important connection with Kassandra.

I hope to show that the association between 'altar + palm-tree' and Artemis is more important than has been realized[16] and is not simply dependent on the Apollo/Delos connection. Thus even an 'altar + palm-tree' on its own can signify 'altar of Artemis', as it does on a krateriskos from Mounichia.[17] The context which gives it the value 'altar of Artemis' here is the ritual shape on which the altar is represented, which belongs to the Attic cult of Artemis, and more especially its facet connected with the *arkteia*.[18] Consequently, the 'altar with palm-tree' must not be presumed to be connected with Apollo and/or Delos and Leto except in cases where this interpretation is excluded by the context—which is only imperfectly understood and can thus rarely disprove an *a priori* hypothesis of this kind.[19] The Mounichia krateriskos is one of a series of scenes belonging to the Attic cult of Artemis connected with the *arkteia*[20] in which the 'altar + palm-tree' denotes a sanctuary and cult of Artemis—since it is represented in scenes depicting, and taking place in, her sanctuaries at Brauron and Mounichia.[21] This series of representations shows that 'altar of Artemis', and in particular 'altar of Artemis Brauronia and Mounichia, Artemis connected with the *arkteia*' was one of the established meanings of 'altar + palm-tree' in fifth-century Attic iconography. I want to argue here that there is a wider association between the sign 'altar + palm-tree' and a very important aspect of the persona of Attic Artemis which comprises her involvement with the *arkteia*: Artemis as protector of *parthenoi* and of their preparation for marriage and transition to womanhood. I will consider a series of iconographical themes in which—I will argue—the 'altar + palm-tree' denotes an altar of Artemis and relates to this aspect of her persona.

The combination 'altar with palm-tree' is found in a representation of the sacrifice of Iphigeneia on a lekythos by Douris in Palermo,[22] that is, a context closely related to Brauron and the *arkteia*;[23] this scene is iconographically related to nuptial imagery, the iconography of the bride.[24] We can be certain that the altar belongs to Artemis, for this is a central element of the story and, moreover, it is inscribed: 'AR' (Artemidos).

2. ALTAR + PALM-TREE, PALM-TREE, AND EROTIC PURSUIT

The second theme in our series is, like the Palermo scene, associated with the nuptial sphere. The 'altar + palm-tree' is represented in many erotic pursuits/abductions, depicting different myths, with different protagonists:[25] the erotic pursuit of a girl by Theseus (which, I argue, connotes also marriage);[26] the erotic pursuit/abduction of Thetis by Peleus[27] (an established paradigm for wedding/marriage),[28] and the erotic pursuit of (probably) Thetis by Zeus.[29] It is also shown in some other pursuits of which the following is a sample: the erotic pursuit of an unidentified girl by Hermes;[30] the pursuit of a girl by Zeus;[31] and an unidentified erotic pursuit of a girl by a man.[32] This appearance of the 'altar + palm-tree' in so many different erotic pursuits/abductions suggests a connection between the latter theme and the sign 'altar + palm-tree' (and the ritual complex which it signified). I shall return to this; but first let us glance at the sign's constituent elements (the altar and the palm-tree) in erotic pursuits/abductions.

The palm-tree on its own (without the altar) occurs in contexts similar to those of the 'altar + palm-tree'; this, as will become clear, confirms the strong connection between the palm-tree and the *parthenos* in contexts pertaining, or related, to the nuptial sphere and also the association with Artemis.[33] Like the altar + palm-tree, the palm-tree on its own is represented in scenes taking place in the sanctuaries of Artemis at Brauron and Mounichia.[34] It is also depicted in several representations of erotic pursuit/abduction involving different protagonists, of which the following is a sample: the pursuit of Thetis;[35] that of the Leukippids;[36] the erotic pursuit of an unidentified girl by Hermes;[37] that of Oreithyia by Boreas.[38] It is also represented on a stamnos showing an erotic pursuit by Theseus as part of its handle decoration:[39] a palm-tree is painted through each handle. It could be argued that since this type of handle decoration is not limited to stamnoi with scenes in which the palm is iconographically relevant, its presence in this pursuit is not significant. In my view, this reading is fallacious, and the palm's inclusion was determined by its appropriateness, and was perceived by fifth-century Athenians to be part of the scene. First, because such a handle treatment is unique in this period[40] (which suggests the use of an earlier schema to represent a significant iconographical element) and second, and more importantly, because this is not an isolated instance of a palm-tree in a scene of this kind, but one which fits within a category of representations of erotic pursuit which include a palm-tree.[41] The altar on its own is a common motif in erotic pursuits/abductions.[42] In some representations of the abduction of Oreithyia a different type of altar is represented, belonging to the same category as our altar + palm-tree: an altar + olive-tree[43] which, of course, represents an altar of Athena. The Leukippids' abduction can be depicted near an altar, in a sanctuary, identified as Aphrodite's by the goddess' presence.[44] I will return to the significance of these representations below.

In my view, the association between 'altar + palm-tree' and *parthenoi* in erotic pursuits/abductions makes best sense in the framework of an association with Artemis and her realm. First, because, we saw, 'altar of Artemis' is an established value of the sign 'altar + palm-tree', and one which is particularly pertinent in a context involving *parthenoi*, their preparation for marriage, and transition to *gynaikes*. Second, because the theme erotic pursuit/abduction has strong associations with Artemis' realm, and particularly with the facet of Artemis' cult to which the *arkteia* belongs. The *arkteia* and the sanctuary of Artemis Brauronia which are connected with the 'altar + palm-tree' are also semantically related to 'erotic pursuit/abduction'. The sanctuary of Artemis Brauronia is connected in myth with girls' abductions: girls and women taking part in the ceremonies are said to have been abducted from there.[45]

Given the Brauronian cult's concern with unmarried girls, and the fact that the *arkteia* was a pre-nuptial rite preparing girls for marriage, and given also the nuptial connotations of the theme 'pursuit/abduction',[46] the association between the Brauronian cult and 'pursuit/abduction' can be seen to be very strong and may even reflect a connection between the Brauronian ritual and the notion 'erotic pursuit/abduction of girls'.[47] The motif 'girls abducted from the Brauron sanctuary' is a version of the general mythological motif 'girls abducted from sanctuaries of, or from choruses of girls dedicated to, Artemis',[48] which is a narrative articulation of the notion 'girl taken away from a place, a realm, belonging to Artemis'. In these circumstances,[49] and since one of the established meanings of 'altar + palm-tree' was 'altar of Artemis Brauronia/Mounichia', fifth-century Athenians would have understood the sign 'altar + palm-tree' juxtaposed to 'unmarried girl', in representations of erotic pursuit/abduction to be an altar of Artemis. This reading was further underpinned by the existence of the established motif 'girl abducted from a sanctuary of Artemis'. For representations of pursuits near an altar of Artemis express pictorially the same notion as that mythological motif: they both reflect, and articulate in narrative terms, the notion 'girl being taken away from a place, a realm, belonging to Artemis'.

One of Artemis' important roles, we saw, pertains to the protection of *parthenoi* and their transformation into gynaikes.[50] Thus, given the nuptial connotations of pursuit/abduction, 'girl abducted from Artemis' sanctuary' clearly pertains to the (pictorial and literary) metaphor which represents wedding/marriage through erotic pursuit/abduction: marriage is—can also be represented as—a girl's forcible removal from the realm of Artemis where she had, as an unmarried girl, belonged. This representation, then, refers to the traumatic aspect of marriage from the girl's viewpoint (as perceived by the established discourse, which is, of course, male) and in general reflects and expresses certain perceptions of wedding/marriage which I discussed above.[51] In the images the pursuit is located in Artemis' realm through the 'altar + palm-tree', which—I am suggesting—represents a ritual complex associated with Artemis' role as protector of *parthenoi* and their preparation for marriage and transition to womanhood.[52]

The altar + palm-tree does not always denote the sanctuary of Artemis in the sense of representing the space from which the girl was abducted, with all its metaphorical meanings. For in the erotic pursuit by Theseus on the volute-krater Boston 33.56[53] the altar + palm-tree is combined with the spatial indicators column and chair which, as I argue in Chapter II.2, could not be combined with the altar + palm-tree in any 'descriptive' representation of space, corresponding to a 'real' reference space. This combination pertains to a conceptual representation in which elements which did not belong together in the 'real' reference space are 'emblematically' juxtaposed to express certain perceptions pertaining to the representation. In this combination the altar + palm-tree connotes Artemis' realm and her role as protector of *parthenoi* (through a double *pars pro toto* trope, by depicting an altar of Artemis, which represents a sanctuary of Artemis, which connotes Artemis' realm).[54] Consequently, the same may be true of other representations in which the altar + palm-tree is not combined with spatial indicators incompatible with it in 'real' space, but may, nevertheless, be representing Artemis' realm rather than Artemis' sanctuary as the mythological location of the pursuit. This suggestion (which does not involve a niggling distinction, but one that will allow us to make sense of some apparent problems) gains support from the consideration of some representations of palm-trees to which we now turn.

It has been suggested[55] that the palm, the altar, and the altar + palm-tree denote the sacred nature of the space in which the pursuit/ abduction took place. This interpretation relies on the assumption (which, I argued above, is incorrect) that the altar + palm-tree does not have specific values and associations with specific deities. More importantly, it disregards the way in which viewers make sense of images:[56] for, given the association of the palm-tree/altar + palm-tree with Artemis (in the ancient iconographical and semantic assumptions) and also the connection between Artemis and her sanctuaries and abductions, even if the palm on its own could indeed denote generic sacred area in other contexts (for which there is no good argument), in the context of pursuit/abduction both the palm and the altar with palm-tree would have been understood by the viewers to refer to the realm of Artemis. As we saw, the one known context in which palm-tree and *parthenos* are juxtaposed is that of the cults of Artemis Brauronia and Mounichia.[57] Thus the palm on its own in pursuits/abductions would also be taken to denote the realm of Artemis. In some representations the palm-tree on its own may denote the sanctuary of Artemis Brauronia/Mounichia and not just 'realm of Artemis'; but this can only be the case when in other versions of the same subject the palm-tree is replaced by an altar + palm-tree—as in the erotic pursuits/abductions involving Theseus, Peleus and Thetis, and Hermes. In other pursuits/abductions the palm-tree represented in some versions is replaced in others by the altar/sanctuary of another deity: Athena's (denoted through the altar + olive-tree) in the case of Oreithyia, Aphrodite's in the case of the Leukippids.[58] The two sets of scenes are not contradictory, nor do they represent different variants of the story. What we are dealing with is two *iconographical* variants of, for example, the abduction of Oreithyia, two different articulations of the theme: one stresses the mythological narrative and situates the myth in the sanctuary in which the story (or one version of it) said it took place, the sanctuary of Athena;[59] the other stresses the character of Oreithyia as a *parthenos* ready for marriage[60] under the general jurisdiction of Artemis, by connoting the realm of Artemis as protector of *parthenoi* and the association palm–*parthenoi* with all the perceptions of the *parthenos* associated with it.[61]

The (simple) altar is likely to have different meanings in the different pursuits/abductions. At one level, given the Greek semantic motif and iconographical schema 'taking refuge at an altar', the representation of the altar in all pursuits/abductions would produce the meaning 'running to an altar for protection', or simply connote the need of such protection, and thus also the fear and trauma of the abducted girl which made the notion of sanctuary at an altar appropriate. At another level, when appropriate, it denoted the sanctuary specified as the location of the abduction in the myth depicted in the scene—in the case of Oreithyia, for example, the sanctuary of Athena. When the altar in an erotic pursuit/abduction was not characterized as belonging to a particular deity,[62] given the relationship between Artemis and the unmarried girl, and the established mythological motif 'abduction from an Artemis sanctuary/chorus', unless the myth or scene specified otherwise, that altar—in so far as it was thought of by the painter and his contemporaries as a specific altar—was probably thought of as an altar of Artemis.[63]

3. PALM-TREES, ALTARS WITH PALM-TREES, AND WOMEN ON ALABASTRA

Another group of scenes in which a girl or woman is juxtaposed to a palm-tree, sometimes in combination with an altar, is found in a series of (mostly white-ground) alabastra.[64] These representations, I will argue, form another major category of images in which the altar + palm-tree is associated with Artemis, *parthenoi*, and the latter's preparation for marriage and transition into womanhood under the former's protection. The scenes on the white-ground alabastra can be divided into three thematic groups: the group I will call 'women's alabastra', the 'Amazon alabastra', and the 'Negro alabastra'; of these only the last two have received much scholarly attention.[65] In my view, the relationship between the three groups and the significance of many of their representations need to be reconsidered.

Since the alabastron was used in Greece as a perfume container,[66] it was associated with beautification, the world of women, and through them also with weddings. These associations can be seen clearly in the contexts in which alabastra are represented on Attic vases,[67] and are also reflected in their iconography: the most important category of scenes depicted on alabastra are scenes from the life of women—including representations of youths and girls.[68] Some vases show 'variants' of the category 'woman': for example Athens 15002 has a Maenad on A and an Amazon on B.[69] Scenes pertaining to *parthenoi* and weddings were most appropriate for the decoration of alabastra,[70] and I shall argue that many of the scenes on the women's alabastra represent, or refer to, aspects of the semantic field 'preparation for marriage/wedding'.[71] At the centre of my investigation is a series of representations on white-ground alabastra by the Syriskos Painter[72] which make best sense if they are seen as related to the realm of Artemis as protector of the *parthenoi* and their transition to womanhood, and if the palm-tree is taken to connote a relationship with that goddess and her realm. First, because 'altar of Artemis/realm of Artemis' is one attested meaning of the altar + palm-tree/palm-tree which fits a women's context; second, and most importantly, because the iconographical analysis of the individual representations leads us to associate the scenes with Artemis' realm and with *parthenoi*/wedding/preparation for wedding, as I shall now try to show.

On Dunedin F 54.78[73] a woman is approaching an altar holding a torch in each hand; there is a palm-tree at the back. The only context which fits the whole scene[74] is that of wedding. For the schema 'woman approaching an altar holding a torch in each hand moving in one direction and turning to look back' is the same as that through which the mothers of the bride and bridegroom are represented in scenes showing the arrival of the wedding procession to the courtyard altar of the bridegroom's house, or—the schema without the altar—in the course of the wedding procession.[75] The woman with a torch in each hand, moving in one direction (towards the altar or in the direction of the procession) and turning back is an established iconographical schema, a sign deployed both in fuller wedding scenes and on its own.[76] Thus the Dunedin image appears to be a reference to, a representation through the *pars pro toto* trope of, the theme 'wedding'.[77] In this scene, then, the altar and palm-tree is represented in a context closely related to that of erotic pursuits (with altar + palm-tree) which were—among other things— paradigms for wedding/marriage. Consequently, and given that one of the established values for 'altar + palm-tree' is 'altar of Artemis', and also the association between Artemis and the *parthenoi* and their preparation for marriage and transition to womanhood, it cannot be doubted that both the painter and his contemporary viewers perceived this altar to be an altar of Artemis and the scene to pertain to her realm. Thus, one frame of reference that came into play in the process through which the viewers made sense of the image[78] was 'wedding', and the other 'Artemis cult in connection with the *parthenoi* and their preparation for marriage and transition to womanhood'. The altar in this scene is clearly polysemic, and helps infuse the whole scene with ambiguity and richness. On the one hand it represents the (house courtyard) altar which the mother with the torches approached in the course of the nuptial rite (and which was not associated with a palm-tree);[79] in this reading the palm-tree refers to Artemis' realm and her jurisdiction over the *parthenoi* and their transition to womanhood through marriage; the combination is similar to that on the volute-krater Boston 33.56[80] in which a courtyard location is conflated with the altar + palm-tree combined into one sign. On the other hand—a difference in emphasis—the altar + palm-tree taken as one sign represented Artemis' altar to which the torches carried by the woman were also appropriate, since the cults of Artemis Brauronia and Mounichia (which were connected with wedding/marriage and especially with the preparation for it) were also associated with torches.[81] In fact, Artemis herself (in Attica) could be visualized with a torch in each hand.[82] Thus, in my view, the female figure with the two torches was also understood to represent a *parthenos* approaching the altar of Artemis Brauronia/Mounichia who presided over her preparation for marriage. This scene, then, when read through fifth-century assumptions (through which the contemporary viewers made sense of it), refers both to the wedding ceremony and to the cultic sphere to which the *parthenos'* pre-nuptial preparation belonged.[83]

On an alabastron by the Syriskos Painter in the Vlastos Collection in Athens[84] a girl/woman is moving to the right towards an altar and looking back; to the right of the altar there is a palm-tree and to the right of the palm-tree a stool. She is holding a flower in her left, and looking at it, and in her right she is carrying (not playing) flutes. Flowers are connected with, among other things, abduction/wedding.[85] Flute-playing is associated with, among other contexts, weddings and the cult of Artemis Brauronia.[86] The stool pertains to the domestic sphere and the world of women—and to some other contexts, like the palaestra, but the other signs make clear that here it is the former that is pertinent.[87] This image then, relates the female figure, first, to the domestic sphere; second to wedding/marriage; and third, to Artemis as protector of the *parthenos* and of her transition to womanhood. Because the flower belongs to the abduction paradigm of wedding, in which the girl is taken from Artemis' realm, and because the flute, we saw, is associated also with the cult of Artemis Brauronia, the emphasis of the image seems to be on the preparation for marriage in the goddess' sphere.

On an alabastron by the Syriskos Painter in a Swiss private collection[88] a woman is standing next to an altar which is on her right and looking back towards a palm-tree to her left (to the right of the altar). An alabastron is hanging in the field. The woman is holding a mirror in one hand and an alabastron from its thong in the other. The mirror and the alabastron pertain to beautification and the erotic sphere;[89] what we must determine is what particular context is referred to here. Contexts in which beautification as represented by the mirror and alabastron was pertinent mostly combine the domestic and the erotic[90]—and thus pertain to respectable women. One of these contexts was that of wedding.[91] Here the juxtaposition of the mirror and alabastron with the altar + palm-tree defines the beautification/erotic context referred to as being that of wedding and its preparation. The comparison of this scene with the very closely related scenes by the same painter on the Dunedin and the Vlastos alabastra, where the references are on the one hand to Artemis and on the other to the domestic and nuptial sphere, confirms this conclusion. Both these scenes, then, are images placing a *parthenos* in a conceptual space which pertains, first, to Artemis the protector of *parthenoi* and supervisor of their preparation for marriage and transition to womanhood, and second, to the nuptial sphere which will bring about that transition, and to the *parthenos*' proper and future concerns, beautification and domesticity.

On the alabastron Brussels R 397[92] a girl/woman (chiton, himation, bonnet) is moving to the right and looking back; she is holding an alabastron hanging from its thong, with another alabastron hanging in the field; to her right is a stool and to its right a palm-tree. The alabastron, we saw, is connected with beautification, women, and weddings; the stool pertains to the domestic sphere. The juxtaposition girl/woman—domestic context—beautification entails that in the eyes of fifth-century Athenians the palm-tree would be read in terms of its association with the *parthenoi* and their protector Artemis. The scene, then, is an image of a *parthenos* concerning herself with her proper task of beautification[93] in the internal space in which she belongs (and to which she will be tied by her marriage for the sake of which the beautification is indulged in); however, through the palm the *parthenos* is also connected with the outside and Artemis' realm.[94] The scene on the alabastron by the Syriskos Painter in Barcelona[95] is the same as on that in Brussels: a girl (chiton, himation, bonnet) is moving forward, holding an alabastron from a thong, with another alabastron hanging in the field; to her right is a stool and to its right a palm-tree. It has the same meanings.

On the alabastron in Athens, National Museum, Stathatos Collection,[96] a girl (chiton, himation, bonnet) is moving to the right and looking back, holding up her chiton with her left hand and with a flower in her right. On her right is a heron/crane[97] and on its right a palm-tree; an alabastron is hanging in the field; on the woman's left is a chair on which stands a kalathos (wool basket). Flowers are connected, we saw, with (among other things) abduction/wedding. The combination flower and *parthenos* here leads us strongly to that pre-nuptial/nuptial sphere. The alabastron is connected with beautification and wedding; the chair denotes the indoor domestic space;[98] the kalathos denotes women's pursuits and, again, the domestic space, but it is also connected with abduction and with the nuptial sphere.[99] In this context, there can be no doubt that the heron/crane, usually represented on Greek vases in connection with domestic scenes and the women's quarters, has a domestic meaning, belongs with the other signs which build up the space 'women's quarters' and the conceptual area 'women's world' in this representation.[100] But there may be an additional connotation to this bird: it is sometimes depicted in representations of erotic pursuit,[101] which links it with the pre-nuptial and nuptial sphere (itself connected with the iconography of domesticity); given the context, this connotation may also have been carried by the heron in the alabastron scene. All these elements, including the palm, are polysemic, and acquire their meaning in context;[102] in this representation their interaction creates the meanings 'domestic context, women's concerns, marriage/wedding, *parthenoi*/Artemis'.

Altars with Palm-Trees

The representation on the red-figure alabastron Kassel 551[103] also fits within the same nexus of meanings and provides additional support for the interpretations put forward here. This alabastron shows a girl (chiton, himation, bonnet) with a phiale in her right and *kanoun* in her left pouring a libation at an altar on the right of which is a palm-tree.[104] The *kanephoros* is a *parthenos* ready for marriage; indeed, there is a close connection between the office of *kanephoros* and marriageability.[105] This places the *kanephoros* within the realm of Artemis the protector of the *parthenoi* and their transition to womanhood through marriage. Of course, the office of *kanephoros* was attached to many cults.[106] But the above suggests a 'conceptual', 'emblematic' connection of the office with Artemis; moreover, there was one occasion of *kanephorein* in a one-to-one relationship with the deity (rather than participation in a procession/sacrifice) involving a *kanephorein* to Artemis by the *parthenos* about to marry,[107] which can be seen as a ritual articulation of the 'conceptual' relationship between *kanephoroi* and Artemis. This is the context which, in my view, fits the alabastron image best. For, as we saw, altar + palm-tree had an established meaning 'altar of Artemis', and connoted Artemis in connection with *parthenoi* and their preparation for marriage. This suggests that, unless there were elements indicating the contrary, the juxtaposition *kanephoros* at altar + palm would lead the fifth-century viewers to read the image as a *parthenos* who *kanephorei* to Artemis before her marriage. Moreover, this representation of a *parthenos* with *kanoun* shows an individual cult act and/or an emblematic representation of a *kanephoros* at the altar, and this, when read through the semantic assumptions which provided the perceptual filters through which the fifth-century Athenians made sense of that representation,[108] fits perfectly the theme '*kanephorein* for Artemis before marriage' and also articulates the conceptual relationship between *parthenoi kanephoroi* and Artemis.

This interpretation gains strong support through the consideration of another scene, on the stemmed dish Copenhagen inv. 6,[109] in which the representation of a girl at an altar holding a *kanoun* in one hand and a branch in the other is combined with the inscription ARTEMIS. Whether it depicts Artemis herself,[110] or, as is most likely given the *kanephorein*,[111] a young woman at an altar of Artemis, this representation shows that the schema 'girl with *kanoun* (on her own) performing a cult act at an altar' certainly pertained to the cult of Artemis—whether or not it also pertained to other cults.

Consequently, since 'altar of Artemis' was an established meaning of 'altar and palm-tree', and since there were no additional elements directing otherwise, the fifth-century viewer would have understood the scene on the Kassel alabastron to be representing an altar of Artemis. Clearly, the two scenes are variants of one basic schema, and the palm-tree combined with the altar on the red-figure alabastron corresponds to the inscription ARTEMIS (which can be described as combined with, qualifying, the altar) on the dish.[112] This, I submit, confirms my interpretation of the 'altar + palm-tree' as 'altar of Artemis' on the alabastra considered above. For the scene on the red-figure alabastron is iconographically so close to the white-ground women series (as well as being painted on the same shape as them) that it cannot be doubted that the altar + palm-tree juxtaposed to a girl in the two cases must have meanings pertaining to the same semantic area.

In these circumstances, we conclude that in this series the palm-tree is associated with the world of women, with special reference to a nuptial/pre-nuptial context pertaining to Artemis' realm. Since the analyses of the individual scenes were conducted independently of each other, the interpretations of each scene gains support from the fact that all the proposed interpretations tally with each other and point to the same semantic/thematic area, an area which itself tallies with my hypothesis concerning the association between altar + palm-tree (or palm-tree) and *parthenos* in other contexts. The scenes on the Syriskos Painter alabastra fall in two categories, based on two (related and alternative) iconographical schemata: the first combines the elements girl + palm-tree + altar, the second girl + palm-tree + stool (or chair). The former places the emphasis on the cultic sphere and the *parthenos*' preparation for marriage within the cultic framework, the latter on the domestic sphere and the *parthenos*' and future *gyne*'s environment and pursuits.

Let us now glance briefly at the Amazon and Negro alabastra, which also depict altars with palm-trees and palm-trees. Wehgartner,[113] who correctly connects the representations on the 'women's alabastra' with the vases' use as perfume containers, takes the (in one form or another) widely held view that Amazons, Negroes, and palm-trees refer to exotic lands famous for their fragrant essences. However, though this may indeed be one of the meanings of these elements, as far as the Amazons, the palm-trees and altars, and palm-trees are concerned, it cannot be the only, or indeed the primary one, for the following reasons. First, this view ignores the Amazons' polysemy in the ancient collective representations and myths, and the fact that they are also—and first of all—a form of 'wild women',[114] which entails that[115] the representations of Amazons, as of Maenads, on the alabastra are 'wild women' variants of the women series.[116] Furthermore, that commonly held view also ignores the Amazons' connection with Artemis,[117] and the association of the altar + palm-tree and the palm-tree with Artemis. In fact, the association 'Amazon—altar + palm-tree representing an altar of Artemis' makes perfect sense in terms of ancient assumptions at a variety of levels, so that the 'altar of Artemis' interpretation produces a series of meanings for these scenes—which are not mutually exclusive, but help construct a rich polysemic reading of a rich polysemic image. First, a representation combining an Amazon and an altar of Artemis can be understood as a mythological scene, representing, as Neils has suggested, an Amazon taking refuge at a sanctuary of Artemis.[118] She may, as Neils' hypothesis suggests, be fleeing from a battlefield and taking refuge at Artemis' altar. Second, these representations combining Amazon + altar + palm-tree can be understood as emblematic representations of the association between the Amazons and Artemis. Third, they may also have been understood in terms of the Amazons' nature as, among other things, *parthenoi*.[119] This produces two different (complementary) readings. The first involves, like Neils' hypothesis, the notion that the Amazon is running to the altar for protection, this time in the course of an erotic pursuit. For fifth-century viewers, the representation of a female moving/running towards an altar + palm-tree called up the schema 'pursued girl and/or her companions running to an altar, especially an altar + palm-tree'; therefore this schema, which was part of the theme 'erotic pursuit/abduction', helped ascribe value to any scene which included a female moving/running towards an altar + palm-tree. Most probably, the images on the Amazon alabastra were ambiguous in this respect also: they were not explicit erotic pursuits, but the notion 'erotic pursuit' was alluded to, and coloured the way in which they were perceived.[120] The second reading involves an emblematic juxtaposition of an Amazon as *parthenos* and the altar of Artemis, similar to that encountered in the women's series (to which this group of Amazon alabastra is closely related), especially on such alabastra as that in a Swiss collection, of which Amazon alabastra such as Cracow 1292 seem to be a 'wild woman version', with the Amazon's weapons replacing the 'proper womanly' mirror and alabastron. The helmet on the ground is the equivalent of a feminine object such as the alabastron hanging in the field: the helmet characterizes the Amazon and her pursuits in the same way as the alabastron characterizes the normal *parthenos*.[121] In these circumstances, and given (*a*) all these associations between the figure of the Amazon and Artemis, and the meaningful ways in which the altar of this goddess can interrelate with that Amazon figure, and (*b*) the established value 'altar combined with palm-tree = altar of Artemis', it was inescapable that a representation in which an Amazon is juxtaposed to an altar and a palm-tree would be taken by the fifth-century Athenians (who shared the above assumptions) to depict an Amazon and an altar of Artemis.

This interpretation is confirmed through a series of iconographical arguments. The representations on the Amazon alabastra involve the following basic iconographical schemata: 1. Amazon + palm-tree + altar;[122] 2. Amazon + palm-tree + stool;[123] 3. Amazon and youth leaning on a stick (with or without stool).[124] The first two are the same as the two variants of the women's alabastra with palm-trees by the Syriskos Painter considered above, with an Amazon replacing a *parthenos*. The third schema is also known from the series representing 'normal' women,[125] is also a variant of the series of alabastra scenes depicting women. Thus, all the schemata on the Amazon alabastra are the same as those on the alabastra representing women and women's pursuits and concerns. The divergences in the iconographical elements deployed in the two variant series are precisely correlated to the differences obtaining between Amazons and 'normal *parthenoi*'.[126] This interpretation, which connects the Amazons with the women's alabastra and with Artemis—as well as with the Negro alabastra and the 'exotic, the foreign'—is strongly supported by the reading of the individual scenes, which, I will suggest, supports my hypothesis that the *parthenos* aspect is an important facet of signification in the Amazon alabastra scenes.[127]

I tried above to indicate how the type of Amazon scene on the Cracow alabastron (based on the schema 'Amazon + altar + palm-tree') would have been made sense of in terms of fifth-century assumptions; I shall now do the same for the other main variants. First, the two variants of the schema 'Amazon + youth'. On New York 21.131[128] an Amazon wearing a panther skin,[129] and with axe and bow (which characterize her as Amazon), is moving to the right (looking back) towards a (folding) stool on which there is a helmet; on the stool's other side is a youth leaning on a stick. The basic schemata to which this scene relates and which help ascribe it meanings through similarities and differences are: (1) the schema 'youth leaning on a stick and girl', seen, for example, on the alabastron Athens Kerameikos HS 107[130] (on which the girl is sitting on the stool); (2) other schemata involving a girl and a youth leaning on a stick in which a stool and a kalathos are included;[131] and (3) the combination 'kalathos on a stool/chair' in scenes such as that by the Syriskos Painter in Athens[132] and the cup Berlin 2289[133] where the kalathos on a seat is represented in a domestic context. We saw above that the stool is associated with a domestic context and the women's quarters. The variant seen here, the 'folding stool', tends not to be associated with women's quarters scenes. It is found in, among others, military contexts, and appears to be more commonly associated with men than with women.[134] Possibly, given its pattern of appearance,[135] the folding stool may have been depicted primarily in contexts carrying connotations of impermanence, with secondary connections such as its appearance in military camps being derivative from it. Raeck takes the helmet on the folding stool to pertain to an arming scene and denote preparations for battle.[136] But this interpretation only covers one facet of this element's meanings; for meaning is created also through the relationships (of similarity and differentiation) with the elements which are (in one way or another) 'alternatives' of the element under consideration, and which replace it in other versions of the representation—and are thus called up by it in the process of meaning-production by the viewer, and help to define it (see Chapter 1). Consequently, the 'helmet on the folding stool' acquires value also through its similarities to, and differences from, the 'kalathos on seat' seen elsewhere in connection with 'normal women'. 'Helmet on the folding stool' is to the Amazon what 'kalathos on stool/chair' is to a 'normal woman'. If the folding stool has the connotations tentatively suggested above, it would be particularly suitable for an Amazon: not, like the other stools, closely connected iconographically with the world of women, but more ambivalent (as the Amazon figure is ambivalent), possibly recalling the residential quarters of a military camp and thus the warlike character and abnormal (for a female) pursuits of this wild woman. In any case, the helmet, which represents war and the warlike pursuits of the wild female who refuses her proper role, replaces the kalathos, implement of the proper, domestic, pursuit of the proper *parthenos* who will become a proper *gyne*.[137]

Thus, this variation on the 'stool/chair + kalathos' is correlative (as are the weapons which replace the feminine attributes in all the Amazon scenes) with the differences between Amazons (both (wild) women and warriors) and 'normal' *parthenoi* and *gynaikes*. In these circumstances, New York 21.131 must be interpreted as an erotic male–female juxtaposition, whether representing an imagined encounter or perceived as an emblematic image. When the image is read in terms of fifth-century assumptions it is impossible to interpret it in any other way. First, because of the schemata and iconographical elements discussed above, and second because elsewhere (both in iconography and in literary myth) youths and Amazons relate in one of two ways (or a combination of both, with one dominant): they are either fighting, or the male is pursuing/abducting the Amazon.[138] Here the war interpretation is excluded: first, the youth is not armed, but leaning on his stick and looking towards the Amazon. Second, the youth and girl schemata which relate to the representation and help the viewer make sense of it have an erotic content (n. 131). The precise significance of the New York scene eludes us; but it is likely to have been ambiguous and ambivalent: has the Amazon (as a refraction of the *parthenos*) abandoned the helmet and war to be subdued by the male and become a proper *gyne*? Will the relationship remain adversarial, with violence being the only way in which the Amazon will be subdued? As with the Theseus–Antiope relationship in the various versions of the myth, I think both 'messages' are to be read in this picture. Certainly, in my view, the scene represented an erotic encounter/relationship between a youth and an Amazon as an image of perceptions pertaining to women, the *parthenos*, and the relationships between the sexes in fifth-century Athens. The notion that the Amazon will be in one way or another ultimately tamed, and that the (distorted) image of the *parthenos* crystallized in the Amazon figure will turn into a proper *gyne*, is probably one of the 'messages' reflected or hinted at in this scene.[139]

This interpretation gains support from the consideration of the variant of the Amazon and youth schema on the Basle alabastron, which represents an Amazon, a youth leaning on a stick, and a heron.[140] This is very similar to the schema on the alabastron

Athens Kerameikos HS 107 which shows a girl (seated on stool), a kalathos, a heron, and a youth leaning on a stick.[141] The schema girl and youth suggests an erotic or potentially erotic situation. The heron, we saw[142] has domestic connotations pertaining to the women's quarter and is also shown in erotic pursuits; this and its representation on Kerameikos HS 107 shows that it is appropriate for 'normal' boy–girl erotic scenes.[143] Thus, this scene is, like the New York one, a transformation of the boy–girl representations, pertaining to the nexus of Amazon scenes which include an erotic content. The Basle scene is closer to, and less sharply differentiated from, the boy–girl scenes, and includes only the heron as an additional element, thus showing the Amazon further down the road to taming and domestication.

The remaining schema of the Amazon alabastra, the combination of Amazon, palm-tree, and stool, stands in a similar relationship to the first (Amazon + altar + palm), as that in which the schema '*parthenos* + palm' stands to '*parthenos* + altar + palm' in the women's alabastra by the Syriskos Painter, and has comparable meanings— pertaining to the Amazon as (among other things) *parthenos*. The fundamental difference lies in the other sets of meanings which in this schema were probably more important than in the other two Amazon schemata: the meanings pertaining to the Amazons as Oriental foreign warriors, with all the symbolism attached to such figures, especially after the Persian Wars. In this other dimension of signification, the palm-tree (which in the *parthenos* set of meanings denotes the association with Artemis and the identity of the Amazon as a *parthenos*) represents also the exotic, and the identity of the Amazon as a foreign warrior, an ally of the Oriental enemies of the Greeks, the Trojans—which made the Amazons adequate symbols for representing the Persians, whom the Trojans and their allies as well as other foreign and/or uncivilized figures symbolized in post-Persian war Greece. The palm, then, like the Amazon and the whole scene, is polysemic.

It is this set of meanings, pertaining to the foreign/Oriental/ Persian aspect of the Amazons, which inspired the creation of another series, another variant on the theme 'exotic foreign warrior/Trojan ally', in which a Negro replaces the Amazon in the iconographical schemata 'figure + altar + palm-tree' and 'figure + stool + palm-tree'.[144] In this hypothesis, the Amazon alabastra are an offshoot/variant of the main women alabastra series,[145] and the Negro alabastra are an offshoot/variant of the Amazon alabastra. The hypothesis that the Negro alabastra developed out of the Amazons series,[146] also suggested by Neils,[147] is supported by chronological as well as iconographical considerations: Negro representations are not found in the earliest group of white-ground alabastra.[148] I am not concerned here with reconstructing the meanings of the representations on the Negro alabastra. I take it that in them the palm denoted above all the exotic; and that the meaning 'altar/realm of Artemis' probably remained attached to the representation of altar combined with palm-tree,[149] and a suitable association created between that realm and the figure of the Negro, on the basis of the fifth-century assumptions.[150]

4. THE HYDRIA GENOA 1155

The altar + palm-tree is juxtaposed to a *parthenos* also in the representation on the hydria Genoa 1155,[151] which shows a girl wearing a chiton and a diadem (apparently) emerging from the ground and being received by two figures: on the left Eros offering her a taenia, towards whom she is turned and to whom she relates through stance and gestures; behind her on the right a woman, the lower part of her body turned towards an altar + palm-tree on the right, the upper part, including the head, turned towards the girl and holding up a cloth. (Another taenia is hanging in the field.) As Bérard[152] has shown, the scene is not mythological, as may at first appear.[153] Bérard took it to be cultic, representing a rite of passage modelled on that of Aphrodite, and involving the covering of the initiate's head; the scene, he thinks, takes place just after the girl's head was uncovered, and the priestess is about to consecrate in the sanctuary the cloth/veil removed from the girl's head. In my view, it is equally possible that the priestess has just removed the cloth from the altar/sanctuary and is about to put it around the girl. I shall return to this. The ritual hypothesis proposed by Bérard is one possible interpretation of this scene; another (not necessarily alternative, but possibly complementary) reading is to suppose that the representation is 'emblematic' rather than descriptive/narrative. Given that there is one important emblematic element in the scene, Eros, the whole representation may be in that emblematic mode, and represent a *parthenos*' emergence into the domain of Eros and

Aphrodite, her maturation into marriageability, through the iconographical schema pertaining to Aphrodite, the goddess who presides over the world of love which the girl is now entering through coming to maturity.[154]

In either case, the fact that the representation is modelled on that of the birth of Aphrodite does not necessarily entail that the altar + palm-tree also pertains to Aphrodite—whether it denotes the sanctuary in which the ritual is taking place[155] or connotes a cultic realm to which the representation and its meanings are pertinent, or both. On the contrary, given Artemis' role as protector of the *parthenoi* and supervisor of their preparation for marriage, we can be certain that the *parthenos*' emergence into the domain of Eros and Aphrodite, and thus her emergence into marriageability[156] which is represented in this scene, took place under the general protection and jurisdiction of Artemis. Thus, if the altar + palm-tree is interpreted as representing an altar/sanctuary of Artemis and/or connoting Artemis' realm and jurisdiction over the *parthenoi*, the representation would correspond perfectly to the cultic/semantic reality of the *parthenoi* as it is reconstructed from other sources. To put the argument differently: if it is correct that one of the major meanings of 'altar + palm-tree' was 'altar of Artemis', and that this meaning was the established value—or one of them—when the 'altar + palm-tree' was juxtaposed to a *parthenos* in a context pertaining to her preparation for marriage/transition to womanhood, the fifth-century Athenians would inescapably interpret the altar + palm-tree in this scene as 'altar/realm of Artemis', unless this meaning was somehow blocked, which it was not. For—despite the birth of Aphrodite schema—the representation of a girl emerging on to Eros, being semantically related to 'erotic pursuit' and the pre-nuptial maturation rite of the *arkteia* which pertain to Artemis, not only does not block the association with Artemis (for those to whom the role of Artemis in connection with *parthenoi* was taken for granted) but on the contrary reinforces it.

If the priestess is going to hand over a himation for the girl to put on, we must imagine that at the end of this scene the latter will be wearing chiton, himation, and diadem; that is, she will be dressed in the same way as—among others—the bride in nuptial scenes, especially if we are to imagine (as the contemporary viewers who shared the painter's assumptions may have done) that she is to wear the himation over her head in the manner of a bride.[157] This hypothesis (that the girl is to be dressed as a bride) would fit the subject of the scene[158] and also my interpretation of the altar + palm-tree. The only other element in the scene besides the altar + palm-tree is the taenia hanging in the field, which is also found in scenes located in the sanctuary of Artemis at Brauron and depicting *arkteia* rites, in erotic pursuits, and in wedding scenes,[159] and so fits perfectly the semantic nexus pertaining to Artemis and the *parthenos* considered here.[160]

5. PALM-TREES AND *PARTHENOI*: VARIA

Space prevents me from undertaking a systematic analysis of all representations in which I consider the palm-tree to pertain to Artemis and/or the *parthenos* and her preparation for marriage and transition to womanhood. Thus I will only mention here a few important and/or characteristic examples of this type of iconographical association.

The hydria Athens N.M. 17469[161] depicts girls around a palm-tree, one holding an alabastron, one a stylized flower; in the field a pyxis/box and twigs. As seen above, the alabastron connotes beautification and is connected with weddings, and the flower is also connected with representations and stories of erotic pursuit/abduction which function also as paradigms for weddings. Pyxides/boxes are found in nuptial representations in association with the bride and have very strong connections with weddings.[162] The twigs also allude to the marriage ceremony in certain representations, namely representations of erotic pursuits.[163] In my view, the only context in which all these elements can be seen to make sense is that of '*parthenoi* in connection with marriage'.[164] If this is correct, this hydria would offer another example of a juxtaposition of *parthenoi* and a palm-tree placed in a context pertaining to womanly pursuits—especially of a nuptial character. The scene may perhaps be interpreted as an emblematic representation of the notion 'marriageable *parthenoi*'.

My second example is on the calyx-krater Bologna Pell. 300,[165] which depicts the race between Atalanta and Hippomenes and includes a palm-tree. If I am right in relating this race to the theme erotic pursuit of which I argue it is a transformation,[166] the presence of the palm-tree in it is comparable to the presence of the

1. Stamnos Oxford A.M. 1911.619. Courtesy of the Visitors of the Ashmolean Museum, Oxford

2. Volute-krater Oxford A.M. 525. Courtesy of the Visitors of the Ashmolean Museum, Oxford

3. Pelike Manchester III.1.41

4. Stamnos London, British Museum E 446. Courtesy of the Trustees of the British Museum

5. Bell-krater Leningrad, Hermitage 777 (St. 1786), side B. Courtesy Hermitage Museum

6. Bell-krater Leningrad, Hermitage 777 (St. 1786), side A. Courtesy Hermitage Museum

7. Pelike Leningrad, Hermitage 728 (St. 1633). Courtesy Hermitage Museum

8. *(left)* Lekane fragment Leningrad, Hermitage. Courtesy Hermitage Museum

9. *(below left)* Neck-amphora Leningrad, Hermitage 709 (St. 1461), side A. Courtesy Hermitage Museum

10. *(below)* Neck-amphora Leningrad, Hermitage 709 (St. 1461), side B. Courtesy Hermitage Museum

11. Cup London, Victoria and Albert Museum, 4807.1901, side A

12. Cup London, Victoria and Albert Museum, 4807.1901, side B, showing altar and palm-tree combined

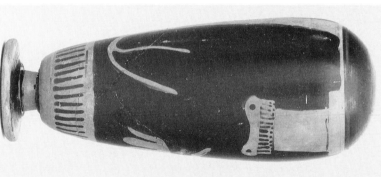

13. Alabastron Kassel 551, showing altar and palm-tree side by side

14. Drawing of Locrian pinax: Hades and Persephone enthroned (after Orsi, *Bolletino d'arte*, 3 (1909), fig. 8)

15. Drawing of Locrian pinax: Offering girl (after Orsi, *Bolletino d'arte*, 3 (1909), fig. 6)

16. Drawing of Locrian pinax: Child in the basket (after Orsi, *Bolletino d'arte*, 3 (1909), fig. 39)

17. Drawing of Locrian pinax: Child in the basket (after Orsi, *Bolletino d'arte*, 3 (1909), fig. 41)

palm-tree in erotic pursuits, where it alludes to the role of Artemis in connection with the *parthenoi*. (Of course, the huntress Atalanta is especially connected with Artemis.)[167]

I will now say something about a group of scenes which form a special category and are again related to Artemis as protector of *parthenoi* and their preparation for marriage. Some scenes pertaining to the *arkteia* show the palm-tree in combination with an animal sacred to Artemis.[168] There are two types of palm + animal combinations in *arkteia*-related scenes: palm-tree and fawn/female deer on a krateriskos fragment from Mounichia;[169] and palm-tree and bear on a red-figure krateriskos which has convincingly been associated with Brauron.[170] The palm-tree and fawn/female deer on the Mounichia krateriskos establishes that this sign has a place in the Attic cult of Artemis concerned with the *parthenoi* and their preparation for, and transition to, marriage. The fawn/female deer without the palm is also associated with girls in the context of the *arkteia*.[171] The palm-tree + fawn/female deer is found in other contexts connected with Artemis[172] and also in contexts involving a girl's transition to womanhood through sexual possession/marriage, such as representations of Io.[173]

6. CONCLUSIONS

I hope to have shown that 'altar of Artemis', especially of Artemis in her role as protector of the *parthenoi* and of their preparation for and transition to marriage, is an important value of 'altar combined with palm-tree' in Attic iconography, especially of the fifth century; and that the palm-tree on its own was also connected with that Attic cult and persona of the goddess, and also, through that, with the marriageable *parthenos*. This interpretation allows us to understand the occurrence of these iconographical elements in scenes which represent different themes but pertain to the same semantic area concerning marriageable *parthenoi* and Artemis in her role as protector of the girls' preparation for marriage and of their transition to womanhood. The representation of these elements refers to this particular facet of the represented subject, stresses this aspect of the theme. In some cases, such as that of Thetis, the relationship with Artemis is to be understood metaphorically—not as 'real' (at the level of myth), for the *parthenos* Thetis did not worship at a sanctuary of Artemis; but as an expression of the perception that the transition from *parthenos* to *gyne* through marriage was an important dimension of this theme (most pertinent because Thetis' abduction was a nuptial paradigm), through the iconographical metaphor 'Artemis' realm' (as, for example, in 'abducted from Artemis' realm') signified through the palm, altar + palm-tree, and palm-tree + fawn.[174] The relative frequency of the palm signs in the representations of the Thetis myth adds some support for this view: for, given this myth's role as a nuptial paradigm, it is precisely what we should expect, if the above interpretations are correct.

It is hardly surprising that, if my conclusions are valid, the *parthenos*' preparation for marriage should have attracted the interest of Attic vase-painters (especially in connection with the decoration of vases destined to be used by women). For marriage was the most important event in a woman's life, and the proper preparation for it was of fundamental importance to her and to society.[175]

7. Epilogue

In a review of the original publication of this essay (in *REG* 102 (1989), 106–7) Metzger accuses me of assuming that iconographical signs had a fixed value—for this is the only possible interpretation of his rhetorical exclamations that imply that I am imagining the vase-painters as using the iconographical signs as printing characters and that I am guilty of a fetishism of the sign, and of his statement that I make the same error as those who ascribe an absolute value to the divine attribute.

But it must be obvious to the most casual reader that far from ascribing a fixed value to iconographical signs, on the contrary, I stress repeatedly and strongly in this chapter (as in the original essay), and in all others, including the methodological Chapter I, that I take signs to be polysemic, and to acquire their meaning in context in interaction with the other elements in the scene and indeed with other semantically related elements which might have been chosen in their place but were not; and that I take the altar + palm-tree to have more than one meaning, and that it is certain types of context which I have tried to determine that, I argued, ascribe it the meaning 'altar of Artemis'.

It is not the case that if we reject monosemic fixed values we have to accept looseness of usage. Nor is it a matter of the painters 'obéissent à des règles impératives et immuables et usent des signes comme on use de caractère d'imprimerie'. The vase-painters' selections, and the viewers' readings, are determined by the parameters of their assumptions—cultural and those pertaining to the conventions of the medium—which we can attempt to reconstruct systematically and thus ultimately reconstruct, as far as possible, the process of meaning—creation and thus also the meanings which the fifth-century vase-painters had inscribed into the scenes and which their contemporaries, who shared their assumptions, constructed out of them.

The only possible explanation of this serious misunderstanding by such a distinguished iconographer (for the hasty reading hypothesis which would have imposed itself in the case of a less conscientious scholar than Professor Metzger must be excluded in this case) is that the schemata of the more common reading models—which involve an implicit dichotomy 'fixed meaning or indeterminacy'—formed a perceptual screen through which my text was filtered. The methodology and readings I am suggesting are very complex and do not fit into simple schemata of either fixed meanings or indeterminacy. I submit that Professor Metzger's misreading of my text (which was the product of the same general cultural, but different epistemological, assumptions as his) proves the point I am making for the necessity of a complex and rigorous methodology for the reading of the ancient cultural artefacts.

NOTES TO CHAPTER II.3

* I would like to thank Dr D. C. Kurtz for extending to me the facilities of the Beazley Archive, and also to acknowledge the help of the Beazley Archive Project, which allowed me to add two vases to my list of altar + palm-tree representations.

1. Altar and palm-tree combined as one sign: here Pl. 12; side by side: here Pl. 13b–c. On the significance of the palm-tree and palm-ornaments on vases: Jacobstahl 1927: 93–102, Deonna 1951a: 162–207; 1951b: 5–58 passim. Deonna's study is unsystematic and speculative; he does not attempt to establish a historical framework and distinguish the secondary associations of the palm-tree, which develop in certain circumscribed contexts: e.g. through the Delian Confederacy which turned the palm-tree into, among other things, a 'political' symbol connoting Athenian power, as in the Eurymedon dedication (on which

see Amandry 1954: 314–15; Gauer 1968: 105–7); through this the palm-tree became associated with Athena, and it is undoubtedly this connection and symbolism that is reflected in Callimachos' lamp of Athena: Paus. 1. 26. 6–7). Also on palm-trees see: A.-B. Follmann, CVA Hannover 1, p. 24; Le Roy 1973: 263–86; Simon 1983: 85; Metzger 1951: 179; Dugas 1910: 236–7; see also below nn. 16 and 19 and the notes in Sect. 3, esp. n. 100. I have been unable to consult H. F. Miller, 'The Iconography of the Palm in Greek Art: Significance and Symbolism' (unpubl. diss., Univ. of Calif., Berkeley, 1979).

2. On this role see especially King 1983: 109–27. As King has shown (*passim* and see p. 122), the transition from *parthenos* to *gyne* is a process beginning with *menarche* and completed with the first *lochia*. Here I use the expression 'transition from *parthenos* to *gyne* through marriage' as shorthand for this. The origins and motivation of the association between the palm-tree and Artemis (and *parthenoi*) are beyond my scope here.

3. See e.g. the oinochoe in the Louvre: Devambez 1962: fig. 136.

4. Bell-krater Harvard 1960.344 (*ARV* 1041.10; CVA (USA 6), pl. 47.2).

5. See Ch. I above.

6. This hypothesis (that it could denote simply 'altar') is implicit in Krieger's view (1975: 77–8, 79, 80) that the altar + palm-tree in Thetis' abduction is the altar of Thetis—for she does not try to show that the palm-tree is otherwise associated with Thetis. A similar implication in Kaempf-Dimitriadou 1979: 5, 42; see also Bérard 1974: 118.

7. See for instance the examples in nn. 10, 26, 52.

8. See Ch. I above.

9. Except in so far as the 'generic sacred' hypothesis has provided the context, and protective screen, for the misreading of several instances of Artemis altars in terms of *ad hoc* explanations which can neither be proved nor invalidated but are nevertheless improbable, especially when set against an interpretation which can account for every aspect of the problem in the framework of a wider investigation.

10. In some cases it is very difficult to determine the context and meaning of this sign (see e.g. the pelike by the Niobid Painter (*Palladion. Antike Kunst. Katalog*, Basle, 1976, no. 34)). The pelike formerly in the New York market (Sotheby) depicting a warrior in front of an altar associated with a palm-tree (Sotheby–Parke Bennett, New York, sale catalogue 1–2 Mar. 1984, no. 69) belongs to a similar category—except that (given the relationship between the battle of Marathon and Artemis Agrotera, and the importance of themes connoting the victories over the Persians at this time) we may

conjecture that the altar here belongs to Artemis—especially since this would allow us to relate A to B (a woman with diadem enveloped in himation): hoplite war under the protection of Artemis (characterizing the male citizen) and a marriageable *parthenos* under the same goddess' protection/supervision. (On the relationship between these two themes see Ch. II.2, end of sect. 6.)

11. In most instances, even when Apollo is not present, there is an Apolline connection in the representations associating Dionysos with the altar + palm-tree. See e.g. the neck-amphora Würzburg H 4533 (CVA (Würzburg 2), pl. 14) which on A shows Apollo with a laurel branch, Artemis, and Leto, and on B an altar + palm-tree.
12. Cf. for instance, *Ilioupersis* hypothesis; Pind., *Paean*, 6. 113; *Ilias Parva*, fr. 15, Kinkel.
13. *ARV* 189.74 ;1632; *Para.* 341; *Add.* 94.
14. See n. 22.
15. See also Boardman 1976: 7.
16. On the association of Artemis with palm-trees: Deonna 1951*a*: 186–7; Kahil 1965: 27; Monbrun 1989: 69–93.
17. Krateriskos from Mounichia: Palaiokrassa 1983: Kk 1, pl. 44a (see p. 186); Kahil 1965: pl. 9.12–14.
18. Kahil 1965: *passim* and esp. 31; 1977: 86–98 *passim*; 1981: 253–5; Palaiokrassa 1983: 68–78; Simon 1983: 83.
19. Beazley (1929: 16 n. 1; see also Cohen 1978: 334) takes the altar with palm-tree on the cup Florence 3 B 3 (*ARV* 41.32, 55.12; *Add.* 76; Cohen 1978: B 41, pl. 70) to be a view of Delos; in this case this is probably correct, because of the proximity of a citharode in the tondo, who may be—and is certainly associated to, and so recalls and calls up—Apollo. But I want to stress that, given the established value 'altar with palm-tree' = 'altar of Artemis' (which, I argue, is more significant than has been realized), it is fallacious to assume (implicitly) that, unless the context excludes it, the value 'Delos/Apollo' should be preferred to 'altar of Artemis'.
20. On the *arkteia* see most recently: Lloyd-Jones 1983: 91–8; Cole 1984: 238–44; Brulé 1987: 177–283; Sourvinou-Inwood 1988*a*; Dowden 1989: 9–47 *passim*. See also above Ch. II.2 sect. 6.
21. See e.g. Kahil 1977: pls. 18.1 and 21.1, p. 87; see also Sourvinou-Inwood 1988*a*: pl. 1.
22. NI 1886, *ARV* 446.266; Gabrici 1927: 331, fig. 142, pl. 94; Kurtz 1975: pl. 10.1. On the inscription: Gabrici 1927: 333, 335.
23. See Sourvinou-Inwood 1971*c*: 339–42; Stinton 1976: 11–13; Lloyd-Jones 1983: 91–6; Osborne 1985: 164, 169; Brulé 1987: 177–238 *passim*; Dowden 1989: 9–47 *passim*.
24. See Jenkins 1983: 141. On marriage and Iphigeneia see Foley 1982: 159–80.

25. Simple altars are represented in many erotic pursuits/abductions; see below on their significance.
26. Volute-krater Boston 33.56 (*ARV* 600.12); I discuss this above in Ch. II.2 sect. 6. On this theme connoting marriage: Ch. II.2 *passim*.
27. For 'altar + palm-tree' in representations of Thetis' pursuit/capture: Krieger 1975: 77–80. See e.g. pyxis Munich, formerly Lugano, Schon (*ARV* 806.93; *Para.* 420; *Add.* 143; Roberts 1978: pl. 66.2); cup London, Victoria and Albert Museum 4807.1901 (*ARV* 89.14; here Pls. 11–12) with the capture/abduction of Thetis on A, and on B the Nereids running towards an altar + palm-tree. Krieger (1975: 77–8, 80) connects the altar with or without the palm-tree and the palm-tree on its own (p. 79) in Thetis' pursuit/capture with the notion that the scene is taking place in a space sacred to Thetis. But though it cannot be excluded that this may have been one of the values attached to the altar in representations of the pursuit/capture of Thetis, as will become clear below, it cannot be the only, or even the main, one.
28. See e.g. on this Roberts 1978: 178–9; see also Ch. II.2, sect. 4 above.
29. Calyx-krater Boston 95.23 (*ARV* 510.3; Kaempf-Dimitriadou 1979: 94, no. 215; CB ii. 68–9, no. 104, pl. 14; Suppl. pl. 13.2). On A Zeus is pursuing Thetis; on B a Nereid is running to Nereus and between them is an altar + palm-tree. On the theme of Zeus pursuing Thetis: Papoutsaki-Serbeti 1983: 106 and 258 n. 441; CB ii. 69. On Nereids and related figures in erotic pursuits see also Benton 1970: 193–4.
30. Hydria Vatican 17882: palm, altar, and column (*ARV* 614.11; Kaempf-Dimitriadou 1979: 104, no. 328).
31. On the kantharos Boston 95.36, side A (*ARV* 381.182; *Para.* 366, 368; *Add.* 112–13; Kaempf-Dimitriadou 1979: no. 204, pl. 1.2).
32. On the phiale Berlin F 2310 (*ARV* 819.50; CVA (Berlin 3), pl. 135); the altar and the palm-tree do not stand next to each other. It is because the altar + palm-tree is closely associated with erotic pursuits and with girls, especially the companions, running towards an altar + palm-tree, that on side B of the column-krater Tübingen 67.5806 (*ARV* 585.27; 1660; *Para.* 393; CVA (Germany 52), pls. 14.5, 15, 16, 17.1–2) Helen's companions are depicted through the schema 'companions fleeing to an altar + palm-tree'. For this schema, which pertains to 'erotic pursuit', is here correlative with, and dependent on, the fact that the Helen–Menelaos encounter at Troy on A is to a great extent modelled on the schema of erotic pursuit—rather than, as other versions of this theme, on that of attack—(Menelaos has already dropped his sword, and Eros is represented). This association, incidentally, confirms the view proposed here, that there is a close association between 'erotic pursuit' and 'girls fleeing to an altar with

Altars with Palm-Trees

palm-tree'. In this context, the scene on B would have been ambivalent and polysemic. One meaning is 'Helen's companions fleeing for protection from the wrath of Menelaos', to an altar which could be understood as either that of Apollo (where Helen is shown taking refuge in some versions of this theme), or of Artemis, with whom the schema altar + palm-tree and fleeing girls was associated—or of both. (In my view, the different altars/sanctuaries at which Helen takes refuge in representations of this theme (Apollo's: see e.g. neck-amphora Vienna 741 (*ARV* 203.101; *CVA*, pls. 55–6; Kahil 1955: pl. 57.1)); Aphrodite's (on this: Papoutsaki-Serbeti 1983: 145 and 265 n. 572); Athena's (oinochoe Vatican H 525 (Kahil 1955: 90–1, no. 72, pl. lxvi) in which Aphrodite and Peitho are present) express different perceptions as well as reflecting different variants.) At another level, especially since A and B are separate scenes, the iconographical schema 'fleeing companions running towards an altar + palm-tree' called up, for the 5th-cent. viewers (who read through their iconographical as well as narrative common assumptions) the 'frame' to which this schema belonged, 'girl abducted from a sanctuary and/or the realm of Artemis', and this in its turn would call up the story of Theseus' abduction of the young Helen from the sanctuary of Artemis. Thus, another (complementary, not alternative) reading of the scene on B would have been as an allusion to Helen's earlier abduction from the sanctuary of Artemis.

33. Prof. M. Lefkowitz pointed out to me that the association 'girl about to be married', palm-tree, and Artemis is also found in *Od.* 6. In ll. 162–8 Odysseus compares Nausika to the palm-tree by the altar of Apollo at Delos; in ll. 150–2 he compares her to Artemis, a comparison also made by the narrator in ll. 101–9; Nausika's wedding is said to be near in l. 27; and since Nausika meets Odysseus when she is with a group of girls outside, the expectation is of a possible sexual encounter, though none takes place. On the comparison of Nausika with Artemis see Burkert 1985: 150. The *Odyssey* passage and Nausika's comparison to the Delian palm are now discussed by Harder 1988: 505–14, who also connects the latter with Nausika's status as marriageable *parthenos*.
34. See e.g. Kahil 1977: pl. 19.1. Figurine of a palm-tree from Mounichia: Palaiokrassa 1983: 53–4, 149, and pl. 31*b*.
35. Krieger 1975: 79.
36. See e.g. volute-krater Halle inv. 211 (KN 63) (*ARV* 599.4) by the Niobid Painter; white-ground bobbin Athens N.M. 2350 (*ARV* 775.3; Wehgartner 1983: pl. 52.2, p. 156): palm-tree towards which a girl is fleeing; B of the same bobbin, which, in my view, depicts an erotic pursuit about to begin, also shows a palm-tree in association with the girls. On the abduction of the Leukippids see Calame 1977: 327–30.

37. Kaempf-Dimitriadou 1979: 104, no. 329.
38. See Kaempf-Dimitriadou 1979: 105, no. 348. On the iconography of Boreas pursuing Oreithyia: ibid. 36–41, bibl.: 69 n. 281, to which add Neuser 1982.
39. Brooklyn 09.3: *ARV* 1084.15; 1682; *Add.* 160; Philippaki 1967: 135, no. 9, pl. 59.2–3.
40. Philippaki 1967: 135 and see 42 n. 3.
41. The scene on the pyxis Athens 1288 (CC1558) (*ARV* 917.198; *Add.* 149; Roberts 1978: 47, no. 5, pl. 21) resembles iconographically and reminds us of pursuit/abduction scenes: it shows a palm-tree and women moving towards it, turning back, in a schema similar to those of our pursued girl and her companions. It is another example of an association between palm-tree and *parthenoi*. The association between palm-trees and girls in abduction-related contexts is not limited to Attic iconography. See e.g. the following non-Attic examples (not hitherto recognized as relating to this context): The Chalcidian cup Würzburg 354 (Rumpf 1927: pls. xl–xlii, see pp. 15–16, no. 20) shows a cruder/wilder version of an erotic pursuit: palm-trees frame naked girls (probably Nymphs); on the other side of one of the palms sexually excited satyrs are about to surprise them. On the Chalcidian psykter-neck-amphora Rome, Villa Giulia, Castellani 47 (Rumpf 1927: pl. cxix and see p. 26, no. 110; Robertson 1975: pl. 41*b*, see p. 138) on one side of a palm-tree a girl is dancing and on the other (hiding behind the tree) a sexually aroused satyr is watching her; the girl is wearing a short garment similar to that sometimes worn by the Brauron bears (see e.g. Kahil 1965: pl. 9.7). I discuss this and some related types of garments worn, or related to those worn, by the Brauronian bears in Sourvinou-Inwood 1988*a*: 119–24. I should add that the fact that, at least in Pausanias' time, there were palm-trees in front of the sanctuary of Artemis at Aulis (Paus. 9. 19. 8) offers another non-Attic connection between palm-trees and a cult of Artemis associated with Iphigeneia and with girls' transition from *parthenos* to *gyne*.
42. See e.g. Krieger 1975: 79 and pl. 3*a*; Kaempf-Dimitriadou 1979: no. 343, pl. 30.1.
43. Altar + olive-tree: hydria Vatican 16553 (*ARV* 594.61; *Add.* 129; Kaempf-Dimitriadou 1979: 106, no. 361, pl. 30.3).
44. Hydria London E 224 (*ARV* 1313.5; 1690; *Para.* 477; *Add.* 180).
45. See e.g. Schol. BT Hom., *Il.* 1. 594; Hdt. 4. 154, 6. 138; Suda s.v. Brauron.
46. See Ch. II.2 *passim*.
47. See Osborne 1985: 161–2, 168.
48. See Calame 1977: 176–7, 189–90. Plut., *Thes.* 31. 2, tells us that Theseus abducted Helen from the sanctuary of Artemis Orthia. On Helen's abduction see Calame 1977: 281–5; Kahil 1955: 305–13.

49. The Brauron girls are shown dancing/running among/towards, palm-trees, altars, altar + palm-trees. Thus there is also a correspondence between the Brauron girls' ritual race and the running of the Leukippids/Nereids'/other companions during the pursuit/abduction.
50. See n. 2. For the Brauronian ritual nexus see also Sourvinou-Inwood 1988a: *passim*.
51. See Ch. II.2 *passim*, esp. sect. 6.
52. Perhaps, given Artemis' connection with ephebes (see Burkert 1985: 407 n. 14; Petsas 1972: 252–4; for association of ephebes with Artemis Mounichia, Palaiokrassa 1983: 21–3, 94; with Artemis Agrotera, Deubner 1969: 209; Parke 1977: 54–5, 194 n. 40; on Artemis' concern with male initiations in Attica see Lloyd-Jones 1983: 96–7), the altar + palm-tree of Artemis was also associated with the ephebic transition. There are representations in which the altar + palm-tree is associated with figures which appear to be ephebes, and may, if this suggestion is correct, represent an altar of Artemis. One such scene is shown on the bell-krater Tübingen E 104 (*ARV* 603.35; *CVA*, pls. 23–4; *Cité*, fig. 62) in which a departing youth is making a libation at an altar + palm-tree. The youth is, as Lissarague (1984: 42) remarks, dressed as a hunter—and, I would add, ephebe—but receives the weapons of a hoplite, which suggests a 'changement de statut, investiture, en quelque sorte, du futur hoplite'. Given the Artemis–ephebes connection, it would make perfect sense for the departure from the status of ephebe to be shown (whether or not reflecting a cultic reality) in connection with an altar of Artemis. It is particularly interesting that the ephebes were associated with Artemis Mounichia, whose altar could be represented through the sign 'altar + palm-tree'. In another series of scenes representing ephebes, as Lissarague has convincingly suggested (1984: 37–9), the altar + palm-tree alternates (as it does in some erotic pursuits) with palm-trees: on the oinochoe Heidelberg Univ. 71/2 (*ARV* 156.54; *Cité*, fig. 57) the ephebes are crouching (according to Lissarague, p. 39 miming an ambush) around an altar + palm-tree for which Lissarague suggests (independently from the hypothesis put forward here) that it may be 'l'indication d'un rapport avec Artemis à qui cet arbre est associé et qui préside au passage de la classe des jeunes gens à celle de guerrier adulte'. On a column-krater formerly in the Hoppin Collection, now in the Fogg Museum Harvard 1925.30.126 (*ARV* 234.11; *CVA* (USA 1), pl. 7), similar figures are shown in ambush in a palm-grove. If the suggestion put forward here is correct, it is possible that, apart from the strong association between Artemis, *parthenoi*, and palm-tree/altar + palm-tree, there was also a similar association involving ephebes. This would suggest an association between Artemis,

the palm-tree/altar + palm-tree, and young people of both sexes in a state of marginality and transition to fully adult status.
53. *ARV* 600.12.
54. Ch. II.2, sect. 6. Spatial indicators build up not a 'photographic', but a conceptual space; 'un espace investi de valeurs symboliques, produit de l'imaginaire social': Durand and Lissarague 1980: 103.
55. See Kaempf-Dimitriadou 1979: 5, 24, 42.
56. On this see Ch. I.
57. As we saw above (n. 33), the 'pre-nuptial' horizon and the Artemis connection is also found in the *Odyssey*.
58. See nn. 53–4.
59. Akousilaos *F Gr. H* 2 F 30: while she was *kanephoros* for Athena Polias (see also Kaempf-Dimitriadou 1979: 39).
60. Articulated also in the Akousilaos story in which she is said to have been a *kanephoros* (for the association between the *kanephoros* and marriageability see below n. 105).
61. On this see Ch. II.2 sect. 6. The same is true of representations of the Leukippids: the palm connotes the realm of Artemis, the sanctuary of Aphrodite the story that this was where the abduction took place and/or the place of Aphrodite in the story as myth and paradigm.
62. For instance, through the presence of that deity e.g. as in the abduction of the Leukippids on London E 224.
63. The scenes located in the Brauronian sanctuary also include simple altars, as well as 'altar + palm-tree' complexes (see e.g. Kahil 1965: pls. 7.4, 8.5). Krieger 1975: 77–8, 79, 80, we saw, connects the altar with or without the palm-tree and the palm-tree in Thetis' pursuit/capture with the notion that the scene is taking place in a space sacred to Thetis. But though this could be one of the values attached to the altar in the pursuit/capture of Thetis, it cannot be the only, or even the main, one. For, we have seen, the altar, the altar + palm-tree, and the palm all appear in a variety of erotic pursuits/abductions, and must thus have a meaning appropriate to all of them, pertaining to the theme erotic pursuit/abduction itself, as it does in the hypothesis put forward here. Moreover, with regard to the 'altar + palm-tree' in particular, Krieger's interpretation relies on mistaken assumptions: there is no evidence to connect the altar + palm-tree, or the palm, with Thetis' cult or Thetis' divine personality, nor does Krieger try to do so, relying instead on the notion that they simply denote the sacred, a notion, we saw, not supported by the pattern of appearance of these signs. The interpretation put forward here makes sense of the appearance of the altar + palm-tree and the palm-tree in all pursuits/abductions in terms of a well-established meaning of the 'altar + palm-tree' and the palm-tree: the meaning 'altar of Artemis' and the association of the palm-tree with Artemis and the unmarried girl. This is also supported

Altars with Palm-Trees

by other connections between pursuits/abductions and Artemis, especially through Brauron, a cult complex closely associated with altar + palm-tree, palms, and unmarried girls. The dolphin which is sometimes juxtaposed to the 'altar + palm-tree' e.g. altar, next to it a palm-tree, over the altar a dolphin, on the pyxis Athens Ceramicus 1008 (*ARV* 806.92; *Add.* 143; Roberts 1978: pl. 64 and p. 96) characterizes the scene as concerning Thetis through the dolphin-allusion (on the use of the dolphin in pursuits/abductions see Ch. II.2 n. 21). In other scenes (see Krieger 1975: 78) a snake is associated with the altar. The combination of the palm-tree with a snake is not confined to Thetis scenes (see e.g. the alabastron with an Amazon in the Brummer Sale 1979—*ARV* 269.3; *Ancient Art: The Ernest Brummer Collection*, ii. auction sale 16–19 Oct. 1979, Zurich, no. 693, p. 329); this suggests that the combination pertains at least in part to the palm-tree; in the case of Thetis it may also be alluding to her metamorphoses, and also, through them, to her divinity.

64. On white-ground alabastra see: Wehgartner 1983: 112–34; Mertens 1977: 128–36.
65. On the Amazon and Negro alabastra (with bibliography): Wehgartner 1983: 116–21, 129–31, and 215 nn. 57 ff.; Neils 1980: 13–15; Raeck 1981: 188–98, 332–4. See also below, nn. 100, 123, 144.
66. Wehgartner 1983: 112. On alabastra see Angermeier 1936.
67. For an example of the nuptial associations of the alabastron see e.g. the lebes gamikos BS 410 (CVA Basle 2, pls. 55–6).
68. See Angermeier 1936: 37–8, 39–41; see also 44. See e.g. *ARV* 99.7–101.30, 363.29–29bis, 624.88–625.94, 661.78–81, 722.1–3; 726.1–727.22, 727.1–2. See also Wehgartner 1983: 128–34 *passim* and pls. 40–44.1.2 *passim*.
69. More on 'variants' below. Athens 15002: *ARV* 98.2.
70. See e.g. the representation of the abduction of Thetis on the alabastron Berlin 4037 (*ARV* 661.78; Furtwängler 1883–7: pl. 54.2) which was a paradigm for marriage and relates the iconography of the alabastra to the nuptial sphere.
71. This is not a systematic analysis of the alabastra series; I am only concerned with showing that scenes involving the altar + palm-tree or the palm-tree are connected with Artemis and the *parthenoi* in the context of a preparation for marriage or marriage itself.
72. *ARV* 264.58–64; discussed by Wehgartner 1983: 117. III and 129–34 *passim* (see her pl. 41.5–6).
73. *ARV* 264.63; Anderson 1955: 45, no. 84, pl. xii.
74. Anderson 1955: 45: 'a maenad?' The woman is not characterized as a Maenad but is wearing the same clothes as the other women on this painter's women's alabastra: bonnet, chiton, and himation.
75. Cf. e.g. the pyxis London D11 (*ARV* 899.146; Wehgartner 1983: pl. 50; Jenkins 1983: pl. 18*a*). See Redfield 1982: 189.

76. For weddings see n. 75. The following are a few examples of this sign depicted on its own. Woman with 2 torches, looking back, moving towards altar: lekythos, formerly in the Basle market (*ARV* 720.23); oinochoe Münster 586 (Stähler 1980: no. 20, fig. 20). Woman wearing peplos and diadem with torches, looking back, no altar: alabastron Cambridge 145 (*ARV* 727.2); lekythos New York 41.162.117 (*ARV* 642.104; Papoutsaki-Serbeti 1983: pl. 38, no. 143, p. 205. The description 'running' (*ARV* and Papoutsaki-Serbeti 1983: 205) is a mistake: she is moving in the normal manner of processional walking; the schema is identical to that of the wedding procession mothers; contrast running women by the same artist as the New York lekythos: e.g. the lekythos Oxford 1890.23 (*ARV* 641.84; Papoutsaki-Serbeti 1983: pl. 22).

77. That the iconographical element 'woman with two torches' was an 'extract' connoting a larger schema through the *pars pro toto* trope is suggested by the combination of, first, the fact that the torches are a specialized, 'narrative' element unlike, say the alabastron, which can characterize more generically the world of women and their concerns; and second, the fact that these 'extracts' were depicted more than once and by different painters. That the woman with torches (at least when juxtaposed to another scene pertaining to nuptial iconography) may have been an 'extract' representing wedding was suggested in *ARV* s.v. 726.12 where the comment on the scenes on the alabastron Exeter 97.1953, which has a woman with torches on A and a woman with wreath on B, reads: 'May have been thought of as extracts from a wedding scene'. Moreover, the pelike Oxford 285 (CVA Oxford 1, pl. 20.5–6), which shows a woman with 2 torches on side B and Eros flying to an altar on side A, connects the 'woman with 2 torches' sign with the nuptial iconography and semantic facet through another route, since Eros has an important place in nuptial iconography and ideology.

78. See Ch. I.

79. See Ch. II.2, sect. 6, and sect. 1 above on the fact that Naples 2422 does not represent an altar of Zeus Herkeios in association with a palm.

80. *ARV* 600.12. This scene is discussed in Ch. II.2, sect. 6.

81. Mounichia: Palaiokrassa 1983: 73–5. Brauron: see e.g. Kahil 1983: fig. 15.7. For the association of the Attic, and especially Brauronian and Mounichian, cult of Artemis with torches: see also Kahil 1965: 27; Palaiokrassa 1983: 40, 73, 74, 75, 95 and 271 n. 298, 166–7,193; Papoutsaki-Serbeti 1983: 267 and n. 608.

82. Soph., *Trach.* 214.

83. On polysemy and ambivalence on Greek vases see Hoffmann 1980*b*: 127–54 *passim* (esp. 128); Bérard 1983: 11. See also Ch. I above.

84. *ARV* 264.58; I was able to consult a photograph in the Beazley Archive.

Altars with Palm-Trees

85. See Ch. II.2 nn. 34–8.
86. Flute-playing at weddings: see e.g. loutrophoros Copenhagen 9080: *ARV* 841.75; *Para.* 423; *Cité*, fig. 139. At Brauron: Kahil 1983: fig. 15.11 and p. 238.
87. See e.g. cup Oxford 1929.464 (stool and kalathos) (*ARV* 379.158; CVA Oxford 2, pl. 51.7); see also *Cité*, p. 29 with examples.
88. *ARV* 264.59; I was able to consult a photograph in the Beazley Archive.
89. See e.g. Eros offering a mirror to a woman holding an alabastron on the hydria Leningrad inv. 4309 (*ARV* 1210.64; *Para.* 518; *Add.* 171).
90. See e.g. pyxis Athens 1585 (*ARV* 1360.2; *Add.* 185); pointed amphoriskos Oxford 537 (*ARV* 1248.10; CVA Oxford 1, pl. 40.3–4). Mirror and stool are very common in domestic contexts; see e.g. lekythos Athens NM 12890 (*ARV* 641.93; Papoutsaki-Serbeti 1983: pl. 10b). Mirror, stool, and kalathos in a domestic context: see e.g. cup Oxford 1925.73 (*ARV* 378.140; CVA Oxford 1, pl. 2.6); white-ground pyxis in Toledo Ohio 63.21 (*ARV* 1675.94bis; *Para.* 434; *Add.* 150). For the combination stool and alabastron see e.g. pyxis Louvre G 605 (*ARV* 943.78; *Add.* 149).
91. See e.g. the pyxis Athens N.M. 1630 (*Cité*, fig. 138).
92. *ARV* 264.60; Wehgartner 1983: pl. 41.5–6.
93. The marriageable girl is associated with beauty in the Greek collective representations, is characterized as beautiful: Calame 1977: 447.
94. On the ambivalence pertaining to the *parthenos* which juxtapositions of this type reflect and express see Ch. II.2, sect. 6.
95. *ARV* 264.62.
96. *ARV* 264.61.
97. It is not always easy to distinguish between herons and cranes in a conventional iconographical system like this. Thus the same bird has been called crane by some and heron by others. For example, Holmberg 1947: 190 calls the bird on this alabastron crane, while Wehgartner 1983: 117 calls it heron. Since this does not affect my argument, I will refer to the birds belonging to this iconographical type as heron/crane.
98. See *Cité*, p. 29.
99. See e.g. the hydria Oxford 531 (*ARV* 520.36; CVA, pl. 32.9). Kalathos and alabastron: hydria Oxford 294 (CVA, pl. 32.3). Alabastron and kalathos in a nuptial context: see e.g. lebes gamikos BS 410 (CVA Basle 2, pls. 55–6); kalathos on a seat in a domestic context: cup Berlin 2289 (*ARV* 435.95; 1653; 1701; *Para.* 375; *Add.* 117). Kalathos held by girl in erotic pursuit: hydria Vatican 16554 (*ARV* 252.47; *Add.* 101; Keuls 1983: 221, fig. 14.28). For the kalathos in the Locrian pinakes see below Ch. III, sect. 2 III (*i*). On the kalathos see also Keuls 1982/3: 31.

100. Hydria Syracuse 36330 (*ARV* 1062.2; CVA, pl. 25); lekanis lid Tübingen s./10 1665 (CVA Tübingen 4, pl. 49.1–2); white-ground alabastron Geneva 20851 (*ARV* 723.3; *Add.* 138): two women, alabastron (and also plemochoe), stool, crane/heron, and underneath the stool what appears to be a kalathos; heron/crane combined with a kalathos in a domestic scene: e.g. white-ground oinochoe Athens 2186 (*ARV* 1562; Wehgartner 1983: 45–6, no. 3; pl. 12.1). Heron associated with a woman and a type of 'wild woman' (Maenad): alabastron London B 668 (*ARV* 98.1; *Para.* 330; *Add.* 85). Heron associated with two 'wild women', a Maenad and an Amazon together: alabastron Athens 15002 (*ARV* 98.2; *Add.* 85). Women and herons on alabastra: e.g. Eleusis 2404 (Wehgartner 1983: 116, no. 5). Holmberg 1947: 190 (and n. 5) acknowledges that all the elements point to a domestic scene, a woman in her boudoir, and that the crane was associated with domestic contexts and women's scenes. But because he begins with the preconceived idea that the palm-tree alludes to Egypt, he decides that the heron/crane also must have the same meaning. This interpretation is based on a methodological fallacy of (*a*) ignoring the polysemy of the palm-tree (and of all signs); and (*b*) basing conclusions on *a priori* notions (that the palm-tree has the same meaning here as it does on the Amazon and Negro alabastra, and that this meaning is Egypt) which ignore inconvenient data (all the contexts in which the palm connected with similar contexts cannot have that meaning). The *a priori* assumption that the palm denotes an exotic environment in the women's scenes is an example of the ways in which our own culturally determined assumptions determine the arrangement and interpretation of the ancient evidence. Because the association Negro–palm = exotic, hence palm = exotic, made sense to us, it has been placed at the centre of the interpretative discourse concerning the women's and the Amazon alabastra as well (helped by the fact that the association 'altar + palm-tree–*parthenoi*' had hitherto not been noticed), even though it did violence to the evidence and produced unsatisfactory interpretations and reconstructions.
101. See the cup Ferrara 44886 (*ARV* 880.11; *Add.* 147).
102. See Ch. I.
103. CVA Kassel 1, pl. 41.5–7; here Pl. 13*a–c*.
104. CVA speaks of 'Baum'; but it is a clear—if somewhat stylized—palm and is correctly identified as such in the Beazley Archive Project.
105. On *kanephoroi*: Schelp 1975: 15–21, 54–5; Brelich 1969: 282–90. The *kanephoros* is a girl ready for marriage: Brelich: 287–90; Calame 1977: 68. See also Sourvinou-Inwood 1988*a*: 54–7, 94–7 nn. 253–66.

106. See Schelp 1975: 20–1.
107. Schol. Theocr. 2. 66: the girls about to marry *ekanephoroun* to Artemis. On the expression *mellousai gamein* see Stinton 1985: 431 n. 50.
108. See Ch. I.
109. *ARV* 787.3; *Add.* 142.
110. As Boardman 1975: caption of fig. 371.
111. And as suggested by CVA Denmark 4, Copenhagen, pl. 159.6 'jeune femme près d'un autel'; *ARV*'s 'woman at altar' and *Cité*, p. 27, 'jeune femme'.
112. Of course it would not necessarily follow that related schemata like these would always represent the same theme. However, given that the 'altar + palm-tree' has the established value 'altar of Artemis', and that this is especially closely connected with *parthenoi* preparing for marriage, we can see that the 'altar + palm-tree' on the red-figure alabastron corresponds to, has the same value as, the altar + the inscription ARTEMIS on the stemmed dish; they both characterize the altar as Artemis' and the cult act as taking place in the cultic realm of Artemis. Other representations of *kanephoroi* at an altar without palm or inscription (see e.g. cup Toledo, Ohio 72.55: Keuls 1982/3: pl. 7.22) may depict an altar of Artemis, if the basic schema was an established schema referring to the pre-nuptial *kanephorein* to Artemis; but it cannot be excluded that additional elements, such as the thymiaterion on the Toledo cup, which seem to us to be 'generically sacred', in fact characterize the scene as belonging to a particular cultic sphere—which may not be that of Artemis.
113. Wehgartner 1983: 131.
114. On the meanings of Amazons: Pembroke 1967: 1–35; 1965: 217–47; Lefkowitz 1986: 17–20, 22–3, 26–7; Tyrrell 1984: *passim*.
115. As Neils, who did not follow up this insight, acknowledged when she correctly categorized the Maenad and the Amazon as 'wild female types' (1980: 21).
116. Most scholars have tried to interpret the Amazon alabastra starting at the wrong end, their relationship with the Negro series. In fact, when they are considered within the broader framework of the totality of (especially early) alabastra scenes, the great majority of which are decorated with scenes pertaining to women, the Amazons (when account is taken of their meanings in myth and collective representations) can be seen to be one variant of that large category, women's scenes. This does not exhaust their meanings, but it provides the foundation and core of such meanings, and a place for us to begin our attempted reconstruction.
117. Tyrrell 1984: 86; Neils 1980: 21 and n. 41; Ridgway 1974: 14 and n. 78; Kahil 1965: 29.

118. Neils 1980: 21. She suggests that the altar may be an altar of Artemis (because of the association of Artemis with Amazons and the stories that the Amazons took refuge at the sanctuary of Artemis at Ephesos), though she does not take account of the established value 'altar of Artemis' for 'altar + palm-tree'. This is interesting, for it shows that the interpretations 'altar of Artemis' for the combination 'altar + palm-tree' on the Amazon alabastra does not depend on the basic hypothesis investigated here, that the altar + palm-tree in women scenes represents an altar of Artemis.
119. On Amazons as, among other things, images of *parthenoi*: Tyrrell 1984: 76, 128.
120. On Amazons represented through the iconography of erotic pursuit/ abduction which is, among other things, a paradigm for marriage reflecting certain perceptions of male–female relationships: Ch. II.2, sect. 9. On the iconography of the abduction of Antiope: Bothmer 1957: 124–30.
121. Neils 1980: 21 speaks of the battlefield represented by the abandoned helmet on Cracow 1292; in the reading focusing on the *parthenos*– Artemis relationship it could represent the abandonment of the Amazon's 'abnormal' pursuits and the preparation for, and espousal of, marriage, which the altar + palm-tree can be taken to refer to.
122. See e.g. Cracow 1292 (*ARV* 270.1).
123. See e.g. the alabastron Virginia Museum of Fine Arts, Richmond, The Williams Fund (78.145) (Shapiro 1981: 88, no. 33; Wehgartner 1983: 117.II, nos. 3–6; 119, nos. 4–5).
124. See New York 21.131 (*ARV* 269.1; *Add.* 102; Wehgartner 1983: 117.II, no. 1, pl. 40.2–3); Basle Kä 403 (*ARV* 269.2; 1641; *Para.* 352; *Add.* 102; Wehgartner 1983: 117.II, no. 2).
125. See e.g. the alabastron Athens Kerameikos HS 107 (Wehgartner 1983: 119, no. 2, pl. 41.1–3) which is very similar to this schema (see below).
126. Of whom the Amazons were, among other things a crystallization (see above, n. 119).
127. In fact it is the primary one; it is in terms of this dimension of signification that the selection of all the elements and their combinations makes best sense.
128. *ARV* 269.1; *Add.* 102; Wehgartner 1983: 117.II, no. 1, pl. 40.2–3.
129. This, among other things, connects her with the other 'wild woman' image, the Maenad.
130. Wehgartner 1983: 119, no. 2, pl. 41.1–3. It is important to note that, according to Wehgartner, p. 119, Kerameikos HS 107 is by the same painter as New York 21.131.
131. See London D 17 (Wehgartner 1983: 124, no. 4).
132. See n. 96.

133. See n. 99.
134. Military contexts (in the residential quarters of military camps): e.g. stamnos in a private collection in Switzerland (*ARV* 361.7; *Add.* 110); skyphos Vienna 3710 (*ARV* 280.171); cup London E 76 (*ARV* 406.1; 1651; *Para.* 371; *Add.* 114). It is not exclusively associated with men, of course: e.g. among the gods, it is used not only by the 'androgynous' (comparable to the Amazons) Athena (e.g. lekythos Münster Arch. Mus. 24, *Cité*, fig. 151) but also by goddesses like Hera (stamnos Paris G 370; *ARV* 639.54) as well as gods, e.g. Dionysos (cup Munich 2645; *ARV* 371.15; 1649; *Para.* 365; *Add.* 111).
135. Other contexts include sport (e.g. the base with cat and dog fight in Athens NM (Richter 1950: fig. 283), and homosexual erotic scenes: *Cité*, p. 116.
136. Raeck 1981: 197 and 308 n. 849; Neils 1980: 21 also takes these elements to denote an arming scene.
137. On the Amazons' refusal of marriage: Tyrrell 1984: 82–3.
138. See on this and the meanings of Amazon's erotic pursuit Ch. II.2, sect. 9.
139. See Ch. II.2, sect. 9.
140. See n. 124.
141. See n. 125.
142. Above and nn. 100–1.
143. I take the 'boy and girl' scenes to be 'emblematic' rather than 'descriptive'; therefore the representation of a youth with a *parthenos* placed (conceptually rather than spatially) in the 'women's quarter' presents no problems. On the problems of identifying women in some representations as *hetairai* or housewives see Raeck 1981: 302 n. 777 with bibliography. In my view, a confusion of the emblematic with the descriptive iconographical mode is one of the reasons why the number of *hetairai* scenes have been overestimated.
144. Amazons and Negroes as allies of Trojans: Neils 1980: 22; Shapiro 1981: 88. Negroes on Negro alabastra as reference to the Persian Wars (and Amazons symbolizing the Persians): Vickers 1981: 546. Both aspects: Wehgartner 1983: 130 and 215 n. 58.
145. Besides the arguments put forward above, this view is also supported by the fact that, in the earliest group of white-ground alabastra, various types of women are common subjects, but, as Wehgartner 1983: 129 notes, what she calls 'exotic motifs' like Amazons are the exception.
146. Under external impulses of which the Persian Wars may have been one. (On the origin of the Negro alabastra see also Wehgartner 1983: 130 and 215 n. 57).
147. Neils 1980: 21–3. In my view, Neils's further reconstructions include iconographical misunderstandings and methodological

fallacies. An example of the former: according to Neils 1980: 21 n. 42, the Syriskos Painter arbitrarily adapted the Amazon/Negro schema to Brussels R 397 'where a woman is depicted on the obverse posed like the archers but carrying an alabastron and moving towards a palm-tree and stool'. In fact, the woman on Brussels R 397 is not posed 'like the archers', but shown in an extremely common schema for women carrying things which we have encountered in other women's alabastra but is also found elsewhere (see e.g. Nike on the lekythos Oxford 1917.58 (CVA Oxford 1, pl. 34.2; *ARV* 309.14; *Add*. 106)). More importantly, Neils's notions of arbitrary transferrals (by the painters), leading to the creation of confused imagery and illogical schemata, and also her theory (p. 23) that 'a mythological subject can evolve into a purely decorative motif exploited for its popularity and exoticness rather than for any intrinsic meaning', are, in my view, untenable. Because, for both artists and viewers, all these signs had certain meanings and connotations (those pertaining to the signs and to the things represented by the signs both in ceramic iconography and in the Athenian semantic universe); and these meanings and connotations, which the artists and their contemporary viewers shared, but we do not, except partly when we reconstruct them deliberately and systematically, (*a*) were in the painters' mind when they created the scenes, and thus helped determine their selections and (*b*) more importantly and concretely, formed the perceptual filters through which the viewers made sense of these images. (On assumptions in the process of making sense of ancient images see Ch. I above.) Thus, whatever the routes and reasons, the iconographical elements were inevitably combined into meaningful schemata, meaningful because (whether or not they seem illogical to us) they were made sense of by 5th-cent. Athenians in terms of the 5th-cent. values and semantic assumptions. It is these 5th-cent. meanings that I have tried to reconstruct above for some of these scenes.

148. See Wehgartner 1983: 114–16, group I.
149. This variant with the altar is rare; this is correlative with the fact that the notion 'altar/realm of Artemis' is not especially pertinent to the Negro figure and its meanings. On the collective representations pertaining to blacks in antiquity see Snowden 1983: 46–9 and *passim*. On the different variants see Wehgartner 1983: 118.IV.
150. Possibly through the category of marginality, marginal warriors, to which the Negroes belonged, and which appears to have been associated with Artemis in Attica (at least in the case of ephebes and Amazons). If this association with marginality is correct, perhaps in spatial terms the palm-tree denotes the marginal as well as the exotic (and, in spatial and/or conceptual terms, the area pertaining to

Artemis). Thus, not only in these alabastra, but also in representations of ephebes in palm-groves (see above, n. 52) the palm may also denote the marginal location of the ephebic activities, as well as the relationship with Artemis' realm.

151. *ARV* 917.206; *Para.* 430; *Add.* 149.
152. Bérard 1974: 118–19; see pp. 117, 119–21, 159.
153. It is fallacious to assume that it is mythological because the schema partly resembles the representation on the pyxis Ancona 3130 (*ARV* 899.144; 1674; *Add.* 148). For this would mean ignoring the differences between the two scenes, and thus disregarding the fact that in a conventional system like that of Attic ceramic iconography small divergences may signify (to the artist's contemporaries who shared his assumptions) significant differences; it is through such divergences that different meanings are created through the manipulation of established schemata (see Ch. I above).
154. The modelling of the scene on the birth of Aphrodite is correlative with the notion which is thus reflected, and expressed, in this scene (whether or not there was the intermediary of a rite, as Bérard's interpretation demands): that the girl has entered the stage of marriageable *parthenos*, in which Eros (for whom she is now ready) will be of paramount importance. (On Eros and marriage see Ch. II.2 n. 67.)
155. So Bérard 1974: 118.
156. See King 1983: *passim*.
157. See (for examples of brides with himation over their head) Ch. II.2 n. 70; for this schema in other scenes with nuptial connotations see Ch. II.2, sect. 5 and n. 69.
158. Especially given the relationship of Eros to nuptial iconography (see Ch. II.2 n. 67).
159. Brauron: see e.g. Kahil 1965: pl. 7. Weddings: see e.g. loutrophoros Copenhagen inv. 9080 (*ARV* 841.75; *Para.* 423; *Cité*, fig. 139). Erotic pursuits: see Ch. II.2, sect. 7.
160. Though it fits other contexts also.
161. Papaspyridi-Karouzou 1945–7: 22–36, additional pls. 1–2.
162. On pyxides (and their iconography): Roberts 1978: *passim*; see also *Cité* p. 96. Pyxides/boxes in wedding scenes in association with the bride: see e.g. lebes gamikos Basle BS 410 (*CVA* Basle 2, pls. 55–6); pyxis Athens NM 1630 (*Cité*, fig. 138).
163. See Ch. II.2, sect. 7.
164. Papaspyridi-Karouzou 1945–7: 22–36 may well be right that these girls are the Pleiades; but whether or not they are, their main character, which makes sense of all the iconographical elements, is 'marriageable *parthenoi*'.
165. *CVA* Bologna 4, pls. 86–7.

166. Ch. II.2, sect. 10.
167. The other tree in the scene may be the miraculous tree which produced the golden apples (and pertains to Aphrodite); if so, the two trees would connote the realm of the two goddesses: Artemis, associated with Atalanta as a *parthenos* and huntress, and Aphrodite, who 'trapped' Atalanta and determined the outcome of the race.
168. Space prevents me from discussing here combinations such as bull and palm-tree (see e.g. the oinochoe in the Basle market, *MuM. Antike Vasen. Sonderliste R.* (Dec. 1977), no. 37, pl. 46) and louterion + palm-tree flanked by bulls (see e.g. black-figure lekythos Athens Agora P 24.067 (Ginouvès 1962: pl. 40, fig. 129)). They pertain to, and define iconographically, the space of the sanctuary. (See Durand and Schnapp 1984: 52; on louteria in sanctuaries: Ginouvès 1962: 299 ff.) The bulls are undoubtedly sacrificial (ibid.). I will only note that the combination bull and palm-tree and winged female figures suggest that the space denoted on the oinochoe in the Basle market is probably the sanctuary/realm of Artemis. (Winged Artemis as Potnia Theron petting a fawn in red figure: see e.g. oinochoe Paris, Petit Palais 315 (*ARV* 307.11; *Add.* 105).) It is, in my view, very likely that, like the altar and palm-tree, the louterion and palm-tree + bulls and the palm-tree + bulls denote the sanctuaries of certain particular deities, one of whom is Artemis: but it is beyond my scope to pursue this here.
169. Palm-tree and fawn/female deer at Mounichia: Palaiokrassa 1983: Kk 46, pl. 50c; I will not go into the well-known association between Artemis and deer/fawn here. (On the iconographical association between Artemis and the deer see most recently Jurriaans-Helle 1986.)
170. Kahil 1977: 90, 91, fig. 4, pl. 19.2. On this scene see also Sourvinou-Inwood 1988a: 62–4.
171. Agora 27342; Acr. 621a (Kahil 1981: pls. 62.9, 62.1); Palaiokrassa 1983: Kk 4, 6, 8. See also Sourvinou-Inwood 1988a: 63, 102 n. 298, 104–5 n. 315.
172. Apollo and Artemis: see e.g. side B of the amphora Oxford 1965.118 (*ABV* 335.1; *Para.* 148; *Add.* 44; CVA Oxford 3, pls. 34–5). Normally the palm-tree and fawn in such scenes are associated with both. Interestingly, in one representation with palm-tree and fawn and altar, the altar and the palm-tree and fawn/deer are represented as more closely associated with Artemis than with Apollo: on the lekythos in Gela (CVA Gela 3, pl. 16) which shows Apollo and Artemis under a portico (represented by columns and an architrave), Artemis is sitting on an altar on the right, turned towards Apollo who is standing on the left; between them, facing in the same direction as Artemis, a female deer in front of a palm-tree; the deer is facing,

as Artemis does, Apollo. Thus the syntax of the scene places Artemis on the altar and the palm-tree and deer in one unit, with Apollo who is facing them constituting another. Artemis sitting on the altar may be a way of articulating the notion that this is a sanctuary of Artemis at which she is receiving Apollo. (For Apollo on his own elsewhere in connection with an altar a fawn/female deer and a palm-tree: e.g. cup Ferrara 2501 (*ARV* 919.1; 1674; *Add.* 149)). I should note in passing that on the pyxis Ferrara 20298 (T27 C VP) (This is the inv. no. given it by *ARV*; another inv. no. also quoted for this vase, in other publications, is Ferrara 12451.) (*ARV* 1277.22; *Add.* 178; Simon 1983: 85, fig. 11) it must not be assumed that the palm-tree and deer is associated with Leto and/or Apollo more closely than with Artemis because it is represented between these two deities; for in this scene Delos is sitting on the omphalos (which belongs to Apollo), and the tripod (also belonging to Apollo) is shown next to Hermes. This scene involves a mixing-up of deities and symbols, probably because the former are represented as, above all, a group, the Delian deities. (On the interpretation of this scene see now Bruneau 1985: 551–6.) But such distinctions do not concern us here; what concerns us is that the combination palm-tree and fawn/deer is associated with Artemis, especially in her role as protector of *parthenoi*, and this is established through the *arkteia* scenes and reinforced through scenes such as that on the Gela lekythos.

173. Io includes a connotation '*parthenos* about to "marry" ' (see e.g. Aesch., *PV* 645 ff.) and this is unambiguously reflected in the red-figure representations. (On the myth of Io and its representation in Attic iconography see now Simon 1985: 265–80; Yalouris 1986: 3–23; see also Schefold 1981: 134–7; Kaempf-Demetriadou 1979: 25.) Sometimes Io is shown as a girl with animal ears and horns pursued by Zeus (see e.g. pelike Naples ex Spinelli 2041 (*ARV* 1122.1; *Add.* 162)), according to the usual 'erotic pursuit' schema. That the erotic dimension is important in this theme can be seen also in the representation of Io and Eros on the cow's head rhyton Ruvo, Jatta 1116 (*ARV* 1551.12; *Para.* 505). Thus, the semantic nexus 'erotic pursuit' (which was also a paradigm for marriage) was associated with Io in the assumptions reflected in 5th-cent. iconography. In my view, this and other evidence (see also Ch. II.1, sect. 2. III) shows that the notion 'marriage'/sexual encounter with Zeus is central to the myth. Consequently, and given that, we saw, the palm was associated with 'erotic pursuit/abduction', the palm-tree + fawn/female deer in Io scenes pertains to the same facet of signification '*parthenos*/Artemis'. A palm-tree + fawn/female deer is represented on the stamnos Vienna Kunsthistorisches Museum 3729 (*ARV* 288.1; 1642; Simon 1985: 273–4, figs. 54–6) which shows the death of Argos. The palm on

its own could be taken—as it usually is—to denote 'Egypt' (another element in the Io myth). Indeed it almost certainly had that meaning also—and it may have been this, the representation of the palm-tree denoting Egypt, that inspired the inclusion of the palm-tree + fawn/female deer sign which was appropriate to another aspect of the theme. But this does not exhaust its significance, as is shown by the fact that the palm is not shown on its own, but in connection with the fawn/female deer, and thus forms a different, if related, complex sign. The palm-tree + deer is shown between Zeus, who is about to touch the animal Io, and the youth on B who is offering a hare to a boy. As Schefold 1981: 135 noted, Zeus' touch here perhaps also alludes to the contact from which Epaphos was conceived and which restored Io to her senses. (On Zeus touching Io: Aesch., *PV* 848; Engelmann 1903: 46 ff.; Simon 1985: 269, 276–7.) The olive-tree shown behind the Hermes–Argos group balances the palm-tree and must allude to the story that Argos had tied up Io at an olive-tree (Apollod. 2. 1. 3); the rope is depicted on the olive-tree's trunk (by that moment of the story presumably cut off), but has been mistaken for decoration comparable to that on the palm-tree, despite its shape, and the fact that it continues alongside the trunk on the lower right-hand side, which cannot fit this interpretation. This provides support for the view that the palm-tree is also a significant part of the scene. If that is correct, given the '*parthenos*–erotic pursuit/contact' meanings of the scene, and the association between the palm-tree and that semantic area discussed here, the palm + fawn would have been understood (by 5th-cent. Athenians) to pertain to Io as a *parthenos* about to become woman. The palm-tree is closely associated with Io in animal form also outside Attica: on the Ionic amphora Munich Staatliche Museen 585 (Stella 1956: ill. on p. 65; Simon 1985: 276, fig. 60) the animal Io is standing in front of a palm-tree, so that the two form a palm-tree and animal combination, and this supports the interpretation proposed here, as does the following consideration. If it is correct that the 'palm-tree + fawn' is represented on the Vienna vase because of its association with *parthenoi* and their preparation for marriage and transition to womanhood, its representation (like Zeus' touch) emphasizes the future sexual encounter between *parthenos* and god. In that case the two sides of the vase are semantically related: both represent an erotic encounter about to take place, one involving a *parthenos* the other a boy (the hare being a homoerotic erotic gift—on this see Koch-Karnack 1983: 63–97). Amazons are also associated with the palm-tree and fawn/female deer—which would fit their persona as *parthenoi* as sketched above: e.g. the black-figure plate Bologna 149 (Callipolitis-Feytmans 1974: pl. 92.3).

174. It is through adjustments of this kind that events pertaining to deities can function as paradigms for mortals.
175. I touch upon this briefly in Ch. II.2, sect. 6, and hope to discuss it more extensively elsewhere.

PART III
Persephone and Aphrodite at Locri:
A Model for Personality Definitions
in Greek Religion

III
*Persephone and Aphrodite at Locri: A Model for Personality Definitions in Greek Religion**

1. THE DEFINITION OF DIVINE PERSONALITIES

Too often in the study of Greek divine personalities assumptions about deities' nature and development have been reflected in the methodology adopted and have thus introduced distortions, forcing the evidence into inflexible interpretative frameworks which may be logical without being correct. I believe we must aim at a 'neutral', bias-free approach which does not allow the operator's convictions to distort the evidence by casting it into a preconceived mould. I shall first set out the factors which, in my opinion, determine the definition and development of Greek divine personalities; these can be established by considering detectable historical developments in these personalities. I then propose an open-ended and flexible methodological framework which will take account of this model but will not depend on its validity.

The first determining factor is clearly the worshipping group and its specific realities and needs as they develop in the course of time. Deities are shaped by the societies that constitute the worshipping group and develop with them. A second factor to be taken into account is the pantheon to which they belong and the spheres of activity of its members. For a pantheon is an articulated religious system within which divine beings catering for the needs of the worshipping group are associated and differentiated; and this nexus of relationships contributes to the definition of each divine personality.[1]

Consequently the study of a divine personality should take account of the other deities of the pantheon to which that personality belonged, and of the (changing) circumstances, economic, social, and other, of the worshipping group. The concepts 'worshipping group' and 'pantheon' bring us to an important aspect of Greek religion: the fact that the Greek deities existed at two levels—the local, *polis* level, and the Panhellenic level.

Too often, in the study of Greek divinities, the local personality of a deity is overshadowed by the Panhellenic one and the individuality of the different local deities is ignored.[2] However, it is extremely unlikely that the establishment and crystallization of a Panhellenic persona for a deity so stamped out the local personalities that only insignificant variations remained. For the parameters affecting their definition differed in different cities, and again at the Panhellenic level. The realities and needs of the worshipping groups differed while some were common to all and also operated at the Panhellenic level. The composition and hierarchy of the pantheon also differed in the different cities, and again at the Panhellenic level. Moreover, of the spheres in which a divine personality manifests itself,[3] that of cult would be especially resistant to change under the impact of Panhellenic religion; for cult operates primarily at the *polis* level, having a function within its structures. These distinctions are based on the assumption that before the emergence in the eighth century of the agents of Panhellenic religion, the Panhellenic sanctuaries and literature such as the Homeric poems, there were significant differences in divine personalities and the composition of the panthea in the different cities. I shall now attempt to justify this assumption by considering briefly the background to the development of historical Greek religion.

Whatever degree of religious uniformity may have existed in the Mycenaean world—and the present evidence does not allow any firm or wide-ranging conclusions—conditions in the Dark Ages favoured radical changes, localized developments, and diversification. The collapse of the Mycenaean palace societies, insecurity, desertion of sites, and movements of population altered the nature of society, and therefore the needs fulfilled by the deities. Moreover, and more tangibly, the collapse of the framework for administering cult controlled by the palaces inevitably affected both cult and, through it, other spheres of religion. Some religious practices previously detected in the archaeological record now disappear. The comparative isolation and inward-looking character of the Dark Age communities, especially in the early part of the era, cannot but have led to separate local developments in religion. Diversification would have been intensified by two factors. First, differences in prosperity, economic activities, and life-patterns between different areas, especially in the later Dark Ages. Secondly, the arrival, in some areas, of intrusive population groups—later glorified into the 'Dorian Invasion'—groups which brought with them new religious concepts and practices expressing the realities and catering for the needs of simpler societies than those of their Mycenaean kinsmen. It is likely that these new practices and beliefs had an important impact precisely because they corresponded more closely to the new realities of the simpler Dark Age societies than did the legacy of the Mycenaean religious tradition. We should envisage, then, a complex process of development and interaction.

These circumstances can explain the important differences between Mycenaean and archaic Greek religion, which cannot be obscured by a few divine names common to both.[4] Even these names are less significant than is sometimes assumed, since identity of name does not entail identity of the other parts of the divine personality, especially when, as here, the panthea to which the deities belong are different.[5]

In the Dark Ages, then, the panthea of the different *poleis* and the personalities of their members were likely to have developed in different ways. Consequently, the divine personalities in local panthea from the eighth century onwards should be envisaged as products of an interplay between on the one hand the Panhellenic personality of each deity, promoted by the great Panhellenic centres and by poetry, and on the other the original local personality, itself developing not only under the influence of this Panhellenic persona, but also in response to changes within the worshipping society. It is clear that in some cases, as in that of Apollo, the Panhellenic dimension is particularly strong, due to the special circumstances of its diffusion.

This model cannot be proved to be correct. However, since it cannot be invalidated (on the contrary, the circumstantial evidence points in its favour) the possibility that it may be correct must be taken into account. Hence the study of Greek divinities must not be based on the assumption that the divine personality of a deity was substantially the same throughout the Greek world. Consequently to avoid the danger of distortions we must study each local divine personality of a deity separately from the Panhellenic one, and not use evidence from the latter to determine the former. Instead, we must recover each local manifestation of the personality, and then relate it to the Panhellenic persona. Moreover, we must not extrapolate from one local cult to another and attempt to interpret an aspect found in one place through another found elsewhere. Nor should we conflate evidence from different parts of the Greek world. The result would be a totally artificial conflation that had no cultic or theological reality. The fact that a given function is, for example, associated with Aphrodite at Sparta only means that this function belongs to her in the context of a particular personality nexus. It is not necessarily found in all, or indeed any, of her other personality nexuses which, I have argued above, had a different profile. Nor is it an inalienable part of an integral complex which included all the aspects of Aphrodite from the whole of the Greek world, and which would be 'the' Aphrodite. For divinities only existed at two different levels of cultic reality: local and Panhellenic. No aspect of a deity has any significance when separated from its organic context. Tendencies and aspects common to a deity throughout the Greek world have to be recovered and tested, not assumed or extrapolated.

To sum up, the study of Greek divine personalities should be based on specific local religious units and rely on internal evidence alone. The Panhellenically consistent traits would then be recovered and tested. It should be clear that this method based on the study of local cults is valid whether or not my initial analysis is correct. For it is a neutral approach that does not introduce any preconceived distortions. Moreover, it allows a circumscribed investigation of the circumstances of the worshipping group, necessary for the study of the development of any one deity, and of the other deities of the pantheon to whom it related.

I now propose to apply this approach to the study of the divine personality of Persephone at Locri Epizephyrii. For reasons of space this investigation cannot be complete. I cannot analyse here the whole material pertinent to Persephone's cult at Locri. I shall only discuss that which illuminates important and/or new aspects of her personality—aspects corroborated by the evidence that will not be discussed. Again I cannot branch off into consideration of the whole Locrian pantheon, on which our information leaves much to be desired. However, I shall consider *en passant* Persephone's consort Hades, and I shall discuss briefly the personality of Aphrodite, with whom Persephone is associated in cult, and of Aphrodite's Locrian *paredros* Hermes. I believe this limitation to be legitimate in so far as it is clear to me, as I hope to show, that Persephone's main concern at Locri was with an area of life and religion that can to some extent be defined as 'circumscribed', in so far as any area of cult can be said to be so—the world of women. I shall only use internal Locrian evidence and I shall not allow its decoding to be contaminated by preconceived ideas based on external evidence, either Panhellenic, or from other local cults, unless the comparison of internal evidence with Persephone's Panhellenic personality shows that a particular Panhellenic aspect was also found at Locri. Since the investigation will use material from one period only, the first half of the fifth century, I shall discuss the personality of the goddess at that period alone. The results may also be valid for other periods—indeed, given the nature of the deity they probably are—but they cannot be assumed to be so. This limitation in time is the reason why I omit discussion of the historical background. The relevant aspect of the worshipping group is the position of women in the city. In general Locri did not differ from any other Greek city in this respect.[6] About the details we are ignorant.[7]

2. PERSEPHONE AT LOCRI

1. Introduction

Our main source of information for the cult of Persephone at Locri is the series of clay relief plaques with religious scenes, from the first half of the fifth century, known as the Locrian pinakes. The great majority of these were found in the Locrian sanctuary of Persephone between the hills Abbadessa and Mannella.[8] Other evidence, albeit not very informative from our point of view, comes from literary sources[9] and a few inscriptions;[10] archaeological finds complete the picture.[11]

Persephone, in whose sanctuary the pinakes were found, is the main deity involved in the cult and myth reflected in the representations. Aphrodite also has a place in that cult, and some types of pinax belong to her.[12] This presence of Aphrodite in a basically Persephonean context is not due to an alleged accidental mixing of two deposits, an untenable hypothesis put forward by Prückner[13] and heavily criticized.[14] The remarkable unity of the whole series[15] shows that we are dealing with one cultic nexus in which the two goddesses are closely associated. The reasons for this association can only be understood after the divine personalities of Persephone and Aphrodite at Locri have been defined. It must be clear from what I said in the first section of this paper that speculation based on the 'general' Greek characteristics of the two deities is worse than useless: it carries a strong danger of distortion.

Given that the series contains types belonging to two different goddesses, and that there is some disagreement about which goddess some of the pinakes belong to,[16] a rigorous methodology needs to be used, which will not depend on preconceptions about the type of scene that we should expect to belong to each goddess. I propose, using internal iconographical evidence alone, to begin with the scenes that indisputably belong to Persephone, and isolate the symbols and cult objects peculiar to her in the local cult. The presence of a symbol in a Persephone-scene is not, clearly, sufficient evidence that it characterizes the goddess and her cultic sphere, and can thus be used for attributing other scenes to Persephone. It must first be ascertained that this symbol does not also appear in contexts associated with Aphrodite. The connection with Persephone can be firmly established only if the same symbol is found in more than one of the types which indisputably belong to her. The symbols and cult objects thus shown to be characteristic of Persephone at Locri will allow us to attribute more scenes to Persephone. Clearly, given the limitations set out above, it is not my intention to divide all the scenes found in the pinakes between the two deities. I am aiming at obtaining firm attributions to Persephone for those scenes that illuminate the important and salient aspects of the goddess' personality.

II. *Scenes indisputably belonging to Persephone's sphere*

The first series of scenes indisputably belonging to Persephone—a fact that has never been doubted—is that showing a bearded and majestic male figure carrying off a girl in a winged chariot.[17] It represents, of course, the Rape of Persephone by Hades. The fact that this is a purely mythological representation, that is, the iconographical expression of a myth, may explain why no symbols are included here, as they are in the context of cult scenes. The only additional feature shown, a chain of flowers,[18] belongs to myth: Persephone had been picking flowers when Hades abducted her.[19]

A Model for Personality Definitions

These scenes confirm that Hades' *Bildvorstellung* at Locri is the same as that elsewhere in Greece: majestic bearded god of mature age, of the same type as those of Zeus and Poseidon who are, of course, characterized by their attributes.[20] They show too that the myth of Persephone's Rape by Hades, associating her with the earth's fertility and with the funerary sphere which was an aspect of her Panhellenic personality, was also an aspect of her personality at Locri.

Another series also indisputably belonging to Persephone is that of 'the Young Abductor', showing a girl being carried off by a beardless youth in a chariot, in one example in the presence of Hades.[21] I have argued elsewhere[22] that these pinakes were wedding dedications; and that they showed an ideal representation of a bride and bridegroom depicted according to the iconographical model of the divine bride and bridegroom of Locri, Persephone and Hades, whose marriage was preceded by an abduction depicted in another series of pinakes. I also argued that these dedications were made by girls who were getting married and seeking Persephone's protection. If this interpretation is correct, it entails that not only was the wedding of Persephone and Hades an important part of the Locrian cult and myth, as Zancani Montuoro has argued,[23] but also, and this is where I go further,[24] that the Locrian Persephone was associated with marriage in a way which suggests that she fulfilled the role of protectress of marriage and weddings. In other cities, this role is usually, though not always, fulfilled by Hera.[25] In this paper, I shall extend my earlier suggestion that Persephone was such a protectress at Locri. First, I shall try to show that there is further evidence connecting this role with her. Secondly, I hope to show that the Locrian Persephone also had a kourotrophic function, a role related to that of protectress of marriage. This would suggest that she was presiding over the world of women and their concerns, that she was a women's goddess like Artemis Brauronia in Attica.

The following cult objects and symbols are found in the 'Young Abductor' series.[26]

(1) The *kalathos* with fruit or flowers (Orsi 1909: 26, figs. 32–3, 27, figs. 34–5; Prückner 1968: pls. 17.1, 21.1; Zancani-Montuoro 1954: pl. xxix; cf. also Quagliati 1908: 165*h*; Prückner 1968: 70–2).

(2) The *cock* (Orsi 1909: 26, fig. 32, 27 fig. 34; Quagliati 1908: 154, fig. 18, 155, fig. 19; Sourvinou-Inwood 1973: pl. 1*b, c*; Prückner 1968: pls. 14.2–4, 15.3, 16.3, 16.8, 17.2, 18.2, and cf. also p. 71).

(3) The *ball* (Zancani-Montuoro 1954: pl. xxx; cf. also Prückner 1968: 71, 72).
(4) The *small chest* (Quagliati 1908: 172, xxv).
(5) *Flowers* (a single flower: Prückner 1968: 71 (type 73), cf. also 72; a wreath of flowers: Prückner 1968, 71 (type 78)).

None of these objects appears in a scene firmly associated with Aphrodite. Note that the only symbols and cult objects that we can definitely associate with Aphrodite, because they appear in scenes which indisputably belong to her (for which cf. below), are the following.

(1) The *dove* (Prückner 1968: pl. 2.1).
(2) The *alabastron* (Prückner 1968: pl. 2.1).
(3) The *'rose-like' large flower* (Orsi 1909: 12, fig. 12; Prückner 1968: pl. 1.1).
(4) The *bowl full of fruit* (Orsi 1909: 13, fig. 13).
(5) The *mirror* (cf. Prückner 1968: 135 n. 107).
(6) The *phiale* is associated with Aphrodite's *paredros*, Hermes (Prückner 1968: 17, fig. 1). As we will see, the phiale is also associated with Persephone and Hades, and the mirror and the alabastron have also gravitated into Persephone's orbit. It will be clear that Persephone, the dominant partner in the cult, has by far the richer nexus of symbols and cult objects.

Two more series, related to each other, also belong indisputably to the sphere of Persephone: the scenes depicting Hades and Persephone enthroned (here Pl. 14), and the 'homage' scenes in which various deities pay homage to an enthroned Persephone or to the enthroned couple.

In the first series[27] we find the following symbols and cult objects.

(1) The *cock* (one is held by Persephone and another is standing under the throne).
(2) *The thymiaterion surmounted by a cock.*
(3) *The stalk of grain* (held by Persephone).
(4) *The phiale* (held by Hades).
(5) *A blooming twig* (held by Hades).
(6) *The throne with a back ending in a goose's head.*

In the homage scenes[28] the divinities paying homage hold attributes which identify them (e.g. Hermes the ram), or offer the cock or other objects appropriate to Persephone or the circumstances.[29] The following deities appear in these scenes: (*a*) *Hermes*, always with a ram, and often presenting a cock to Persephone;[30] he is sometimes accompanied by a female figure to whom I shall return below; (*b*) *Dionysos*, holding a kantharos and a vine, sometimes also accompanied by a female figure; (*c*) *Apollo*, in one type with a lyre, in the other with a lyre and a bow;[31] (*d*) *Triptolemos*, holding a stalk of grain in one hand and with the other guiding the winged serpents of his chariot; (*e*) the *Dioskouroi*, who are represented as horsemen, sometimes followed by a female figure; they hold a cup or a kantharos and a shield or a lyre. Prückner[32] also recognizes Athena in one of the types. The identification of the female figure with the mantle drawn over her head who sometimes accompanies Hermes, Dionysos, or the Dioskouroi, and never appears alone[33] is difficult. She is shown offering a cock, a ball, a small chest, and an alabastron. Of these the cock has already been connected with Persephone, and we shall also find that Persephone herself is holding a cock in some homage scenes. The ball, offered together with cock by the woman accompanying Dionysos, has been associated with Persephone through the 'Young Abductor' scenes. Moreover, as we will see later on, the ball and the cock appear, again as joint offerings to Persephone, in the hands of mortal girls. The small chest, already found in the Young Abductor series, is also, we shall see, held by Persephone herself in some homage scenes. The alabastron, we saw, is associated with Aphrodite; we have not, so far, found it in the realm of Persephone. Three out of the four objects connected with this problematic goddess then belong to the realm of Persephone and do not characterize the enigmatic figure. The alabastron, which only appears once, held by the goddess accompanying Hermes, belongs to Aphrodite, though, we shall see, it also seems to have been attracted into Persephone's sphere. Therefore, the identification of the unknown goddess must depend primarily on context, the associations with the male companions. I am inclined to agree with Zancani Montuoro[34] that a different goddess is shown in each of the three contexts: Ariadne with Dionysos, a married couple; Aphrodite with Hermes, an illicit pair important in Locrian cult; Helen with the *Dioskouroi*[35] (Helen's marital affairs had been presented in a favourable light by Stesichoros, a poet connected with Locri,[36] in the *Palinode*, a poem associated with Locri).[37]

With regard to the significance of the homage scenes, there can be little doubt that, as Zancani Montuoro has argued,[38] they represent various deities paying homage and offering gifts to Persephone or Persephone and Hades on the occasion of their wedding. Prückner,[39] who has rejected this interpretation in favour of a much less convincing one of his own,[40] has not produced any cogent arguments against it. Before considering his objections, I want to suggest that Zancani Montuoro's interpretation can be extended, and that the meaning which she has established for these scenes can be given another level of significance. Persephone's wedding had an important place in Locrian cult and myth, and there are strong reasons for thinking that she was the patron goddess of weddings and marriage. This nexus of ideas present in the minds of the Locrians would inevitably affect their perception of mythological events involving Persephone as a bride. In this context, the paying of homage and offering of gifts by various deities to Persephone the bride or to the bridal couple would, I think, also be seen as expressing the importance of weddings, and of the institution of marriage, in the divine, and therefore also in the human, world. This double point of reference, mythological-narrative and conceptual, may have been an important reason for the adoption of the iconographical schema of the homage for the representation of this event, instead of the usual procession of deities, a choice which Zancani Montuoro[41] attributed to the restrictions of space. Alternatively, the choice of schema may indeed have been dictated by the restrictions of space, and this choice may have facilitated the perception of the representation in 'symbolic-conceptual' terms. But I am convinced that, given the nexus of ideas about Persephone in the Locrians' minds, whatever the exact mechanics, these representations could hardly have failed to be perceived also as an expression of the importance of weddings and marriage.

If this is correct, the fact that these scenes were not only representations of a mythological event, but also expressions of an idea in representational language, and that at the conceptual level Persephone the divine bride expressed and 'symbolized' the concept of marriage, may explain why she could be shown alone, where in a strictly narrative scene in this context we would have always expected the bridal couple. And this answers the first objection put forward by Prückner[42] against Zancani Montuoro's interpretation, the objection that, if that interpretation were correct, Hades would have always been present. His second objection is that, if the scenes were indeed showing the paying of homage and the offering of gifts at Persephone's wedding, all the gods, or at least all the most important gods, would have been present, and certainly Zeus would not have been absent. This objection takes no account of the fact that all the deities involved in the scenes are important divinities in the Locrian pantheon,[43] and that there is clearly a local emphasis in the selection. The series does not represent Panhellenic myth, like the representations Prückner has in mind; it does not even, if I am right, represent only myth, but also an evaluation of a human social institution in mythological-narrative terms drawing on local myth and cult. But the absence of Zeus does indeed require explanation, since his cult was important at Locri.[44] In my opinion, it is due to the fact that it would be inappropriate to have Zeus, the highest divinity, paying homage to Persephone and Hades; it would be particularly inappropriate if it is correct that this iconographical schema also expressed a 'symbolic' evaluation of marriage in which Persephone, representing the concept of marriage, played the dominant role; for in that case, the connotation of homage would have been emphatically present in the perception of the scene.

The following symbols and cult objects are associated with Persephone in this series.

(1) The *phiale* (Zancani Montuoro 1954: pls. xv–xvi and p. 81; Orsi 1909: 11, fig. 9; Prückner, 1968: pl. 24.3).

(2) The *cock* (Orsi 1909: 9, fig. 7, 11, fig. 9; cf. Zancani Montuoro 1954: 86).

(3) The *stalk of grain* (Orsi 1909: 9, fig. 7; cf. Zancani Montuoro 1954: 81).

(4) The *small chest* (Zancani Montuoro 1954: pl. xviii; Prückner pl. 24.3).

(5) The *kalathos* (Zancani Montuoro 1954: pl. xviii).

The following objects are held by Hades.

(1) The *kantharos* (Zancani Montuoro 1954: pls. xiii–xvi, cf. p. 81).

(2) The *phiale* (Orsi 1909: 11, figs. 9–10).

(3) The *pomegranate* (Orsi 1909: 11, fig. 9).

(4) The *goose* (cf. Zancani Montuoro 1954: 79–80).

(5) The *sceptre ending in a figurine of a sphinx* (cf. Zancani Montuoro 1954: 80).

The *throne with a back ending in a goose's head* is associated with both divinities.

As I mentioned earlier, among the offerings to Persephone, the cock, the ball, and the small chest should be taken as appropriate to the receiving goddess; the alabastron, only offered in Prückner's type 105[45] by a woman who is probably Aphrodite is more dubious; it may characterize the offering goddess.

Of these objects and symbols, the following also occur in the other series belonging to Persephone's sphere which I have considered, and therefore emerge as firmly connected with that sphere. The *cock* (Young Abductor, enthroned couple, homage series); the *stalk of grain* (enthroned couple, homage); the *small chest* (Young Abductor, homage); the *kalathos* (Young Abductor, homage); the *ball* (Young Abductor, and in the homage scenes as an offering to Persephone in association with the cock). The *phiale* is associated with Hades in the enthroned couple series and with Persephone in some homage scenes; it belongs then to this sphere, but not exclusively, since, we saw, it is also held by Hermes. The *throne with a back ending in a goose's head* firmly belongs here (it is found in the enthroned couple and the homage series). The *flowers* are found in the Young Abductor and in the Rape of Persephone series. There are also some other symbols and cult objects which are only found in one series, but are nevertheless likely to belong to Persephone's sphere. For they are never found in types that can certainly be attributed to Aphrodite and are frequently associated firmly with Persephone's realm either through context (objects consistently held by Hades in one series, like the kantharos) or through their nature (for example, the thymiaterion surmounted by a cock which incorporates an element firmly connected with the goddess). These are, firstly, the thymiaterion decorated with a cock, and secondly, the following objects held by Hades: the kantharos, which also belongs to Dionysos; the pomegranate; the goose; the blooming twig.

I now propose to consider these symbols and cult objects connected with Persephone's sphere and examine their associations and significance. This will allow us to understand and determine their connection with Persephone more precisely.

III. Symbols and cult objects

(*a*) The *stalk of grain* is associated with Persephone everywhere in the Greek world, especially in her character as Demeter's daughter.

The connection is particularly close in Eleusinian contexts.[46] This symbol denotes fertility, especially fertility of the earth, which was, of course, conceptually connected with other kinds of fertility. The *blooming twig* held by Hades undoubtedly has the same connotation of fertility.

(*b*) The *cock* appears to have been very closely connected with Persephone at Locri. Apart from its frequent appearance in connection with the goddess on the pinakes, the cock is also found in the hands of some female terracotta figurines found in Persephone's sanctuary, where terracotta figurines of cocks, mostly in low relief, more rarely moulded in the round, have also been discovered.[47] Because of the close connections between Locri and Sparta,[48] it may be of interest to note that the cock is associated with the enthroned Underworld deities in the problematic Laconian reliefs.[49] Porphyry, *De abstinentia* 4. 16, suggests that the cock was sacred to Persephone not only at Eleusis, but in the Greek world in general. For the Greeks in general the cock had two main symbolic connotations. Firstly, maleness, in the double sense of bravery and fighting spirit, and of male sexuality;[50] and secondly, the terrifying things of this world and especially death[51]—from which aspect is derived the bird's apotropaic character. This second aspect explains the cock's widespread association with Persephone. The first makes it appropriate to a wedding/marriage context and therefore also to Persephone in her persona as divine bride and patron of marriage. The cock in a specifically nuptial context can be seen as symbolizing maleness, including male sexuality and aggressiveness, harnessed into marriage, an institution necessary for society.

(*c*) The terracotta *balls* with incised decoration, very similar to those shown on the pinakes, which have been found at Locri[52] were probably dedications to Persephone made in circumstances similar to those of the dedications depicted on the pinakes (see below). For the Greek mentality the ball, like the cock, had, it appears, a double connotation, one chthonic and funerary, the other nuptial and pre-nuptial.[53] The second of the two associations may be at least partly derived from the Greek practice of girls dedicating toys to divinities on marriage as symbols of childhood left behind,[54] a dedication marking their passage to the new status which was the fulfilment of their destiny. I wonder whether the funerary connotations of the ball may not have derived from the cases of girls who died unmarried, and for this reason had a ball buried with them as symbol of their unfulfilled destiny.[55] In any case, it is clear that the ball was an appropriate object for Persephone and a particularly appropriate offering to her in a nuptial context.

(d) Terracotta *pomegranates* predominate among the terracotta fruit discovered in the sanctuary of Persephone.[56] Of the many female figurines found near the Marasà temple, some are holding a pomegranate while others hold a dove,[57] the former, in my opinion, representing Persephone or a votary of hers, the latter Aphrodite or a votary of Aphrodite. In the Laconian reliefs the pomegranate is associated with the Underworld couple.[58] In the Greek world in general the pomegranate had both chthonic and fertility connotations.[59] It was widely associated with Persephone,[60] and this association was expressed mythologically in a most important episode of her myth, her final binding to Hades and his realm through the consumption of a pomegranate given to her by Hades.[61] Consequently, it is a most appropriate symbol for the bridal couple Hades and Persephone.

(e) The *goose* is held by Hades in some homage scenes, but it also appears in many other types in the motif of the throne decorated with a goose's head with which Hades and Persephone are firmly associated in the pinakes. This firm association suggests that this throne of the pinakes reflects a real throne in the sanctuary or a throne in which the cult-statues of the deities were seated. Terracotta geese were included among the finds of the Manella sanctuary of Persephone.[62] Zancani Montuoro[63] has pointed out that a goose appears to be associated with Hades (inscribed 'Eumolpos') on side B of the skyphos London E 140 by Makron.[64] Outside Locri the goose is generally associated with Aphrodite, being considered an erotic bird,[65] but in Lebadeia it is associated with Persephone.[66] This confirms the need for establishing attributes and associations for each local cult unit, instead of importing those that seem more widespread throughout Greece. It is possible that it was the erotic aspect of the bird that here attracted it to the sphere of the marriage deities, but this is by no means certain.

(f) The *kantharos* is associated with both Hades and Dionysos in the pinakes. This causes no ambiguity, for context and especially iconography make clear the identity of each of the gods. The vessel is, of course, a Panhellenic attribute of Dionysos. It is associated with the seated couple of Underworld deities in the Laconian reliefs.[67]

(g) The *phiale*, in the pinakes associated with Hades and Persephone, but also with Hermes, is elsewhere widely associated with many different deities, denoting the performance or the reception of libations.[68]

(*h*) *Flowers* and flower-picking are associated with Persephone in the myth of her abduction: she was picking flowers when Hades carried her off, and this was reflected in one of the pinakes types depicting the Rape, in which Persephone is holding a chain of flowers. For this reason flowers were appropriate to wedding contexts at Locri[69]—hence also the flower or chain of flowers held by the girl in some Young Abductor types. Terracotta flowers, and especially, it seems, flowers of lotus, have been found in the Locrian sanctuary of the goddess, where one lotus flower made of ivory was also discovered.[70] Many of the female terracotta figurines found in the sanctuary and one 'maschera' hold flowers in their hands.[71] In the Locrian colony of Hipponion a legendary flower-picking of Persephone in the area was the *aetion* for cultic *anthologein* and *stephaneplokein* by the women.[72] Pollux[73] mentions a festival of Persephone, or rather Kore, called Anthesphoria which was celebrated in Sicily: *Kores para Sikeliotais Theogamia kai Anthesphoria*; Zancani Montuoro[74] argued that this Anthesphoria was part of the celebration of Persephone's wedding, commemorating Persephone's flower-picking before the abduction. It is clear then that the flowers and flower-picking were firmly associated with Persephone at Locri, but also elsewhere,[75] and especially with Persephone in her bridal aspect.

(*i*) The *kalathos*, held by Persephone in some homage scenes, also appears in the Young Abductor series where it contains fruit or flowers—it is difficult to distinguish which. We have the same difficulty in another series of pinakes[76] in which girls are picking fruit or flowers from a tree and placing them in the kalathos. Since these scenes are shown on pinakes they must have a religious significance; the presence of a cock in some of the types indicates that we are in the realm of Persephone; this is confirmed by the firmly Persephonean connections of the kalathos with or without the fruit or flowers. The appearance of the kalathos with the fruit or flowers in the Young Abductor series may suggest that it is connected with weddings and that the garden scenes also involve a nuptial context. According to Zancani Montuoro,[77] the nuptial context involved in the garden series is mythological, that of Persephone's wedding. But it is equally possible that the context is cultic and that we are dealing with Locrian nuptial rites. If the objects in the kalathos are flowers, as their appearance in the garden series seems to suggest, we would be dealing again with the flower-picking motif in a nuptial context. If they are fruit, and specifically, as Zancani Montuoro suggests, pomegranates and quinces, we would still have connotations of wedding and fertility.[78] We have seen that terracotta fruit, and especially pomegranates, were dedicated at the sanctuary of Persephone at Locri. In any case, it would appear that Persephone's connection with the kalathos derives from its use as a receptacle for fruit or flowers in the course of nuptial rites celebrated in Persephone's cultic sphere.

(*j*) The *small chest* probably became associated with Persephone and nuptial contexts because of its place in the life of women; if it is seen as a miniature version of a household chest it belongs to the sphere of house management, one of the main functions of a Greek wife; if it is seen as a small container for jewellery and the like it belongs to the beautification activities appropriate to brides and women in general. Small chests and other small replicas of items of furniture appear to have been dedicated at the sanctuary of Persephone, to judge from the ivory plaques found there which had been inlaid in them.[79] These dedications may perhaps provide some confirmation for my argument that Persephone at Locri was closely associated with women and their concerns.

(*k*) With regard to the *sceptre ending in a figurine of a sphinx*, the firm chthonic and funerary associations of this monster hardly need mentioning.

Now I propose to consider briefly the associations and values of these symbols, when these are known independently of their context in the pinakes, and see what these can tell us about the connotations of the deity with whom they are associated; and whether these connotations derived from the symbols confirm or invalidate the conclusions about the Locrian Persephone's personality and functions reached so far on the basis of the content of the scenes, conclusions which will, I hope, be confirmed and extended below through the examination of further scenes.

The small chest, we saw, indicates the sphere of women and their concerns in general. The usual function of the kalathos in ancient Greece is as a wool-basket:[80] it belongs therefore to the sphere of female activities. It is undoubtedly for this reason that at Locri it was transferred to the cultic activity of flower- (or fruit-)picking which is widely associated with women even outside Persephone's cult.[81] It is also generally associated with marriage rites.[82] The flowers, and the blooming twig, carry the connotation of the blooming, rejuvenation, and fertility of nature. The stalk of grain also denotes fertility of the earth and prosperous agriculture. The cock has a double connotation, a funerary one and one of maleness and aggression. The ball belongs to the sphere of marriage and prenuptial rites and also to the funerary realm. The pomegranate has connections with fertility and with the funerary sphere. The sphinx has funerary connections.

Even if we knew nothing else about the goddess with whom these symbols are associated, we could still deduce from the associations that she was concerned with the life of women and their pursuits, with special reference to marriage; she was associated with fertility,[83] especially of the earth; and she was also involved with the funerary sphere. This confirms the conclusions reached through consideration of the scenes indisputably belonging to her. We may note that her connections with fertility and the funerary sphere are aspects found also in the Panhellenic personality of the goddess, while her connections with marriage and the world of women are peculiar to Locri.

IV. *Further scenes from the cult of Persephone*

I shall now consider some cycles of scenes which can be attributed to the cultic realm of Persephone on the basis of the symbols and cult objects in them which have been firmly connected with this goddess in the previous sections.

I shall begin with a nexus of scenes involving young girls, and take as the centre of the investigation a group of scenes showing young girls bringing offerings to a seated goddess. I take these to be either 'ideal representations' in which the goddess is represented as 'ideally' present at the moment of the offering, in the manner of votive reliefs, or depictions of actual cultic scenes in which Persephone's priestess receives the offerings on behalf of the goddess.

Before discussing this series, I want to mention a type which is related to it but does not belong in it.[84] It shows Hades and Persephone seated, Hades holding a phiale and pomegranate, Persephone a cock. A kantharos stands on a small table in front of them. A girl wearing chiton and himation is pouring a libation.

The context is clearly different from that of the presentation of offerings to Persephone. The girl may be a priestess offering a libation either in front of the cult-statues, or to the two deities perceived and represented as 'ideally present'.

The series includes the following types.

1. Prückner's type 89.[85] Persephone is seated, holding a stalk of grain together with poppy-flowers or poppy-heads;[86] on a table in front of her there is a cup; a girl wearing chiton and himation is offering a *kanoun* with pastry and a pomegranate. This scene is very similar to another one, illustrated in Orsi 1909: 15, fig. 16,[87] in which two girls wearing chiton and himation are shown in front of an altar; the first is holding a similar *kanoun* with pastry and a small phiale and stooping over the altar on which she is about to place something; the other is offering a cock and a kalathos; between the first girl and the altar stands a thymiaterion surmounted by a cock; a kantharos and two phialai are hanging on the wall. There can be no doubt, given all the objects firmly connected with Persephone, that this scene, like the related type of the offering girl with which I started, belongs to Persephone's cultic sphere, and that therefore a *kanoun* with pastry was an appropriate offering in this goddess' cult in at least two different rituals. What may appear slightly surprising is that in the scene with the two girls a mirror is also hanging from the wall, an object which appeared in connection with Aphrodite and her cultic realm: I will return to this later. To return to the offering of the single girl, we have already noted the double connotation of the pomegranate. The *kanoun* had definite nuptial associations in ancient Greece, for it had an important place in wedding rites.[88]

Prückner suggests, I think rightly, that the girls bringing offerings here and in his types 90 and 91—my nos. 2 and 3—are brides worshipping Persephone the heavenly bride, but he does not extend this interpretation to a second group of scenes showing girls bringing offerings, my nos. 4–7, which he arbitrarily attributes to Aphrodite. He does not draw any conclusions about Persephone's personality and functions from this suggestion. Presumably, he takes the (few) scenes which he interprets in this way as *ad hoc* acts of worship rather than part of a cultic cycle connected with a well-established aspect and role of the Locrian Persephone.

2. The fragmentary type illustrated in Orsi 1909: 14, fig. 14 (Prückner's type 91).[89] Persephone on the left holds a stalk of

grain, the girl, who is wearing a 'spotted' peplos, is offering a cock; one of her forearms and hands is missing and so is most of the space between goddess and girl.

3. The type illustrated in Quagliati 1908: 211, fig. 59 (Prückner's type 90).[90] Persephone is seated on the left; her attributes are the stalk of grain together with poppy-heads, the cock, and the phiale. The girl is offering a toilet box and a mirror. The mirror thus appears again in Persephone's sphere—for the goddess' attributes leave no doubt as to her identity. We have to accept then that the mirror and the toilet box, both belonging to cosmetic activities, were considered appropriate offerings to the Locrian Persephone, and that the mirror was associated with that goddess in other contexts, although it was also connected with Aphrodite. The explanation of this phenomenon clearly lies in Persephone's close association with girls of marriageable age and with weddings, an event above all requiring beautification. It must have been this that brought the mirror, and perhaps to a lesser extent the toilet box, into Persephone's orbit. Many bronze mirrors and some other bronze toilet articles were found in the Mannella sanctuary,[91] but, given the joint cult, the possibility cannot be excluded that these were dedications to Aphrodite.

4. The type illustrated in Plate 15; Orsi 1909: 8, fig. 6 and Quagliati 1908: 197, fig. 47. Persephone (or her priestess) is seated to the right, on a throne the back of which ends in a goose's head. She has her himation pulled over her head and is holding a deep phiale, while she has just dropped a wand.[92] A small table stands in front of her under which a goose is shown—it is not clear whether it is to be understood as a real bird or as a decorative motif in relief. On the table lies a folded 'spotted' peplos and on top of it a box decorated with rosettes. Two phialai and an oinochoe, clearly metallic, are hanging on the wall. The girl is standing on the other side of the table; like the companion of Dionysos in the homage scenes, she is offering a cock and a ball. These, I argued, are very appropriate offerings in a wedding context. The various symbols and offerings make clear beyond any doubt that the goddess concerned is Persephone. The folded peplos is, I believe, a bridal peplos placed in front of the goddess, or her priestess, so that she can bless it, perhaps, if the woman is the priestess, by sprinkling it with the wand with water from the phiale. If the woman is Persephone we would be dealing with an 'ideal representation' of

the blessing modelled on actual rituals. Zancani Montuoro's reading[93] is not dissimilar from mine at this point, but our interpretations of the precise circumstances differ, as she is thinking in terms of Persephone's own wedding and bridal peplos.

5. The type illustrated in Quagliati 1908: 199, fig. 48 (Prückner's type 32).[94] Here too, Persephone (or her priestess), dressed as in 4 and seated to the right, holds a deep phiale and has just dropped the wand. There is an altar in front of her, on the side of which there is a cock, probably to be understood as sculpted in relief. On the altar lies a peplos similar to the one in 4 and on top of it a very ornate crown which, if the peplos is indeed a bridal peplos, should be the bridal crown. The girl is holding a ball in her right hand, while with her left she holds up the end of her apoptygma, stretching it out towards Persephone.

6. A type closely related to 5, and preserved in a fragmentary form, is that illustrated in Quagliati 1908: 224, fig. 73, Prückner, pl. 7. 1, and especially in Allard Pierson Museum, *Algemeene Gids* (Amsterdam, 1937), 2077, pl. xcviii.[95] It differs from 5 in two respects. Firstly, the girl is here wearing a 'spotted' peplos, and secondly, her outstretched apoptygma contains flowers or, perhaps less probably, fruit. It is important to note that in one of the types of the garden series, that illustrated in Prückner 58, fig. 10, a girl, who is wearing 'spotted' garments, is holding the flowers (or fruit) which she is picking in the folds of a similarly outstretched himation. The offering of the flowers (or fruit) and the connection with the flower-picking (or *karpologia*) of the garden scenes give nuptial connotations to the scene discussed here, as does the ball held in the girl's right hand.

7. Prückner[96] mentions and describes a seventh type, his type 33, which appears to be unpublished. His use of the word 'Truhe' for both table and altar elsewhere does not allow the reader to understand which of the two is depicted here. In any case, a Siren is shown under it, and on top of it just a box, no peplos. The wand is also missing. This may suggest that the use of the wand was connected with the peplos, and that this is a different moment in the ritual. With regard to the Siren, a being of chthonic, funerary nature,[97] it is worth mentioning that Sirens are associated with cocks in another type of scene, illustrated in Orsi 1909: 19, fig. 21.

There are then clear nuptial connotations in this series, and a less pronounced fertility connection, itself appropriate to a nuptial context. A nuptial context is suggested by the *kanoun*, the flowers (or fruit) and the connection with flower-picking or *karpologia*, and the offerings of cock and ball. The fertility aspect is denoted by the stalk of grain and the pomegranate. In my opinion, the combination of several symbols and offerings suggestive (even if perhaps not always conclusively in each case) of a nuptial context, the presence of the peplos and crown and especially their place in the syntax of the scene, and the fertility connotations, leave no doubt that these scenes show a cult act which was part of the Locrian wedding rites. Specifically, they appear to represent young girls making offerings to Persephone on the occasion of their wedding, and getting their bridal peplos blessed by the deity.

This series relates, we saw, to the garden series. It also relates to another cycle of scenes, to which I now turn, which involves six groups of representations. This cycle is connected with the 'offering girls' series through two elements, found in nos. 4–7 of the latter series: first the peplos, and secondly the woman wearing chiton and himation and holding the deep phiale and wand (whom I will call, for brevity's sake, the *phialophoros*). I shall argue that in this cycle of six groups of scenes involving a *peplophoria*, the *phialophoros* is the priestess of Persephone. That the *peplophoria* cycle belongs to Persephone is shown by the various objects and symbols present and also by the connection with the 'offering girls' series.

Of the six groups two show the *phialophoros* and four girls carrying an unfolded spotted peplos.[98] A third group[99] shows a girl carrying a folded peplos on a tray resting on her head, followed by the *phialophoros*; other elements that appear are the thymiaterion surmounted by a cock, a small boy, and a cock walking in the procession. A fourth group[100] represents a woman, himation drawn over her head, carrying a cock and with a tray containing a folded peplos resting on her head. In the fifth group[101] a girl carrying the peplos on a tray and followed by the *phialophoros* arrives in front of a seated deity who has her himation drawn over her head and is holding a cock; under her seat there is a hydria. This type would suggest that in this cycle at least the *phialophoros* figure is a priestess, while the goddess, shown as 'ideally present', is meeting the peplos-carrying procession. The final group[102] shows a girl putting the peplos away in a chest which stands in front of a throne with a back ending in a goose's head. Since this simple act is shown on pinakes, it must have had a religious significance. The peplos, the type of throne, and the kalathos and kantharos hanging on the wall indicate that we are still in Persephone's cultic sphere. A mirror is also hanging on the wall: we saw that this object had entered Persephone's orbit. There is also a lekythos on the wall, but this conveys no information to us. The context indicates a sacred garment kept in a sanctuary; this, in combination with the *peplophoria* scenes, suggests an occasion of garment presentation to a goddess, a well-known ritual act in Greek religion. Zancani Montuoro suggested that we are dealing with the presentation of Persephone's bridal peplos; she considered the whole nexus of scenes involving the peplos as part of Persephone's *theogamia*, but was undecided as to whether these are cultic scenes taking place in the Locrian sanctuary or mythological ones, though she is inclined towards the latter view.[103] I think that she is right about the garment being Persephone's bridal peplos. For, first, Persephone in her character as bride was most important in the Locrian cult, and secondly, the 'offering girls' series showed us the peplos and the *phialophoros* figure in a nuptial context. However, I do not think that we are dealing with the same peplos and the same nuptial context in the two cases. In my opinion, the *peplophoria* is part of a festival celebrating Persephone's wedding, while the 'offering girls' series represents ritual activities that were part of the Locrian marriage rites in which Persephone played a role.

There are differences between the two series of scenes which, I think, justify their attribution to different cultic occasions. Firstly, in the 'offering girls' series the *phialophoros*, deity or priestess, is the person who is receiving the offerings, and in front of whom the peplos has been deposited; while in the *peplophoria* the *phialophoros*, who must be a priestess, is differentiated from the seated deity who will receive the peplos—she is part of the offering procession. Even if the *phialophoros* is a priestess in the 'offering girls' series, her role is to stand in for the goddess, while in the *peplophoria* she is acting as an officiating priestess and comes in front of the 'ideally present' Persephone. This difference in the role of the *phialophoros* may indicate that, although the general cultic context is the same, involving the bridal aspect of Persephone, the specific occasion is different in each case. A second such indication lies in the fact that in the 'offering girls' series there are only two participants in the action: the offering girl and the receiving goddess (or priestess). This

A Model for Personality Definitions 169

contrasts sharply with the *peplophoria* scenes (always involving more than one mortal person) and also, and especially, with the gift-offering procession depicted in the type illustrated in Zancani Montuoro 1960: pl. vii, which also forms part of the celebration of the *theogamia*. The one-to-one relationship suggests offerings by individual girls, not part of the celebration of a festival in honour of a divinity. Another consideration which strengthens the cumulative evidence in favour of this hypothesis is that Persephone (or her priestess standing in for her) who is receiving the offerings from the girls, appears to be 'blessing' the bridal peplos lying in front of her. It is unlikely that the goddess, or her priestess standing in for her, would have been shown doing this to Persephone's own bridal peplos. Finally, according to Zancani Montuoro,[104] the 'offering girls' series has comparatively few types, but is represented by a large number of specimens. This situation would fit perfectly the hypothesis that these pinakes were votive offerings for a standard occasion, weddings, at which girls dedicated 'ideal' representations of themselves performing a cult activity that was part of the marriage ritual; an activity through which they obtained the blessing and protection of the goddess and which also symbolized their change of status. Some at least of the different types may represent different stages in the ritual. If my interpretation is correct, the 'offering girls' series confirms Persephone's role as protectress of marriage at Locri.

I shall now consider another group of scenes which I shall call the 'children in the basket'.[105] The following scenes from this group are published: 1. Quagliati 1908: 193, fig. 44; 2. Quagliati 1908: 194, fig. 45; 3. Quagliati 1908: 195, fig. 46; 4. Plate 16 and Orsi 1909: 29, fig. 39; 5. Orsi 1909: 30, fig. 40 and Prückner, 1968: 31, fig. 3, 6. Plate 17 and Orsi 1909: 30, fig. 41. This cycle of representations can be divided[106] into two basic groups: one with a simple iconographical schema, and one more elaborate. The scenes following the simple schema all show a majestic woman, wearing chiton and himation, seated on a throne with a back ending in a goose's head, in front of an ornate table or chest on which there is a large elaborate basket; the woman has drawn the basket towards herself and opened it, and is holding the lid up with her right hand; in the basket lies a small child, its back propped up against a cushion, dressed in a himation, with long hair falling on its shoulders and arms around its knees. In (1), which is preserved in a fragmentary form, there is nothing in the space under the throne, although this is preserved. In (2) very similar, and also fragmentary, a mirror hangs on the wall above the basket. (3), also fragmentary, has a hydria under the table or chest. In (4) there is a bird under the throne and an alabastron as well as a mirror on the wall; under the table there is a hydria. (5) shows a table and a basket of a slightly different type, a similar throne and a bird under it similar to the one in 6, a mirror on the wall and a kantharos under the table; on the table, behind the basket, there is a round flask.

There is only one illustration of the elaborate iconographical schema. (6). This is a fragmentary representation in which the basket, held open by the woman, faces away from her and towards the approaching figure of a young woman of whom little is preserved, the right arm and hand raised in a gesture of adoration, and the left hand holding a kalathos upright on the palm. The child is standing in the basket. It is clearly female. This is shown first by the hair-style, which can be found in young men, but not, to my knowledge, in small male children, and secondly, by the aggressively female coquettish stance in which the child is shown. Moreover, the 'spotted' himation worn by the child also appears to have female connotations, for we found it worn by girls in a nuptial context; furthermore, the bridal peplos was also 'spotted'. The full scene, as well as other types involving the elaborate iconographic schema, are described by Prückner.[107] The full version of 6 includes a door on the left; in some cases the kalathos is filled with painted flowers or fruit.

In another group of scenes with the enlarged schema the enthroned woman has taken the child on her lap, while the basket again faces away from her and towards the young woman. In one of these types the approaching young woman has a taenia, a mirror, and an alabastron.

In my opinion, the different iconographical arrangements and positions of basket and child correspond to different moments in the action. First, the woman opens the basket and finds the child; then she picks the child up, turns it round and places it on her lap; finally she puts it back in the basket standing on its feet. This is surely the only way in which the differences between the iconographical schemes can be explained. I will argue below that this explanation also makes sense in terms both of iconography and of content.

Before attempting to interpret this cycle of scenes, we should first try to determine the cultic sphere to which it belongs. This cycle does not tie up with any other; nor is a clearly identifiable male god present, such as Hades or Hermes. For this reason, we have to rely exclusively on the symbols and cult objects for the identification of the cultic sphere. It would be very dangerous to begin, as Prückner does, with a hypothesis about the significance of the scene, and then try to fit the facts and symbols into it.[108]

The following symbols and objects are found in this cycle: mirror, alabastron, hydria, kantharos, bird (on which see below), kalathos, spherical flask, taenia, throne with a back ending in a goose's head. Of these, the spherical flask and the taenia can offer no help.[109] The kalathos is firmly fixed in Persephone's cultic sphere, and so is the throne with a back ending in a goose's head. The mirror, we saw, had entered Persephone's sphere at Locri, although it was also associated with Aphrodite. The alabastron belonged to Aphrodite; it is offered to Persephone by a goddess that is probably Aphrodite in one type of homage scene, but it also appears associated with objects and symbols belonging to Persephone in two 'furniture' pinakes[110] (representations of furniture, vessels, and other objects). It would appear then that the alabastron, although primarily characteristic of Aphrodite, had also been attracted into the orbit of the dominant Persephone, probably because of her association with women and their pursuits. The hydria and the kantharos are alternatives in the 'children in the basket' series. The kantharos at Locri was associated with both Hades and Dionysos. The hydria we have found only once and in connection with Persephone.[111] However, it also appears on a furniture-type pinax[112] which shows two cocks and a hydria standing on a very ornate chest under which there are two Sirens; the cocks and the Sirens from Persephone's sphere surely place the whole representation in her realm, giving us another instance of the association of the hydria with Persephone. The fact that kantharos and hydria are alternatives in the scenes with the children in the basket suggests that the kantharos is present here in virtue of its association with Hades rather than Dionysos, since, if this is right, both vessels would belong to the same cultic sphere, that of Persephone–Hades. Prückner assumes that the vessels are associated with the child. But even if this *a priori* assumption is correct, it cannot be excluded that the cultic sphere was also relevant. I argue below that there was a connection both with the child and with the cultic sphere.

It appears then that all the symbols considered so far which are fixed in one cultic sphere belong to Persephone. No symbol exclusively associated with Aphrodite is present. Now we turn to the bird under the throne. Prückner takes it to be a dove, a bird firmly connected with Aphrodite. However, in all the drawings published, the bird looks definitely different from the dove held by Eros in the scene illustrated in Prückner, 1968: pl. 2, and from the doves shown in the scene illustrated in Orsi 1909: 13, fig. 13, and Prückner, 1968: pl. 11.2. I do not know what this bird is, nor do I know whether it is a naturalistic representation of the bird it is meant to portray. But it seems to me significant that the artists have represented it differently from Aphrodite's dove. This suggests to me that it is meant to be a different bird. Consequently, it cannot, I believe, be argued that this bird brings a discordant Aphroditic note into the scenes.

Scholars have interpreted this cycle of scenes in various ways, but there is general agreement that we are dealing with mythological representations, and that both woman and child are divine.[113] There is no compelling reason for this view; indeed, there is at least one strong argument against it. The child in the basket cannot be divine, for its sex can be either male or female. No divine female infant—other than the new-born Athena in totally different circumstances—has a role in myth; but even if one did, the variation of male or female infant in the same myth and mythological representation would be, in my view, highly implausible; moreover (given the modalities of operation of Greek mythopoea), it would need to presuppose closely connected mythico-semantic profiles for the two figures for the alternation to be, first, possible, and second, meaningful. I do not believe that such a scenario fits the modes of operation of Greek religion; but even if this is wrong, the persona of Dionysos (the candidate proposed for the male child) (cf. n. 113 and also n. 114) is both too important and too idiosyncratic for him to be alternating with an (unknown) minor local figure (cf. n. 113) even as a baby—by which time he already had a significant history behind him.

That the sex of the child is female in at least one case cannot, I believe, be doubted. It is true that the sex of the children shown still lying in the basket is difficult to determine. But I would like to suggest that it was indicated through the presence of either the kantharos or the hydria under the chest; that the hydria, associated with fetching water and other female occupations, denoted a girl, the kantharos, a drinking vessel, belonging to the sphere of the symposion and male pursuits, a boy. The fact that we are dealing with Persephone's cultic sphere would have determined which vessel associated with women (hydria rather than, say, alabastron) and which drinking vessel (kantharos rather than, say, kylix) was chosen. I shall argue that these pinakes were dedicated by parents of children on a certain occasion in the child's life. The vessel may have been added by the artist as a means of differentiating the sexes and thus making, with little effort, some pinakes appropriate for girls and others for boys. Alternatively, a hydria or a kantharos, depending on the sex of the child, may have been actually placed under the chest during the cultic act which, I believe, is represented in the scenes. This must then have been considered sufficient indication of sex for the artist not to have otherwise differentiated children of different sexes. It is interesting to note that the one (published) case in which the sex of the child is clearly shown to be female depicts what I argue is the culminating point of the ritual. In any case, I want to stress that the fact that the child in the basket is female at least once shows that this is not a mythological scene involving a divine male infant as is generally believed, but a representation of a cultic event involving different (mortal) children. There is one cultic occasion that suggests itself: the 'presentation' of children to a goddess, an act through which the child is placed under the protection of the divinity.[114] This interpretation would make sense of the whole iconographical nexus; it introduces the most important cultic occasion in which Greek children were involved, and, moreover, it offers iconographical parallels for our scenes. I should note that even if we disregard the argument based on the sex of the children, the existence of the iconographical parallels still indicates that this is the most likely interpretation of the scenes showing children in a basket.

The most important parallel is provided by the so-called Ino-Leukothea or Albani relief[115] which is very closely connected stylistically with the Locrian pinakes and is likely to have been made at Locri.[116] A seated goddess has a female child standing on her lap; two children and a woman are shown standing, the woman offering a taenia. Another close parallel is found on the Ikaria stele[117] which shows an enthroned woman holding a child on her lap while two more children and two men are approaching, the latter making gestures of adoration. The third close iconographical parallel, or rather cycle of parallels, is provided by a series of Corinthian vases depicting the presentation of small boys and small girls to Demeter.[118] The following representations offer some close parallels with our scenes. (i) Pyxis Paris Bibliothèque Nationale, Cabinet des médailles 94 (Payne 1931: no. 878; *BCH* 94, 51, fig. 3): Demeter, seated between two women, holds a small naked boy on her knees. (ii) Pyxis Munich 7741 (*BCH* 94, 49, fig. 2): the scene on the shoulder shows Demeter seated with a girl standing on her lap. (iii) Bottle Béziers 22 (*BCH* 94, 54, no. 5; *Ant. K.* 6, pl. 17.1, 3, 5 and 6): representation similar to (ii). (iv) Alabastron Berlin 4285 (Payne 1931: no. 1203; *BCH* 94, 53, fig. 5): Demeter is shown seated with a small boy on her knees. The following monuments also depict presentations of children: Acropolis reliefs nos. 581[119] and 3030;[120] Xenokrateia relief, Athens, National Museum, no. 2756.[121]

In these circumstances, there can be little doubt that the scenes on the pinakes show the presentation of Locrian children to the kourotrophic deity, a cult act designed to ensure the goddess' blessing for the children. All these scenes depicting children's presentations, at Locri and elsewhere, show the deepest meaning of this cult act, by representing the goddess, understood as 'ideally present', in physical contact with the child. At Locri, the presentation would seem to have involved the placing of the child in a basket, a ritual which can be explained when it is remembered that in myth divine children who were 'adopted', as it were, by kourotrophic goddesses were similarly placed in baskets. Such was the case with Erichthonios, who was 'adopted' by Athena,[122] and, most importantly, with Adonis who was 'adopted' by Persephone, to whom he was handed over by Aphrodite hidden in a basket.[123] In reality, it would be the priestess of Persephone who 'received' the child in the basket on behalf of the goddess in the course of a ritual which would have involved the three stages which are represented in the pinakes. Alternatively, the Locrian ritual may not have involved an actual placing of the child in the basket. In this case, the scenes on the pinakes should be understood as 'symbolic', emblematic, representations of the ritual, modelled upon the myth involving Persephone in a kourotrophic role, the myth of Adonis. The pinakes were probably dedicated by the parents on the occasion of their child's presentation.

In any case, these scenes show that the Locrian Persephone was a kourotrophic goddess, the protectress of children. This aspect of the goddess is another manifestation of her involvement in the world of women. The role of protectress of women, marriage, and children

A Model for Personality Definitions 175

is peculiar to the Locrian personality of the divinity. This Locrian personality also included aspects found in her Panhellenic persona, that is her involvement with the fertility of the earth and with the funerary sphere. However, at Locri her character as Demeter's daughter, so prominent in Panhellenic religion, is hardly noticeable; Demeter's place in the Locrian cult of Persephone is minimal.[124] This phenomenon is undoubtedly related to the fact that at Locri the emphasis was on Persephone the bride and spouse.

It is of some interest to the problem of the definition of divine personalities that the personality and functions of Persephone at Locri as recovered here resemble in many ways those of Hera at the sanctuary of Foce del Sele.[125]

3. APHRODITE AT LOCRI

It is impossible to conduct here an exhaustive investigation into the Locrian Aphrodite's personality. This would involve, among other things, a discussion of the attempts to identify her Locrian sanctuary. It would also involve a lengthy examination of some problematic scenes[126] that could belong to her but cannot conclusively be shown to do so, because of the paucity of the symbols and cult objects that can firmly and exclusively be associated with this goddess. I should, however, point out that this uncertainty in no way affects the conclusions about Aphrodite's personality which can be drawn from those series that definitely belong to her and from some fragments of literary evidence which are relevant to this cult. The problematic scenes would neither add anything to, nor change anything in Aphrodite's personality as it can be recovered from the evidence available.

The following types of scenes indisputably belong to Aphrodite.

1. Aphrodite, Hermes, and Eros.[127] Eros, holding a lyre, is standing on his mother's arm. Hermes has the kerykeion, Aphrodite holds a 'rose-like' flower;[128] between the two there is a simple thymiaterion. Prückner[129] thinks that the scene shows the cult-statues of the deities.

2. Eros holding a dove and a winged girl holding an alabastron are drawing a chariot; Aphrodite is standing in it while Hermes is just climbing on.[130] The precise meaning of the scene is problematic.[131]

3. A woman hands a statue of Eros over to another woman in the presence of a third who is holding a taenia.[132]

4. The birth of Aphrodite.[133] Aphrodite has just emerged from the sea and is being received by two female figures, one of whom is about to envelop her in a garment. The goddess is shown in a smaller scale.

5. The temple of Aphrodite and Hermes.[134] The scene shows a temple with open doors through which can be seen the cult-statues of Aphrodite and Hermes, the latter holding a phiale. An altar stands in front of the temple, decorated with a relief representation of a satyr copulating with a hind. A youth and a girl are standing by the altar, the former pouring a libation on it. Alternatively,[135] only a girl is shown and a mirror hanging in the air.

6. A fragmentary type showing a temple with mixed Doric and Ionic elements like the one in 5 and with a pediment decorated with two doves; in the temple there is a goddess holding a bowl full of fruit, and in front of her a girl playing the double flute.[136]

We can deduce the following about Aphrodite's personality. First, the myth of her birth from the sea, which was part of the goddess' Panhellenic myth, was included in her personality-nexus at Locri. Secondly, she is associated in cult with Eros.[137] The fact that Eros had a place in cult is shown by types (1) and especially (3), which depicts a ceremony involving a statue of Eros. I must stress that the Eros with whom Aphrodite is associated here is not the funerary Eros of later South Italian iconography. This last cosmic and funerary Eros is iconographically very different from the one on the Locrian pinakes. The funerary type is effeminate, and this is not a casual characteristic, but an important aspect of his iconography which expresses important concepts about his personality.[138] Moreover, the Locrian Eros is not associated with the funerary one through any common attributes. On the pinakes Eros is shown as a boy, according to the usual iconographical type of Eros-Love.[139] This is important. For it must be clear that there is nothing in Aphrodite's personality or associations at Locri that would connect her with the funerary sphere. Aphrodite does have a funerary aspect in some places,[140] but not at Locri. Therefore her cultic association with Persephone is *not* due to a common funerary aspect.

Aphrodite's cultic association with her son, Eros, in his persona as Love indicates that her Locrian personality involved her Panhellenic aspect as goddess of love. This is confirmed by the symbols and cult objects that are associated with her. The attribute most closely connected with the goddess at Locri, the dove, was considered by the Greeks a love-bird;[141] it is, of course, widely associated with Aphrodite throughout Greece.[142] The mirror and the alabastron, connected with beautification, point in the same direction.

The last piece of information provided by the pinakes about Aphrodite's personality will confirm the above considerations. It is that she was cultically associated with Hermes who was her *paredros* at Locri. A similar cultic association between the two is also found in local cult-units in other parts of the Greek world.[143] This cultic pairing must reflect a connection between the two divine personalities: in the local cults in which such an association is found, each deity's personality must have important aspects that relate to important aspects of the personality of the other. Failing an analysis of the relevant cults, we can make an *a priori* supposition about which aspects of the two deities are likely to have been related in a way that could be expressed by pairing them. If we construct an artificial conflation by adding up all the aspects and functions of the two deities in all the different Greek cults, we will recover all the aspects and functions that each can 'cover', in the Greek world. We will then find that there are firmly established aspects of Hermes, his ithyphallic powers and connections with wild nature, which relate to Aphrodite's persona as goddess of love, sex, and the creative force which operates in nature through love.[144] It is likely that it is this connection which led to the pairing of the two deities in some cults where their divine personalities were emphatically characterized by these aspects. The evidence for Hermes' personality at Locri is too scant to test this hypothesis. Aphrodite on the other hand does appear to be primarily the goddess of love. Further consideration of the functions of the two deities as a pair in the Locrian cult will, in my opinion, confirm this hypothesis.

First we may note that, as Prückner points out,[145] Aphrodite and Hermes are not a married couple, they are a pair of lovers. The argument for this is not so much that they are not attested as spouses anywhere (Prückner's argument) but that the Panhellenic religious dimension, particularly through the Homeric poems, would have had an especially powerful influence in matters such as the relationships between the twelve gods. Even if originally in some cults Aphrodite and Hermes had been a married couple their relationship would have been transformed into an illicit one under the impact of Panhellenic religion.

Secondly, we may consider again the scene depicting the temple of Hermes and Aphrodite, and especially the relief decorating the altar. It shows a satyr copulating with a hind, an act of bestiality, involving on the one hand a satyr, the embodiment of the wildest pole of human nature, especially of human sexuality, and on the other hand an animal. The fact that this scene was judged suitable for the decoration of an altar to Hermes and Aphrodite—whether in reality or only on the pinakes—is significant.[146] It shows that bestiality too belongs to the cultic sphere of Hermes and Aphrodite, the illicit lovers. In other words, Aphrodite and Hermes preside over love and sex in its entirety, as a cosmic principle, which includes manifestations that society may classify as perverse. The demiurgic aspect of the love represented by Aphrodite may conceivably be hinted at through the bowl of fruit in scene (6), which may be taken to denote fertility. In the same scene, the presence of the flute-playing girl in front of Aphrodite may perhaps illustrate the goddess' connection with the illicit aspects of love, since the flute is associated with *hetairai*.

4. Persephone and Aphrodite at Locri

We can now see how Aphrodite relates to, and is differentiated from, Persephone at Locri: she relates to Persephone first through Persephone's connections with fertility, a concept associated with that of love as a demiurgic force represented by Aphrodite; secondly, through Persephone's role as protectress of marriage, because Aphrodite represents and presides over the demiurgic aspects of love and sex which are very important for marriage. But there is also a clear differentiation between the types of love which concern the two goddesses. Persephone, and Persephone and Hades as a bridal pair and married couple, presided over its institutionalized forms operating within the polis and harnessed to the needs of society. Aphrodite—and her lover Hermes—stands for love and sex as a cosmic principle, which includes all its manifestations, that is also its illicit and 'aberrant' forms which do not serve society, love unconfined by institutions.

The story about the *votum* made by the Locrians to Aphrodite in 477/6[147] tends to confirm these remarks. The inhabitants of

Locri had vowed to prostitute their virgin daughters at the festival of Aphrodite if they managed to defeat the tyrant Leophron of Rhegion who was attacking them. I have expressed elsewhere my agreement with the view that this *votum* had nothing to do with an alleged institution of sacred prostitution.[148] The point I want to make here is that this inversion of the structures and values of the city, this strikingly 'outside' manifestation of sexual activity imposed on the virgin daughters of citizens, was a vow to Aphrodite not because Aphrodite was a warrior goddess at Locri—there is not the slightest hint of evidence to that effect—but because she had jurisdiction over love and sex as a principle, unconfined by social institutions, and therefore also over the 'outside' manifestations of love and sex. However, by this vow—which was never fulfilled—Persephone's realm was impinged upon. For in the normal course of events, the virgin daughters of citizens, destined to marriage and the 'social' forms of love and sex, came under the jurisdiction of Persephone. There may be some evidence suggesting special dedications to Persephone aimed at propitiating her for the prospective encroachment on her realm. There is one type of pinax[149] which shows a seated Persephone holding a phiale and a very elaborate stylized flower, and in front of her a girl, much younger than those in the 'offering girls' series, offering a ball, accompanied by a warrior offering a cock. This is a variant of the scenes with the dedication of a ball and a cock to Persephone by girls about to get married. The important variations are three: the younger age of the girl, the fact that she is only offering a ball, and the presence of the warrior by whom the cock is offered. These similarities and divergences suggest that the occasion portrayed here is related to the ritual of the wedding dedications but has a different significance and function. The fact that this scene is only found in one type of pinax represented by many specimens[150] may suggest that we are dealing with a special dedication, and therefore also with a special occasion. In my opinion, the occasion that would best meet these requirements would be a special dedication to Persephone at the time of the *votum* aiming at propitiating the goddess in the face of the prospective (*even if only symbolic*) encroachment on her realm; a dedication that would mark the end of girlhood for the girls who were to be prostituted, in a way similar to that taking place before marriage, and with Persephone's blessing. The very young age of the girl on the pinakes I take to be an iconographical convention for expressing 'premature abandonment of girlhood, before the normal occasion of marriage'. The girl is dedicating the ball as a symbol of the childhood (and virginity) that was to be left behind. The cock, symbol of maleness, aggression, and male sexuality (in this case not to be harnessed into marriage), as well as of the fighting spirit, is offered by the figure who embodies the concepts expressed by the cock at the symbolic level: the warrior, a representative of the male citizens and soldiers of the city who were instrumental figures in the *votum*. If this interpretation is correct we would have gained an interesting insight into a special manifestation of the relationship and interaction between the cultic spheres of Aphrodite and Persephone.[151]

The emphasis in the Locrian Aphrodite's personality on the cosmic and demiurgic aspect of love, which includes also the sexual activities situated 'outside' the city structures, and the cultic association with Hermes and Eros, make Aphrodite's personality at Locri distinct. It is closely related to the Panhellenic personality of the goddess, but has also been shaped by its interactions with the personalities of Persephone and Hermes. Both, the former primarily by differentiation, the latter by association, tended to reinforce the emphasis on the cosmic and demiurgic aspect of love and sex which is not confined by civic institutions. Persephone's personality at Locri includes some of the aspects which characterize her Panhellenic personality, but without the close association with Demeter. Moreover, it contains some other functions not associated with her elsewhere: she presided over the world of women, with special reference to the protection of marriage and the rearing of children, that is of those female activities that were most important for the life of the polis.

I submit that these conclusions confirm the observations made in the first section of this chapter on the definition of divine personalities, and demonstrate the validity of an approach based on the study of local cults.[152]

Notes to Chapter III

* This chapter was first published in *JHS* 98 (1978), 101–21. An abridged version was published in Italian under the title 'Due protettrici della donna a Locri Epizefirii: Persefone e Afrodite', in G. Arrigoni, ed., *Le donne in Grecia* (Rome and Bari, 1985), 203–21.

1. Cf. also Vernant 1974: 110.
2. Contemporary scholars do allow significant variations in some divine personalities in some of the colonies, where they attribute them, usually unsupported by any evidence, to the notorious indigenous influences. Against the approach that assumes many take-overs of indigenous cults by Greek colonists in cases where there is no explicit evidence for such a phenomenon cf. Pugliese-Carratelli 1964: 19–28; Pembroke 1970: 1255–8. For recent discussions of various aspects of religion in the Greek colonies see Malkin 1987: *passim*; and Polignac 1984: 93–126, cf. 166–8 for bibliography on the question of the relations between Greeks and the indigenous population.
3. The (interrelated) spheres in which a divine personality manifests itself are the following. The sphere of divine name with its subordinate sphere of epithet, that of *Bildvorstellung* including the attributes, the sphere of myth, the sphere of cult, involving a deity as a recipient of worship, that of theology in the sense of sets of beliefs about the functions and areas of activity of the deity, and finally, the sphere of 'ideology', derivative from the previous one, primarily through the agency of literature, involving the deity as an embodiment of certain ideas and concepts.
4. Cf. Brelich 1968: 922.
5. Cf. ibid.
6. Cf. Pembroke 1970: *passim*.
7. On the polis of Locri Epizephyrii and its culture see now also *Locri Epizefirii: Atti del sedicesimo convegno di studi sulla Magna Grecia. Taranto 3-8 ottobre 1976* (Naples, 1977).
8. Publication of pinakes: Orsi 1909: 1–42; Quagliati 1908: 136–234; Zancani Montuoro 1940: 205–24; 1954a: 71–106; 1954b: 79–86; 1955, 283–308; 1960: 37–50; 1964: 386–95; Prückner 1968. Cf. also Oldfather 1910: 114–25; 1912: 321–31. I have briefly discussed the circumstances of the pinakes in Sourvinou-Inwood 1974a: 132–4.
9. Cf. Giannelli 1963: 187–204; Prückner 1968: 4–7; Zancani Montuoro 1959: 227 n. 5.
10. Zancani Montuoro 1959: 227.
11. Orsi 1909: *passim*; P. Orsi, NSc. 1909, 321–2; id., NSc. 1911; Suppl. 67–76; Zancani Montuoro 1959: 225–32; de Franciscis 1971: 75–9.
12. I have discussed this problem briefly in Sourvinou-Inwood 1974a: 133.
13. Prückner 1968: 63, 68.
14. Cf. J. Boardman, CR 21 (1971), 144–5; G. Zuntz, *Gnomon*, 42 (1971), 492–4.
15. Cf. Boardman, cit. n. 14.
16. Cf. Sourvinou-Inwood 1974a: 133.
17. Orsi 1909: figs. 36–7; Prückner 1968: 75, fig. 13, pl. 12.
18. Cf. Prückner 1968: 70, type 59.

19. Hom. Hymn. Dem. 6 ff.
20. On Hades' *Bildvorstellung* cf. P. E. Arias in *EAA* iii. 1081–2; Despinis 1971: 139–40.
21. Orsi 1909: figs. 30–5; Quagliati 1908: figs. 18–23; Prückner 1968: 71, fig. 12 and pls. 13–21.1–3; Zancani Montuoro 1954*b*: pl. viii.
22. Sourvinou-Inwood 1973: 12–21.
23. In the articles cit. in n. 8.
24. I should note that Zancani Montuoro (1955: 299 n. 2), while talking about a different type of scene, makes a tentative remark which to some extent points in the same direction: 'Può darsi per altro, ma poco giova l'indovinare, che il quadretto fosse offerto da una sposa locrese, che le sue nozze intendeva assimilare a quelle della dea, per invocarne la protezione.' However, she is clearly thinking in terms of an *ad hoc* dedication rather than of an established cult function of the goddess.
25. Farnell 1896: i. 188–92.
26. Since the objects held by the girl were appropriate to this ritual occasion which belonged to Persephone's sphere, it follows that they were connected with Persephone's cult.
27. Orsi 1909: fig. 8; Zancani Montuoro 1954*a*: pl. xxiii; Prückner 1968: 76, fig. 14, pl. 22, and cf. also pp. 75–6.
28. Orsi 1909: figs. 5, 7, 9–11; Zancani Montuoro 1954*a*: pl. xiii–xxii; Prückner 1968: pls. 23–30.4. Cf. Zancani Montuoro 1954*a*: 79–90; Prückner 1968: 77–81.
29. Zancani Montuoro 1954*a*: 83.
30. Ibid. 83–4.
31. Zancani Montuoro (ibid. 86) interprets the scene illustrated in Prückner 1968: 47, fig. 7, as Ares paying homage to Persephone. However, the presence of a young girl in the scene suggests that this is a representation of a very different kind (cf. below). Her suggestion concerning a presence of Artemis (in the scene illustrated in Orsi 1909: fig. 14) is also unlikely to be correct. For the 'spotted' peplos worn by the offering figure connects this scene with another series, that of the 'offering girls', see below.
32. Prückner 1968: 79.
33. Cf. Zancani Montuoro 1954*a*: 89; cf. also ibid. 85 and Prückner 1968: 155 n. 592, with reference to the scene illustrated in Orsi 1909: fig. 5.
34. Zancani Montuoro 1954*a*: 89.
35. As Zancani Montuoro suggested (ibid. 85).
36. Cf. Dunbabin 1948: 168–9.
37. Cf. Vallet 1958: 310–11.
38. Zancani Montuoro 1954*a*: 90.
39. Prückner 1968: 80.
40. Ibid. 80–1. He suggested that the scenes show the presentation of

deities newly arrived at Locri to the old established deities Persephone and Hades; Aphrodite is acting as their patron in this presentation because their cults were annexed to hers. I need hardly point out that this interpretation depends on a series of wholly unsupported assumptions. But it may be worth mentioning that there is no iconographical parallel for such a situation; while there are parallels for deities paying homage to a divine bridal couple, albeit shown in a different iconographical schema, that of procession. On this point cf. also Zancani Montuoro 1954a: 90.

41. Ibid. 90.
42. Prückner 1968: 80.
43. Cf. Giannelli 1963: 187–210.
44. For the cult of Zeus at Locri cf. Giannelli 1963: 187–210; de Franciscis 1972: 143–58.
45. Cf. Prückner 1968: 79.
46. On Persephone's association with the stalk of grain see B. Conticello, *EAA* iv. 386–94.
47. Orsi 1909: 14; cf. also Lissi 1961: 92, no. 91.
48. I have discussed this matter in Sourvinou-Inwood 1974b: 189–90, where a bibliography can also be found.
49. Cf. Andronikos 1956: 253–314 *passim* and esp. p. 305.
50. See Hoffmann 1974: 195–220; cf. esp. 204–6.
51. Cf. Hoffmann 1974: 213–14. For the cock as offering to the dead cf. Stengel 1910: 142, 152, 192; Keller 1920: 140.
52. Cf. Lissi 1961: 96, no. 111, pl. xli.
53. On the significance of the ball cf. Schneider-Herrmann 1971: 123–33. Cf. also Brendel 1936: 80–9. Cf. also the Attic black-figure ball belonging to a *hetaira* which bears a funerary inscription published by Hoffmann 1963: 20–2. On the pinakes depicting the dedication of a ball cf. also Torelli 1977: 163–4.
54. Cf. *Anth. Pal.* 6. 280.
55. Balls have been discovered in tombs; cf. n. 53 and Schneider-Herrmann 1971: 131 n. 138; Torelli 1977:164 and n. 27. It is not impossible that balls made of perishable materials were also placed in tombs and have not survived.
56. Orsi, *NSc.* 1911, Suppl. 71; cf. also Lissi 1961: 96, no. 112, pl. xli.
57. Orsi, *NSc.* 1890: 262. Prückner is mistaken in thinking (1968: 30) that the same figurines hold both dove and pomegranate.
58. Cf. Andronikos 1956: 305.
59. Cf. Studniczka 1911: 129, 141.
60. Cf. B. Conticello, *EAA* iv. 390. Because of its fertility aspect it can also, in some places, be associated with other goddesses (cf. e.g. for Hera: Paus. 2.17.4); the argument that the same attribute can

characterize iconographically different deities has now been set out by Metzger (1985: 173–8).
61. Cf. *Hom.H.Dem.* 371–4, 411–13. It is interesting to note that this hymn was known at Locri and is reflected in one of the scenes on the pinakes (cf. Prückner 1968: 82–4 and 82, fig. 15; cf. also Richardson 1974: 168–9).
62. Orsi 1909: 15.
63. Zancani Montuoro 1954*a*: 80 n. 2.
64. *ARV* 459.3; Pfuhl 1923: 111, fig. 437.
65. Cf. Thompson 1936: 329.
66. Paus. 9. 39. 2.
67. Cf. Andronikos 1956: 305. Andronikos (1956: *passim*) has argued convincingly that the seated figures are Underworld deities and not heroised dead. On these reliefs see also, most recently, Parker 1989: 147–8 and 166 nn. 28–9.
68. Cf. Simon 1953: 7.
69. On flower-picking associated with Persephone's myth and cult, especially in connection with her bridal aspect, see Blech 1982: 349. Cf. also Gernet 1968: 43 on the general connection between marriage rites and flowers and flower-picking.
70. Orsi 1909: 15, 36.
71. For figurines: Orsi 1909: 14; for the 'maschera' ibid. 13.
72. Strabo 6. 256.
73. Pollux 1. 1.37.
74. Zancani Montuoro 1954*a*: 102.
75. A flower is also associated with the couple of the Underworld deities on the Laconian reliefs (cf. Andronikos 1956: 305).
76. Orsi 1909: 31, fig. 42; Quagliati 1908: fig. 72; Prückner 1968: 58, fig. 10 (and pp. 58–60); cf. also Zancani Montuoro 1954*a*: 98–9, where she reads the objects as fruit, pomegranates, and quinces; and cf. Neutsch 1953/4: 62–74.
77. Zancani Montuoro 1954*a*: 98–9.
78. Cf. ibid.; for the pomegranate see also above.
79. Orsi 1909: 37.
80. Cf. Richter and Milne 1935: 14, 15, and illustration on p. 17; Simon 1959: 64 and 68, fig. 40. On the kalathos see also Keuls 1982/3: 31; 1983: 221. For a kalathos in a scene of an erotic pursuit by a god see e.g. Keuls 1983: 221, fig. 14.28. The kalathos also had an important place in some cults of Demeter (cf. Nilsson 1906: 350–2; cf. also A. Longo, *EAA* iv. 295).
81. Cf. Hesych. s.v. Ἡροσάνθεια; Phot. s.v. Ἡροάνθια.
82. Cf. n. 69.
83. It is conceivable that the naked female figurines with pronounced erect breasts found in the sanctuary of Persephone (Orsi 1909: 14) may also have connotations of fertility.

84. Prückner, who describes it (1968: 77, type 88), lumps it together with one of his two series of offering girls. His division of the cycle of the offering girls into two series, one of which he attributes to Aphrodite and the other to Persephone, is purely arbitrary. As will be apparent from the discussion, there is a fundamental unity in the series, and the symbols and cult objects firmly establish that the goddess involved is always Persephone.
85. Described in Prückner 1968: 77.
86. The poppy is associated with Demeter (cf. Steier in *RE* s.v. Mohn 2445).
87. Prückner 1968: 65, type 51.
88. Cf. Schelp 1975: 11, 25–6, and *passim*; cf. also Deubner 1925: 210–23.
89. Prückner 1968: 77.
90. Described ibid.
91. Orsi 1909: 39.
92. A deep phiale and wand are held by a priestess in a series of scenes showing processions involving girls which I will discuss below.
93. Zancani Montuoro 1954a: 94–5.
94. Prückner 1968: 49.
95. Cf. also ibid.; he does not distinguish this from the previous type.
96. Ibid.
97. Cf. Buschor 1944: *passim*; Vermeule 1979: 75, 169–71.
98. In one (Orsi 1909: 21, fig. 25; Prückner 1968: 43, fig. 5, Prückner's type 16, p. 42), the *phialophoros* precedes the girls, in the other (Orsi 1909: 22, fig. 26; Zancani Montuoro 1960: pl. II; Prückner's type 17, p. 42), she follows. The peplos is held differently in each case.
99. Orsi 1909: 16, fig. 17; Zancani Montuoro 1960: pls. I, III; Prückner 1968: 43–4.
100. Orsi 1909: 17, fig. 18; Zancani Montuoro 1960: pls. IV.3, V.1–2.
101. Described in Zancani Montuoro 1954a: 93 and Prückner 1968: 45–6.
102. Prückner 1968: pl. 4.4.
103. Cf. Zancani Montuoro 1954a: 90–102; 1960: 40–5; cf. also 1955: 283–308.
104. Zancani Montuoro 1954a: 93.
105. On this group cf. Prückner 1968: 31–6.
106. Cf. ibid.
107. Ibid. 32.
108. Whether we understand the symbols and objects as present in the sanctuary in the course of this cult activity (or in the location of the mythological event if the scenes show a myth), or added by the artist as 'emblematic' iconographical elements, they must have belonged to the same cultic sphere as the ritual (or myth) depicted, since the

artists' selections were inevitably determined by the parameters of their ritual knowledge, by the nexus of associations between cultic sphere and cult objects and symbols.

109. The taenia does appear elsewhere sporadically, but the discussion of these scenes is beyond by present scope. I believe that the taenia is not cultically fixed in one or the other sphere, for it seems to occur in both —not, perhaps, surprisingly when we consider its wide cultic usage.
110. Orsi 1909: 19, fig. 22, 20, fig. 23.
111. Zancani Montuoro 1954a: 93; Prückner 1968: 45–6.
112. Orsi 1909: 19, fig. 21.
113. Quagliati 1908: 195–6 was reminded of Erichthonios and thought in terms of a myth parallel to his. Oldfather 1910: 121–2 though of Dionysos, but considered it also possible that the scenes show symbolically the birth of a child, considered as an offering of the gods as well as the product of the parents. Later he opted for Iakchos (Oldfather 1912: 325–6), a view also favoured by Orsi 1909: 30–1 and others. Studniczka 1911: 143 identified the child as Adonis, a view embraced by Ashmole 1936: 17 and Higgins 1954–9: i, no. 1219. Prückner 1968: 32–4 argued against all these theories and suggested that the scenes show Aphrodite and Dionysos. For Simon 1977: 19 the scenes involve two different divine children, Dionysos and an unknown and unattested daughter or adopted daughter of Persephone.
114. On the presentation of children to deities cf. Kontoleon 1970: 1–21, *passim*; 1974: 13–25; Lambrinudakis 1971: 218–28; 1976: 108–19; Walter 1923: 32; 1937a: 103; 1937b: 59–61; cf. also Guarducci 1937–8: 37–8. That the child may have been a mortal baby has also been suggested independently, in the year this essay was first published, by Hadzisteliou Price 1978: 172, who puts this forward as one of two alternatives, the other being that two different goddess–child couples are involved, Demeter with Persephone and Persephone with Iakchos or Dionysos: 'Otherwise it is Persephone with a newly or not yet born mortal baby put under her protection by the mother who dedicated the votive Pinax.'
115. Cf. Kontoleon 1970: pl. vi; 1974: pl. 5γ; cf. also 1970: 10 n. 1 and 17.
116. Cf. Ashmole 1922: 248, 252.
117. Kontoleon 1970: pl. i; 1974: pl. i. Kontoleon has argued very convincingly that this relief represents a presentation of children (1970: 1–21), and again, refuting the objections of Rühfel 1974: 42–9 in Kontoleon 1974: 13–25. Lambrinudakis 1976: 108–19 has completed Kontoleon's case.
118. Callipolitis-Feytmans 1970: 45–65; Jucker 1963: 47–61; cf. also Lambrinudakis 1971: 218–28, who argued convincingly in favour

of interpreting the scenes as presentations of children to the kourotrophic deity.
119. Kontoleon 1970: pl. iv; 1974: pl. 6 and p. 20.
120. Walter 1923: 32, no. 3030; Kontoleon 1974: pl. 4.
121. Walter 1937a: 102 fig. 1 and 101–3.
122. Nilsson 1967: 317; cf. also G. Becatti, *EAA* iii. 419–20.
123. Cf. Apollod. 3. 14. 4.
124. Cf. also Zancani Montuoro 1959: 227–8; Demeter appears on a type of pinax which reflects the *Hom.H.Dem.* (illustrated in Prückner 1968: 82, fig. 15; cf. also n. 61 above).
125. On Hera at Foce del Sele cf. Zanotti Bianco in Zancani Montuoro and Zanotti Bianco 1951: 14–18.
126. Cf. e.g. Orsi 1909: 24, fig. 28; Prückner 1968; 53, fig. 9.
127. Orsi 1909: 12, fig. 12; Prückner 1968: pl. 1.1.
128. On the flower cf. Prückner 1968: 16.
129. Ibid.
130. Quagliati 1908: 189, fig. 41; Prückner 1968; pl. 2.
131. Cf. a discussion in Prückner 1968: 22–7.
132. Quagliati 1908: 191, fig. 42; cf. Prückner 1968: 67–8.
133. Quagliati 1908: 212, fig. 60, 213, fig. 61; Prückner 1968: 37, fig. 4; cf. Zancani Montuoro 1964: 386–95 (and figs. 4–7); cf. also Prückner, pp. 36–8.
134. Zancani Montuoro 1940: pl. i and fig. 2; Prückner 1968: 17, fig. 1, and pl. 1.2; cf. Zancani Montuoro, pp. 205–24; Prückner, pp. 17–19.
135. Prückner 1968: type 2A, p. 135 n. 107.
136. Orsi 1909: 13, fig. 13.
137. On Eros cults cf. Farnell 1909: 445–6, 476–7; Schneider-Herrmann 1970: 86–117 *passim* and esp. pp. 87–8.
138. Cf. Schneider-Herrmann 1970: 86–117 *passim* and esp. p. 105.
139. On the iconography of Eros see Greifenhagen 1957; Richter 1958: 255–7; E. Speier in *EAA* iii. 426–33; Bérard and Durand 1984: 32–3; *LIMC* iii. 850–952, esp. pp. 850–942.
140. Cf. Farnell 1896: ii. 652–3.
141. Cf. *EAA* i. 116; *RE* s.v. 'Taube'.
142. Cf. *EAA* i. 116 and 119. The wild dove, *phassa*, was associated with Persephone (cf. Keller 1920: 123).
143. Cf. bibliography in Prückner 1968: 134 n. 114; to this should be added Nilsson 1967: 503, and especially a recently discovered sanctuary of Hermes and Aphrodite at Kato Syme Viannou in Crete, which is in the process of being excavated, and on which see now Lebessi 1985 and Lebessi and Muhly 1987: 102–13. In this sanctuary Hermes is closely connected with trees, hunting, and animals, especially wild goats. On Hermes' personality at Kato Syme see esp. Lebessi 1985: 163–87.

144. On this last demiurgic aspect of Aphrodite cf. Aesch., *Tr. GF* iii F 44 (= fr. 44 N²); Soph., *Tr. GF* iv F 941 (= fr. 855 N²); Eurip. fr. 898 N².
145. Prückner 1968: 22.
146. Prückner (1968: 29) has misunderstood this significance. He talks as though the representation involved two animals copulating and misses the element of bestiality when he takes the relief to be a reference to Aphrodite's power, through love, in the animal as well as the human world.
147. I have discussed this *votum* in some detail in Sourvinou-Inwood 1974*b*: 186–98.
148. Cf. n. 147.
149. Prückner 1968: 47, fig. 7.
150. Zancani Montuoro 1954*a*: 86.
151. I should note that in other cities Aphrodite is sometimes associated with marriage, the family, and the polis structures (cf. Farnell 1896: ii. 655–7; cf. also Detienne 1972: 120–1; Friedrich 1978: 84–5). Nilsson (1967: 524), who denies this, has conflated evidence from different parts of the Greek world and treated it as though it came from one cultic unit.
152. Torelli 1977 was published in a volume which, despite its apparent date of publication, appeared after the present essay was first published in *JHS* 98 for 1978 (based on a paper delivered at the Triennial Conference of the Greek and Roman Societies in Aug. 1975). Torelli and I agree in the conclusion that many of the pinakes pertain to the rites which transform *parthenoi* into *gynaikes*, which was an important aspect of the cult of the Locrian Persephone—though several of our interpretations of particular scenes on different types of pinax differ. As Torelli correctly states in an addendum written in 1979 (Torelli 1977: 184), our methodologies also differ. It is true that my analyses do not encompass all pinakes types, but in these analyses the types considered have been studied in the context of, and their values were reconstructed through the reconstruction of, their relationships to the whole series and to the sanctuary in which they were found, seen in its social and historical context. The procedure adopted here is based on the conviction that (for reasons that have, I hope, become apparent through the methodological discourse in Ch. I) it is necessary to conduct exhaustive analysis of the individual iconographical types, starting with those whose meanings are more immediately accessible to us without the need to import any external assumptions, and working systematically outwards.

PART IV
Myth and History

INTRODUCTION

The three chapters in this section are concerned with the study of narratives that purport to be recounting past historical events but are in fact mythological creations, myths which present themselves as 'history'.*

Chapters IV.1 and IV.2 discuss myths which pertain to, and purport to give information about, the early history of the Delphic oracle. The myth of the 'Previous Owners of the Delphic Oracle' which is investigated in Chapter IV.2 has been widely believed by many modern scholars to give correct information about that early history and has significantly affected the investigation, scholarly discourse, and understanding of this most important oracle and Panhellenic sanctuary.

Chapter IV.3 investigates a story in Herodotos which purports to narrate past historical events situated in the not too remote past, but which in fact can be proved, through a detailed investigation, to be a mythological narrative.

These stories are not, of course, 'just lies' or 'sheer inventions'. The myths discussed in the first two chapters express, and are articulated by, significant perceptions about, among other things, the Delphic oracle. The Herodotean narrative articulates in the mythological idiom, which was an established type of discourse for talking about the past, the notion 'disordered succession, interruption of a dynasty'.

NOTE

*Ch. IV.1 is a version of an essay published in *CQ* 29 (1979), 231–51, modified to take account of recent developments in archaeological research (cf. Ch. IV.1 n. 21), and also to make it more accessible to the Greekless reader. Ch. IV.2 was first published in J. Bremmer, ed., *Interpretations of Greek Mythology* (London and Sydney, 1987), 215–41; and Ch. IV.3 in *Opuscula Atheniensia*, 17 (1988), 167–82.

IV.1

The Myth of the First Temples at Delphi

The intriguing myth of the first temples at Delphi is first attested in Pindar's fragmentary eighth *Paean*.[1] This text, and Pausanias 10. 5. 9–13, are the only two sources that actually tell the story of the first temples, while a few others simply mention, *en passant*, one or more—but not all—of these legendary temples, without setting out to give an account of the myth.[2]

The version found in Pausanias runs as follows. The first temple of Apollo at Delphi was made of laurel and had the shape of a hut. The laurel branches used in its construction came from the Tempe valley in Thessaly. The second temple had been built by bees who used as building materials their own wax and birds' feathers. It is said, Pausanias adds, that Apollo sent this temple to the land of the Hyperboreans—a land, it should be remembered, to which the Greek *logos* ascribed now a real historical identity, now the character of a Paradise. At this point Pausanias inserts a rationalizing euhemeristic alternative version of the myth, according to which this second temple had been built by a man, a Delphian called Pteras, and took its name (*pterinos*) from the builder. The third temple was made of bronze—and there is nothing extraordinary in that, Pausanias hastens to add; he backs up this statement by mentioning other known bronze buildings, some actually existing in his own day. Then he proceeds to reject the version told by Pindar, according to which this bronze temple had been built by gods[3] and had golden singers singing above the pediment. Clearly then, Pausanias is rationalizing again here, rejecting the miraculous elements, and pulling the bronze temple into the realm of reality and everyday life. He concludes his account of the third temple by giving two alternative versions of its end: it was said either to have disappeared in an opening of the earth, or to have been melted down by fire. The fourth temple was made of stone; it was the work of Trophonios and Agamedes. Pausanias identifies this fourth temple with the one which was burnt down in 548 BC.

Trophonios and Agamedes, the legendary architects who lived before the Trojan War, were credited with the construction of many splendid buildings believed to have been erected in the heroic past. They were usually considered to be the sons of Erginos, a king of Orchomenos, but Trophonios was sometimes said to have been Apollo's son.[4]

Pindar's account of the myth survives in fragmentary form. The first three mythical temples are mentioned in the surviving text[5] which breaks off after the discussion of the bronze temple. However, there can be no doubt that Pindar went on to talk about the fourth temple, built by Trophonios and Agamedes. For after the break the text appears to be concerned with Erginos, their father.[6] Moreover, Pindar elsewhere tells of Trophonios and Agamedes building a temple to Apollo at Delphi.[7]

The following points are worth noting with regard to Pindar's account. First, as remarked by Lobel,[8] while we were told by Pausanias that Apollo had sent the second temple to the Hyperboreans, 'the "mighty rushing wind" is a new detail'. Secondly, the discussion of the bronze temple contains several interesting points. This temple is said to be the work of Hephaistos and Athena, and this divine craftsmanship goes together with the generally supernatural character that Pindar's account attributes to this temple—and which, we saw, Pausanias rejected. We are told that above the pediment six Charmers (Keledones) made of gold were singing. I shall discuss the significance of the Keledones later on, when I come to consider the third temple in detail. With regard to the manner of the disappearance of this bronze temple, Pindar tells us that the gods hid it inside the earth which they had opened up with a thunderbolt. Lobel[9] notes that Pindar's account 'covers' both versions given by Pausanias. I take this cautious expression to mean 'could have given rise to' both versions, which is likely to be correct.

The sources that refer to one or more of the mythological first temples *en passant*, without dwelling on the story, are the following. Aristotle, *De Philosophia*, fr. 3R, mentions the temple made of feathers (*pterinon*) and the bronze temple. The Scholia to *Paean* viii, P. Oxy. 841, fr. 107, mention the laurel temple, and may have included the information that the laurel used in its construction had come from the Tempe valley.[10] Strabo 9. 421 distinguishes three temples: the one made of feathers (*pterinon*), which he ascribes to the mythical sphere; the 'second' temple, built by Trophonios and Agamedes;

and 'the present' temple, constructed by the Amphiktyones. Clearly, then, Strabo has constructed here a tripartite model, one pole of which is occupied by reality and the present, and the other by the sphere of myth and supernatural things. Between the two poles, and mediating in a way between them, lies the world of the heroic, legendary past, to which Trophonios and Agamedes belong. Eratosthenes, *Catast.* 29, mentions the temple of feathers which he locates in the land of the Hyperboreans. Plutarch, *De Pyth. orac.* 402 D, quotes a verse, which he presents as an ancient oracle, in which the birds are told to contribute their feathers and the bees their wax. Finally, Philostratos, *Vit. Apollon.* 6. 10–11 first mentions the temple made of wax contributed by bees and feathers contributed by birds and then quotes the oracle mentioned by Plutarch; next comes a passage referring to the (bronze) temple with the golden Keledones which I will discuss later.[11] I shall attempt to explore the significance of this myth by first investigating each of its component elements, each temple, separately, and then examining the relationship between the component elements and the structure of the myth.

First, the laurel temple. Three nuclei of religious–mythological reality need to be investigated here, in order to provide a framework for the assessment of this mytheme, proceeding from the generic to the specific: (i) the significance and role, if any, of the laurel in a Delphic Apolline context; (ii) the significance and role, if any, of the Tempe laurel in particular in a Delphic Apolline context, and (iii) the laurel temple.

The numerous and close connections of Apollo in general and the Delphic Apollo in particular with the laurel are too well known to need discussion.[12] It may, however, be worth stressing that the laurel at Delphi is associated with the act of divination itself. It was used in the ritual preparation of the Pythia, who also wore a laurel crown, and it adorned the prophetic tripod.[13] More importantly, the evidence indicates that the Pythia shook a laurel while prophesying.[14] The laurel from the Tempe valley had a special place and significance in Delphic myth, cult, and ritual.[15] Apart from our myth, the Tempe laurel appears in three mythological and/or cultic contexts, which are associated into one nexus in the version in which our sources record them, though they have been shown to have had different origins, and to have been fused into one complex, probably in the archaic period.[16] The first context is mythological: after Apollo had killed the dragon he had to be purified; he did this at the Tempe valley, and afterwards he cut some of the laurel that grew there, crowned himself with it, and carrying some more of it in his right hand he returned to Delphi.[17] The second context involving the Tempe laurel, the boys' *theoria* to Tempe, was supposed to take place in memory of Apollo's purification and its aftermath, which became the *aition* of this institution—this in the version of the institution which belongs to the 'unified nexus', the earlier role and significance of this *theoria* being another matter. In commemoration of Apollo's purification, every eight years the Delphians sent to Tempe boys from noble families, headed by one of their number who was *architheoros*.[18] On arrival at Tempe the boys performed a sacrifice and then cut laurel branches from the trees growing there and crowned themselves. They then returned to Delphi following a definite itinerary, the route called Pythias. Wherever they went festivities took place and the boys were escorted with great honour. Finally, they returned to Delphi; the laurel they brought with them from the Tempe valley was used to make the laurel crowns for the victors in the Pythian Games.[19] And this use of the Tempe laurel provides the third context in which the Tempe laurel had a special significance and role.

Two conclusions can be drawn from all this. First, that the Tempe laurel was special, had a particular holiness in the cult of Apollo at Delphi. Secondly, that the act of carrying the laurel (*daphnephorein*) was of importance at the level of myth and cult. It is difficult to doubt that this was also reflected at the theological level, and that Apollo was Daphnephoros at Delphi, despite the fact that the epithet is not attested here. I will be arguing elsewhere that the persona of Apollo-of-the-laurel/Apollo Daphnephoros was of great importance in the earlier stages of the history of the Delphic sanctuary, but became overshadowed by the god's character of law-giver and guide, establisher of order, and authority on purification, which developed as a result of the role which the Delphic oracle was called upon to fulfil in the archaic Greek world.

Let us now consider the concept of the laurel temple.[20] A temple made of laurel is a perfectly feasible proposition. And we have just seen that the laurel had an all-important place in the Delphic cult of Apollo, with the Tempe laurel possessing a special holiness. In these circumstances, the story of the first temple of Apollo at Delphi, made of laurel, can theoretically be explained in two alternative ways. It may have arisen as an expression, and perhaps even partial explanation in mythological spatial terms, of the laurel's important role in Delphic cult. Or, it could be reflecting a true historical memory, a memory of a shrine made of laurel. The former interpretation is almost certain to be correct.[21]

Now, let us turn to the consideration of the mytheme of the second mythical temple, the one made of wax and feathers.

This mytheme contains 'miraculous' elements, which were absent from that about the laurel temple. A temple made of laurel is a perfectly possible thing, one made of wax and feathers is only feasible as a model, not at the scale necessary for an actual place of worship. Then, the wax-and-feathers temple is said to have actually been constructed by bees, with the bees themselves contributing the wax, and the birds adding their feathers. It was also said to have been sent by Apollo to the land of the Hyperboreans, through a mighty rushing wind, according to Pindar. So, its miraculous elements place the temple made of wax and feathers firmly in the realm of myth. Obviously, this does not mean that the mytheme may not contain reflections of cultic elements and phenomena. I shall now therefore consider the place, if any, of birds and bees in Delphic cult and myth.

It is well known that Apollo had connections with many birds, at Delphi as well as elsewhere, and that he was particularly closely associated with the raven, which is believed to be a prophetic bird, the swan, the kirkos (a type of hawk), and to a lesser extent, the crow; doves appear to have had special rights at Delphi.[22]

There were probably specific reasons for his connection with some of these birds; the close connection with the swans, for example, could be due to the fact that the singing swan was a topos in Greek (and Latin) literature and probably also folklore.[23] But the god's extensive and wide-ranging connections with all kinds of birds was probably determined by the fact that he had become the prophetic god *par excellence*, and birds were involved with prophecy through the practice of *oionoskopeia*, in which their flight or cries were taken to be omens.[24] Amandry has argued[25] that the flight of birds was observed at Delphi as a preliminary to the main consultation.

Bees have an extensive place in Greek cult and myth.[26] Some priestesses, mostly of Demeter, were called 'Melissai',[27] some priests in the cult of Artemis at Ephesos were called 'Essenes',[28] and other holders of a cult office in the same cult were called 'Melissonomoi'.[29] These belong to the wider category of Greek priests, priestesses, and cult personnel or worshippers bearing the names of animals.³⁰ And this leads us back to Delphi. For the first connection between bees and the Delphic cult of Apollo is found in a verse of Pindar (*Pyth*. 4. 60–1) where the Pythia is called 'Delphic Bee', *Melissas Delphidos*. It could be argued that this expression which is attested nowhere else need not be a cult title, but simply a poetic expression. However, this is not likely to be the case. First, the scholia (Schol. Pind., *Pyth*. 4. 106b) do not interpret this Pindaric expression as an *ad hoc* metaphor; they explain that it denotes the Delphic priestess and add that the priestesses were called bees, primarily those of Demeter, but by extension also all priestesses, because of the bee's purity. Second, and more importantly, the fact that the word *Melissa* was established as a cultic title, a name for some priestesses, would, in my opinion, make it rather unlikely that Pindar should have used it for a priestess to whom this did not cultically belong as a simple metaphor.

The second connection between bees and the Delphic oracle and cult is rather indirect, though significant. Pausanias (9. 40. 1) tells us that the Pythia had advised some Boeotians to consult the oracle of Trophonios. The Boeotians could not locate that oracle, until one of them, Saon, followed a swarm of bees which led him to it. So here we have an oracle, belonging to a hero involved with building a temple to Apollo at Delphi, and associated with Delphi,³¹ revealed by bees.

The third connection between bees and the Delphic Apollo is very important, but also somewhat problematic. The relevant text is the *Homeric Hymn to Hermes*, vv. 552–66:

> σεμναὶ γάρ τινες εἰσὶ κασίγνηται γεγαυῖαι
> παρθένοι ὠκείῃσιν ἀγαλλόμεναι πτερύγεσσι
> τρεῖς· κατὰ δὲ κρατὸς πεπαλαγμέναι ἄλφιτα λευκὰ
> οἰκία ναιετάουσιν ὑπὸ πτυχὶ Παρνησοῖο,
> μαντείης ἀπάνευθε διδάσκαλοι ἦν ἐπὶ βουσὶ
> παῖς ἔτ' ἐὼν μελέτησα· πατὴρ δ' ἐμὸς οὐκ ἀλέγιζεν.
> ἐντεῦθεν δὴ ἔπειτα ποτώμεναι ἄλλοτε ἄλλῃ
> κηρία βόσκονται καί τε κραίνουσιν ἕκαστα.
> αἱ δ' ὅτε μὲν θυίωσιν ἐδηδυῖαι μέλι χλωρὸν
> προφρονέως ἐθέλουσιν ἀληθείην ἀγορεύειν·
> ἢν δ' ἀπονοσφισθῶσι θεῶν ἡδεῖαν ἐδωδήν,
> ψεύδονται δὴ ἔπειτα δι' ἀλλήλων δονέουσαι.
> τάς τοι ἔπειτα δίδωμι, σὺ δ' ἀτρεκέως ἐρεείνων
> σὴν αὐτοῦ φρένα τέρπε, καὶ εἰ βροτὸν ἄνδρα δαείης
> πολλάκι σῆς ὀμφῆς ἐπακούσεται αἴ κε τύχῃσι.

There are certain august virgins, born sisters, glorying in their fast wings, three in number; with their heads besprinkled with white meal they dwell under a fold of Parnassos, teachers of divination apart, divination which I practised while I was still a boy minding the cattle; and my father did not mind. Flying from this place now here now there, they feed on honeycombs and accomplish all things. And when they are inspired through having eaten yellow honey, they want willingly to speak the truth; but if they are deprived of the sweet food of the gods, then they lie, as they buzz together. These I give to you; gladden your heart questioning them precisely, and if you teach a mortal man, many times he will hear your voice, if he is lucky.

It is quite clear[32] that Apollo is talking of three winged bee-women virgins, who were sisters and lived under a ridge of Parnassus, and who had taught Apollo the art of divination.[33] When they have eaten honey they deliver true prophecies, if they are deprived of it, false ones. This oracle of the women-bees Apollo offered to Hermes.

The first problem is how these bee-women are understood in the hymn to be delivering prophecies. Amandry[34] suggests that it is their flight, different in each case, that reveals their decision. Latte[35] and Jacoby,[36] on the other hand, are thinking of prophecies derived from the buzzing of the bees (women-bees in the mythological image presented in the hymn). Perhaps what the hymn is articulating is a combination of observation both of the way of flying of the women-bees and of their buzzing. For on the one hand, as implied by Amandry,[37] vv. 558–9 suggest that they reveal things through their flight, and on the other, as Jacoby notes,[38] 'the *agoreuein* alongside of *doneousai* can only mean the buzzing of the bees'.

Amandry[39] comments: 'Sous une forme allégorique, il y a là une allusion à un ancien rite agraire où la farine et le miel constituaient l'essentiel des offrandes.' And this is surely correct as far as it goes—except that, instead of speaking of an allegorical form and allusion, I would prefer to say that our passage offers the mythological expression and crystallization of a rite of this type. As for the kind of old agrarian rite involved, it is surely likely to have been a rite of divination through the observation and interpretation of the flight and/or buzzing of bees:[40] bees which would be special sacred bees kept somewhere within the sanctuary.[41]

It cannot be proved conclusively that divination by bees had been practised at or around Delphi (cf. n. 33) and that this is what the mythological image of the *Hymn to Hermes* reflects. But apart from the fact that this is the most likely explanation for the emergence of this mythological image, other considerations make this hypothesis likely. Prophetic bees at Delphi would help make better sense of the myth mentioned above in which bees revealed the oracle of Trophonios to Saon, when he and other Boeotians were trying to locate it, on the advice of the Delphic Pythia. For the bees in this story have supplemented the directions of the Pythia, taking on, as it were, her role, from a certain point onwards. This would make better sense if there had existed a conceptual framework in which a divination pattern through prophetic bees was included. To put it differently: if we attempt to recover the perceptions shaping, and expressed in, this myth we find, that, above all, the myth ascribes special prestige and holiness to the institution of the oracle of Trophonios, which is presented as having taken place under Delphic authority. Clearly, one representation shaping this myth is this 'glorification' of the Trophonios oracle. Another is the particular modality of the operation of the guidance given by the Delphic god. The consultation of the Pythia obviously represents the chief established mode of Delphic divination. In this myth the mytheme 'bees reveal the oracle', which is a version of the schema 'animals discovering an oracle',[42] has the role of 'extending', taking over, the Pythia's role. This structural relationship of the Pythia and the bees in the myth suggests the possibility that the two may have been functionally correlative in the Delphic cultic nexus; that there was a Delphic divination mode involving prophetic bees, and that this helped determine the creation of the mytheme 'bees reveal the oracle' in this myth, that the modality of the operation of the divine guidance was here shaped by two Delphic divination patterns. If the motif of 'animals discovering an oracle' had already been extant at that time, its influence may have also operated, helping to determine and shape the selections that led to the creation of the 'bees reveal the oracle of Trophonios' mytheme. Otherwise, it would be conceivable that the story about the oracle of Trophonios could have given rise to the motif of animals discovering oracles.

The hypothesis of the existence of prophetic bees may also explain why the Pythia should be called *Melissa*. She would have fulfilled a role similar to that of the bees and this similarity of role could easily lead to a parallelism being made between the two, resulting in the Pythia being called 'bee', especially given the existence of the mythological image of the prophesying women-bees which could serve as intermediary.

Moreover, if we accept that divination through prophetic bees had been practised at Delphi, we obtain for this oracle a nexus of interconnected divination patterns[43] parallel to that found at Dodona. It is irrelevant whether or not these patterns coexisted in time, since we can detect the influence of the two marginal patterns on the central one, and it is this contact that matters.

Delphi: bees—Pythia who could be called 'bee'—laurel-tree (cf. above: especially *Homeric Hymn to Apollo*, vv. 394–6).
Dodona: doves—priestesses who were called 'doves' (cf. above n. 30)—oak-tree (cf. Homer, *Odyssey* 14. 327–8, 19. 296–7).

Whether this similarity was due to the influence of one oracle over the other, or to a common 'mentality' operating at a particular stage in the history of Greek religion—a stage in which physical objects and animals were of paramount importance, and which in historical terms I would identify with the so-called Dark Ages, with cultic elements continuing, and being reinterpreted in, the early historical age—it could be argued that the close correspondence between the nexus known from Dodona and that emerging from Delphi, if we include the element of the prophetic bees, may offer some additional support for the validity of this reconstruction of the Delphic model.

As we have seen, the miraculous elements contained in the story of the second temple place it in the realm of myth. It is now clear that this myth contains some reflections of cultic phenomena. Specifically, the association of both birds and bees with the Delphic Apollo and the Delphic sanctuary, and the concept of these creatures 'serving' Apollo inherent in the story, corresponds to a cultic association at Delphi between birds and Apollo and divination, and bees and Apollo and divination—an association in which both types of creatures 'serve' Apolline divination. But these cultic facts do not, of course, fully explain the significance of the story of the second temple. We can see why it was the birds and the bees that were chosen for the role of animal builders of a Delphic temple to Apollo, especially since both are builder-animals. But where did the idea of such a temple built by animal creatures come from in the first place, and what meanings did this mytheme have in the eyes of the ancients? As we saw in Chapter I, these questions cannot be answered except in the context of a consideration of the whole myth, in which each mytheme acquires meanings. Some mythemes, like that of the laurel temple, make sense to us even in isolation, but of course this 'sense' is a very partial one, for we do not know yet what was its full meaning and significance in the context of the whole myth about the first Delphic temples.[44]

Let us now turn to the third temple, the one made of bronze. As we have seen, in the Pindaric account we are told that above the pediment of this temple six Charmers (Keledones) made of gold were singing. Pausanias dismisses this, arguing that Pindar invented the Charmers in imitation of the Homeric Sirens. This last is undoubtedly true, and the surviving text of the eighth *Paean* includes[45] the passage referred to by Athenaeus (7. 290E) which speaks of the Keledones in Pindar, who, in the same way as the Sirens, made those who listened to them starve through pleasure by making them lose thought of food. Another comparison between the Keledones and the Sirens is found in Philostratos,[46] so it seems that, at least in later antiquity, the relationship between the Keledones and the Sirens had become something of a topos. But the Keledones do not only relate to the Sirens, although it is with these that they are more closely connected. They belong, like the Sirens, to a wider series of destructive female monsters, part-human,[47] who lure or lead people to destruction, and which also includes the Harpies, the Sphinx, and the Lamia.[48] And this image of dangerous female monsters bringing about death is surely the mythological expression of two fundamental fears, the fear of death and the fear of the alien female nature, as presented in the male discourse and seen through the eyes of the 'establishment world' of men—two fears which have been fused into one, into a model of death caused by a polarized version of the female.

There is a second level of meaning in Pindar's Keledones; another model apart from that of the 'dangerous female monster' has also operated upon their conception. The meaning of vv. 82–6 is reconstructed by Lobel[49] as follows: 'Pallas put (enchantment) into their voice and Mnemosyne (or her daughters, the Muses) revealed to them the present, past (and future).' In other words, the Keledones, thanks to a divine gift, possessed knowledge of present, past, and future. This implies that they had the gift of prophecy. As there was a close connection between poetry and prophecy in early Greek literature[50] it could be argued that this prophetic gift of the Keledones was a simple corollary of their identity as singers. But even if it was that connection that had triggered off the endowment of these creatures by the mythological imagination with prophetic powers, in my opinion, these part-human female singers of past, present, and future, associated with a temple of Apollo at Delphi, could have hardly failed to be perceived as mythological prefigurations of the human female prophetess who operated in the historical temple of Apollo at Delphi. In fact, I am inclined to believe that this 'mythological prefiguration of the Pythia' aspect was the primary model that inspired the creation of the Keledones; that the mythological imagination—Pindar's or whoever else's—was inspired to create the Keledones by the desire to invent mythological prefigurations of the Pythia. In other words, I think that, in the process of mythologizing about a mythical temple of Apollo at Delphi, created by divine craft, the model of the historical temple, created by human craft, in which a woman delivered prophecies, operated on the mythological imagination and prompted the creation of corresponding prophesying female beings of a supernatural character—to match the supernatural, divine craft that created the temple. In my opinion, two factors determined the modelling of these prophesying females on the type of the 'destructive female monster' in general and the Sirens in particular. The mythopoeic imagination was first pulled towards that general area of 'dangerous female' under the influence of the 'different', and in a way scaring, aspect of prophetic powers, which in a mythopoeic process can easily slide into its polarization[51] 'dangerous female prophet'. This, under the influence of the already established type of 'destructive female monster', was polarized further, so that the supernatural prophesying females took the form of destructive female monsters. This last polarization, and the crystallization of the supernatural prophesying females into the Keledones of the Pindaric description, was undoubtedly prompted by the fact that within the sphere of the dangerous female creatures the Sirens, singing destructive female monsters, provided an excellent model for supernatural females singing (prophecies) at the temple of Apollo, the god par excellence associated with music and song.

The choice of the wryneck for the animal element of the Siren-like Charmers (cf. above, n. 47) was probably determined by the fact that in ancient Greece wrynecks were associated with love charms.[52]

Like the second temple, the third temple as described in the Pindaric version is placed in the sphere of myth by the miraculous elements it contains. It was constructed by Hephaistos and Athena, the two craftsmen gods, it had the singing golden Charmers, and the gods made it disappear into the earth which they had opened up with a thunderbolt. I have argued that the Keledones who possessed the gift of prophecy were visualized as a mythological prefiguration of the Pythia, shaped on the model of the attractive but destructive female monster who brings death.

As we have seen, Pausanias attempted to rationalize the story, remove the bronze temple from the sphere of myth, and place it in the world of reality. Those elements of the story that can be explained in terms of everyday reality—like the concept of a bronze building—he accepts as true, those that cannot, he dismisses as false. The implication is that the story contains a core of historical truth, to which have been added other 'false' elements, the miraculous ones. Clearly, this a priori assumption—which I shall argue is false—can only be relevant to the question of the origin of the story; it is of no relevance to the understanding of the myth as told by Pindar. For this version, found in Pindar, whatever its origin, forms one interlocking organic whole, in which each element has a significance with reference to all the other elements and to the whole, and no part of the story can be considered in isolation. For this reason, in this version, the material, bronze, and the miraculous character are inextricably bound. With regard to the origin of the myth, there is no reason to attach any authority to Pausanias' attempt to make sense of the myth on the basis of the a priori assumption that it originated in historical fact—which inevitably entails arbitrarily separating the elements that can be made to fit the world of reality from those which cannot, and interpret the former as reflecting historical fact and the latter as later accretions.[53] Indeed, I consider it very unlikely that this myth started life as a factual report about a real temple made of bronze, and then miraculous elements became attracted to this core and transformed the historical account into the story of a miracle. For Pausanias was correct in claiming that bronze buildings belonged to everyday reality. Bronze was indeed used in Greek architecture, mostly for revetment, metal plaques covering a surface of different material;[54] architectural members made entirely of bronze were rare.[55] This surely suggests that bronze was not a sufficiently exotic building material in archaic Greece to attract around it miraculous elements. So it is difficult to see why and how the account of a historical temple made of bronze should be transformed into a myth about a magical bronze temple built by gods; while, we shall see, the mytheme 'miraculous bronze temple built by gods' makes perfect sense as an organic whole in the context of the whole myth of the first Delphic temples. This does not prove that Pausanias' implied theory about the origin of this myth is fallacious. But it does indicate that, not only it has no basis—other than its unconscious appeal to the implicitly rationalizing modern scholars' mental cast—it also is, when taken on its own terms, less likely to be correct than its alternative. Consequently, the attempt to recover the significance and circumstances of the creation of this story must not be based on the assumption that the 'bronze temple' element is primary, and reflects a historical reality, and the other elements secondary.

The fourth temple was made of stone and had been built by Trophonios and Agamedes. The involvement of Trophonios and Agamedes in the construction of a temple of Apollo at Delphi is first attested in the *Homeric Hymn to Apollo*, vv. 294–9, where we are told that Apollo laid down the foundations of the temple, upon which Trophonios and Agamedes laid *lainon oudon*, a footing of stone, and then men built the rest of the temple of wrought stones.[56] From a logical, rationalizing point of view, this complex process of collaboration in the construction of this temple appears odd. When we attempt to reconstruct the story's ancient meanings by making sense of it in terms of the ancient perceptions we find that what it says is that three types of beings participated in the construction of the temple, in this chronological order: a divine being, Apollo, heroic beings, Trophonios and Agamedes, and human beings, the tribes of men. The three stages and the three types of participants in the construction of the temple of Apollo as told by the poet of the hymn, can be seen as standing for three different orders, three different spheres which are included, and play a role in the Greek notion of a cult-unit, a sanctuary, or temple. The divine, in virtue of whom, and in the name of whom, the cult-unit functions; the human, which operates the cult-unit at the practical everyday level; and the heroic, a sphere which partakes of both the divine and the human, and which has the function of bringing close, in terms of projecting into the mythological past, the god and the man—a function which accounts for the frequent phenomenon of various Greek sanctuaries and cults tracing their foundation to legendary, heroic figures, as for example, the sanctuary of Artemis Brauronia in Attica, believed to have been founded by Iphigeneia, or the cult of Zeus Hellanios at Aegina whose founder was believed to be Aiakos. The mythological image of temple-building also expresses the concept of cult-foundation, and this is why the hymn shows the three orders of being involved in cult participating in the building of the temple.

The position of Trophonios and Agamedes as heroes, lying between the divine and the human, is reflected in their role in the construction of the temple described by the hymn, which comes between that of Apollo and that of the tribes of men. The same situation is surely found in the myth of the first temples. In this myth, the human element lies outside the nexus of the four mythological temples: it is provided by the actual, physical temple visible in the Delphic sanctuary, at whichever period one considered the myth. The temple of Trophonios and Agamedes, the heroic craftsmen, again provides the heroic element,[57] while the temple preceding theirs, the one made of bronze, was made by gods, Hephaistos and Athena. The tripartite model[58] then found in the hymn is also present here, but analysed into three different temples when in the hymn it was combined into one. In the hymn the divine element is represented by Apollo, the owner of the temple, and this is particularly appropriate, given that the context is that of the foundation of the sanctuary, oracle, and temple. In the myth of the first temples, it is Hephaistos and Athena, the craftsmen gods, who represent the divine element. But, of course, the myth of the first temples contains more than the 'human–heroic–divine' tripartite model. It also includes the temple of wax and feathers and the laurel temple. So we now have to consider whether these two mythological structures have any significance in terms of the scale of values which, I argued, lies behind the nexus of the third and fourth mythological temples considered together with the real, historical temple at the Delphic sanctuary.

Bees and birds belong to the animal world. They thus take us out of the anthropomorphic human–heroic–divine sphere, and into nature—here contrasted with the world of man and culture. A nature which is not devoid of craft (albeit non-human, non-culture craft); for both birds and bees are builders.

Moving backwards from the wax-and-feathers temple, we reach the first temple, the one made of laurel. From the point of view of the values that we have so far decoded, the important thing about the laurel is that it is a wild plant that can grow spontaneously. So the laurel, and the laurel temple, take us one more move away from the anthropomorphic sphere and from culture and into wild nature. It is perhaps significant in this respect that nothing is said about who constructed this temple, or how it was done.

We can tabulate the successive temples as follows:

First temple: laurel: plant—no craft: wild nature
Second temple: wax and feathers: animals (animal craft): nature

The third and fourth and historical temples belong to the *anthropomorphic* sphere and to culture, and this nexus can be analysed as follows:

Third temple: bronze: Hephaistos and Athena (divine craft): divine sphere
Fourth temple: stone: Trophonios and Agamedes (heroic craft): heroic sphere
Historical temple(s): stone: men (human craft): humanity

If we take the anthropomorphic sphere as one unit, the order within the myth, from the first temple onwards, is ascending, from the lower and wilder, to the higher and more civilized. But within the anthropomorphic sphere itself, the order is descending, from the divine, to the heroic, to the human. This descending order is determined by the fact that the 'chronological/historical' framework of the story makes it necessary that the last stage should involve the actual historical temple, representing humanity and everyday reality. This dictated the order: human, before that heroic, before that divine. The same chronological framework dictated the ascending order from the laurel temple to the anthropomorphic unit, ascending order which gives an evolutionary dimension to the myth and expresses the Greek way of structuring the universe.

The anthropomorphic unit, we saw, is found on its own, independently of the nature-unit, in the *Homeric Hymn to Apollo*, where, through the image of temple-building, it expressed in mythological language the concept that three interlocking spheres are involved in, and operate within, a sanctuary. This significance of the anthropomorphic unit was extended when, in the myth of the first temples, the nature-unit was also brought into play. In this myth, the nexus of the two units expressed in mythological language, through the image of temple-building, two groups of ideas. First, it set out all the different categories of things and beings which were involved in the Delphic sanctuary; in this way it also expressed the fact that Apollo's might extended over all categories of life, plant, animal, human, and heroic. Secondly, by means of this first group of ideas, the myth also expressed in mythological language a structured view of the cosmos. This is obviously the result of the fact that the evolutionary model expressed in this myth was ordered according to the Greek conception of world-order.

With regard to the creation of the myth of the first temples and the stimulus that led to it I would like to suggest that there are two possibilities. The first is that this myth was created as a result of the juxtaposition in the mythological imagination of, on the one hand the tripartite model articulated in the *Homeric Hymn*, and on the other the god's close connection with the laurel and the birds and bees; the latter situation inspired the extension of the tripartite model to take in all aspects of the cosmos, and express the god's power over all forms of life, and his close association with some particular life forms, the laurel among plants and the birds and bees among animals.

The second possibility is that, before the myth of the first temples was articulated in the form in which we know it from Pindar, an earlier myth had already articulated Apollo's intimate connection with the laurel, and the laurel's potent physical presence in the very adyton, into the story that Apollo's first temple at Delphi had been made of laurel. Such a mythological articulation could have taken place in a variety of contexts, especially cultic contexts in which the laurel played an important part or even was the symbolic focus. If this hypothesis is correct, and the mytheme of the laurel temple had been created before, and not in the context of, the myth of the first Delphic temples of Apollo, the latter may have been shaped as a result of the juxtaposition on the one hand of a story about the existence of an early temple of Apollo made of laurel and on the other of the tripartite model of the *Homeric Hymn*, in which the image of temple-building, through being broken down into stages, was used to express the concept of the three interlocking spheres of cultic and cosmic reality. This juxtaposition stimulated the mythological imagination and inspired further myth-making. We can tentatively suggest the parameters which determined this process. The tripartite model of the hymn provided a conceptual framework in which different stages in temple-building expressed different orders of beings and of cultic and cosmic reality. The story of the laurel temple, associated with a cultic reality, the central role of the laurel in the Delphic cult, and at the same time containing as dominant element a plant, a thing clearly belonging outside the divine–heroic–human anthropomorphic nexus, offered a model for representing the different stages of temple-building—which were used for expressing different orders or beings and of cultic and cosmic reality—through different structures, successive temples. And it provided the stimulus for extending the orders of cultic and cosmic reality represented in the myth beyond the anthropomorphic sphere.

Within the framework thus outlined the laurel temple provided one pole, that of wild nature, while the historical temples, products of human craft, and representing humanity and historical reality, fixed the other. Beyond this last—going backwards—unfold the remaining two-thirds of the 'human–heroic–divine tripartite model, each sphere now represented by a separate structure. So the model now runs: temple made of laurel; temple made by gods; temple made by heroes; end of 'myth': temple made by men, i.e. real, historical temple. This as it stands is clearly unbalanced: the animal world is not represented and the transition from the laurel temple to the rest is abrupt: as an evolutionary model this would be incomplete, and as a narrative pattern unsatisfactory. Thus this mythological structure created the space for, demanded the gap be filled by, a temple involving the animal world. As for the type of animals involved, given that the parameters operating on the mythoplastic imagination were (a) craftsmanship and (b) cultic connections with the Delphic sanctuary, the selection of the birds and the bees can be said to have been inevitable.[59] The choice of material for the temple was also inevitable: wax and feathers, the bodily products of the animal creatures themselves, and material which they could, and did, naturally handle.

In terms of narrative structure, the bronze temple is the element that provides the linking points between the 'nature' unit and the anthropomorphic one. First, this temple belongs to the latter, but is built of a 'characteristic' material, like those of the former. Secondly, it helps to link the two units through its complex relationship, through sets of similarities and contrasts, to the last temple of the previous unit, the wax-and-feathers one. Both these temples contain miraculous elements, which are lacking from the other two. A contrast is provided by the fact that the second temple involves the lowest form of craft, animal craft, and the third the highest, divine craft (represented by the craftsmen-gods Hephaistos and Athena). Both temples 'disappeared' in a supernatural way. The second temple was sent to the land of the Hyperboreans by means of a mighty wind—an element belonging to nature. In the case of the bronze temple, a thunderbolt was involved, an element of nature, but one believed to be the product of divine craft, like the bronze temple itself.

This set of similarities and contrasts between the two temples clearly enhances strongly the narrative unity of the whole myth. It also suggests that the present forms of both mythemes are likely to have been created at the same time in the context of the same mythopoeic activity.

The choice of bronze as the 'characteristic' material of the third temple may have been a chance one. But it is also possible, indeed more likely, that this choice was the result of the influence of another historical/evolutionary model involving materials: that of Hesiod's generations.[60] The Hesiodic model may have influenced the mythopoeic process of the creation of the Delphic myth in the following way. In Hesiod's scheme, the 'present-day' generation is that of iron; it is preceded by the generation of heroes, in its turn preceded by the generation of bronze. So there are already two correspondences between the Delphic and the Hesiodic sequence: the last two stages are the same: present-day, preceded by heroic; and in both different materials stand for different stages in the evolution. Given these already existing similarities, the creator of the Delphic mythological model, poet or theologian that he was, may have slipped naturally into the choice of bronze for the material characteristic of the evolutionary stage preceding the heroic one, under the influence of the fact that bronze was the material characteristic of the stage preceding the heroic one in Hesiod. As we saw, a 'characteristic' material was desirable for the creation of an effective link-up with the laurel and the wax-and-feathers temples. And thus the bronze temple, built by Athena and Hephaistos, representing divine craft, was invented.

The last question to be considered is, when and by whom was this myth created. The *Homeric Hymn to Apollo* gives us, in my opinion, a *terminus post quem*, in so far as the version of the tripartite model human—heroic—divine contained there appears (to me) to be primary, and likely to have at least partly stimulated the creation of our myth.[61] I would be tempted to think that it was Pindar's mythological and poetic imagination which invented the myth. Another possibility would be to suppose that Pindar had not invented the myth, but was responsible for its elaboration, for example, for an elaboration of the motifs of the wax-and-feathers and the bronze temples, and the creation of the complex set of relationships between the two. In any case, the creator of this myth used pre-existing motifs, modifying and reinterpreting them, and invented new ones, mostly using established cultic and mythological material. The result was a myth which gave mythological expression to some aspects of the Delphic cultic reality and through it also presented an articulated and evaluative view of the world.

Notes to Chapter IV.1

1. Vv. 58–99 (Snell 1964); cf. Lobel 1961: 45–50. I should note that an extraordinarily bad study has been written on three-quarters of this myth: Elderkin 1962.
2. Cf. below.
3. Pindar says (cf. below) that it was built by Hephaistos and Athena, but Pausanias only mentions Hephaistos as the divine builder in the version which he mentions in order to reject.
4. Cf. Paus. 9. 37. 3.
5. All that survives of the account of the laurel temple is δάφν, but cf. also the mention of the laurel temple in the Scholia to Pind., *Paean* 8; P. Oxy. 841, fr. 107.
6. Cf. Snell 1964: 43; Lobel 1961: 49–50.
7. *Isthm.*, fr. 2 Snell; cf. also Plut., *Consol. ad Apoll.* 109A–B.
8. Lobel 1961: 46 on fr. 22.
9. Lobel 1961: 46.
10. The text reads [ἐποιήθη ὁ π]ρῶτος ἀπ [ὁ] δάφνης, δάφνη [and Snell 1938: 435, supplements [δὲ ἐκομίσθη ἐκ τῶν Τεμπῶν vel sim.
11. For the temple of Trophonios and Agamedes cf. also the following sources: Steph. Byz. s.v. *Delphoi*; Paus. 9. 37. 4; Schol. Arist., *Nub.* 508, ed. Dindorf; Cic., *Tusc.* 1. 47. 114 (and cf. also above, n. 7). The account of their involvement in the building of a temple to Apollo given in the *Homeric Hymn to Apollo* will be discussed below.
12. For the important role of the laurel in Delphic cult and ritual cf. Amandry 1950: 126–9 (cf. also Eur., *Ion*, vv. 112–24).
13. Cf. Plut., *De Pyth. orac.* 397a, *De E apud Delph.* 385C; Luc., *Bis acc.* 1; Aristoph., *Ploutos*, 39; Schol. Aristoph., *Ploutos*, 39; Callim., *Iamb.* 4 fr. 194. 26–7 Pfeiffer; cf. also Amandry 1950: 126–9.

14. Cf. Aristoph., *Ploutos*, 213; Schol. Aristoph. *Ploutos*, 213: Callim., *Hymn to Delos*, 94; cf. Aristonoos, *Paean to Apollo*, vv. 9–11 (Colin 1909–13: no. 191), where Apollo is said to prophesy shaking a laurel branch; (the date of this paean is the third quarter of the 4th cent. BC). And cf. Amandry 1950: 129–34. The *Hom. H. Ap.*, v. 396 (cf. 394–6) makes clear that the laurel played a fundamental role in the act of divination. I intend to discuss elsewhere the problems associated with this passage, and some other aspects of the laurel's place and significance in Apolline cult and ritual. Bousquet's suggestion (*Gnomon*, 32 (1960), 260) that the laurel tree shaken by the Pythia and situated in the adyton was artificial, made of bronze, is very unconvincing, especially when three factors are taken into account: (i) we know from the texts, and especially the *Ion*, that real laurel was used a lot, and grew in the sanctuary, (ii) laurels grew in the adyton-courtyard of the temple (and oracle) of Apollo at Didyma, (iii) several inscriptions mention the δάφνη ... παρὰ τοῦ θεοῦ or δάφνη ... παρὰ τοῦ Ἀπόλλωνος, which is likely to be referring to the adyton laurel, and which is a real laurel from which crowns were made.
15. It is interesting to note that according to Hesychius, the laurel growing at Tempe had a special name; it was called *dyareia* (cf. Hes. s.v. *dyareia*).
16. Cf. Brelich 1969: 398–412 and 427–8, cf. also pp. 397–8, 412–27.
17. Theopompos, *F Gr. H* 115 F 80; Callim., *Aet.* 4, fr. 86; according to Hypoth. Pind., *Pyth.* c., Apollo was purified in Crete, but he then went to Tempe and brought the laurel to Delphi from there; in Artistonoos' *Paean* (cf. n. 14) Apollo is said to be *seion* a laurel branch when prophesying in vv. 9–11 and to have been purified at Tempe in v. 15, but no connection is made by the poet between laurel and Tempe purification. Cf. also Nicander, *Alexipharmaca*, vv. 198–200; Steph. Byz. s.v. *Deipnias*; Plut., *Quaest. Graec.* 293C; and Tert., *De cor.* 7. 5.
18. Theopompos, *F Gr. H* 115 F 80; Steph. Byz. s.v. *Deipnias*; cf. also Callim., *Iamb.* 4, fr. 194, Pfeiffer 34–6; Hypoth. Pind., *Pyth.* c; Plut., *De mus.* 1136A, *De def. orac.* 418A–B, *Quaest. Graec.* 293C.
19. Theopompos, *F Gr. H* 115 F 80; Hypoth. Pind., *Pyth.* c; cf. also Callim., *Iamb.* 4, fr. 194, Pfeiffer 34–6.
20. Cf. Schaefer 1939: 54 ff.; Bérard 1971: 59–73.
21. In the original version of this essay, published in 1979, I had accepted Bérard's interpretation of the 8th-cent. so-called Daphnephoreion in the sanctuary of Apollo Daphnephoros at Eretria, according to which this structure had walls made of laurel branches (see Bérard 1971; Auberson 1974: 60–8; cf. also on this building Auberson and Schefold 1972: 118–19 and 116, fig. 22). I have now become convinced, as a result of Drerup's study of this building (1986: 3–21) which has been corrected and refined by Coulton (1988: 60–2), that

interpretation is wrong. Of course, this does not exclude the theoretical possibility that a temple made of laurel may have existed at some point in time, but we do not now have any grounds for thinking that the laurel temple had a cultic reality. The disappearance of the laurel temple does not alter the overall picture presented in this essay except in so far as the constituent elements that went into the making of the first temple is concerned: it is not the case, as had appeared in 1979, that this mytheme reflects directly cultic history; it does, of course, refract elements of cultic reality, for Apollo is closely connected with the laurel in cult, as well as in myth; but this relationship had not been articulated into a laurel temple in cult, only in myth.

22. On Apollo's many associations with birds cf. the following, where references to ancient texts and representations as well as further bibliography can be found: Jessen 1955: 281–309; Keller 1920: *passim*; Amandry 1950: 57–8; Metzger 1977: 422 n. 13, 425, 426, 428; cf. also Metzger 1951: 171–5. On doves at the Delphic sanctuary cf. Eur., *Ion*, vv. 1196–8; Diod. Sic. 16. 27. 2.
23. Cf. e.g. Eur. fr. 773 N² 33–4; Aristoph., *Av.* 769–84; Callim., *HAp.* 4–5, *Hymn Del.* 249–54.
24. Cf. Aristophanes' comparisons between birds and Apollo in *Av.* 716–22. On birds and divination see also Bodson 1978: 94–6.
25. Amandry 1950: 57–9.
26. Cf. Cook 1895: 1–24; Ransome 1937: 91–111, with several errors; Picard 1922: 183–4, 228–9; 1940: 279–83 (very speculative); Feyel 1946: 5–22; Keller 1920: 421–31; Bodson 1978: 20–43. Cf. also Willetts 1962: 216–18, 257. Apart from the sources quoted in these discussions, cf. also Pind., *Encom.*, fr. 123, Snell³ 10–11.
27. Pindar fr. 158 Snell³; Hesych. s.v. *Melissai*; Porphyr., *Antr. Nymph.* 18; Grenfell and Hunt 1922: 1802, col. ii, 29–35; cf. also pp. 155–60; Callim., *HAp.* 110–12; Schol. Pind., *Pyth.* 4. 106c; Serv. Schol. Virgil, *Aen.* 1. 430; Lactant., *Div. Inst.* 1. 22.
28. Paus. 8. 13. 1; Dittenberger 1915–24: 352.6, 363. 16.
29. Aesch., *Hieriai Tr. GF* iii, F 87 (= fr. 87 N²).
30. Cf. e.g. the bears at Brauron (cf. Aristoph., *Lys.* 645; Sourvinou-Inwood 1988a: *passim*); the *tauroi* of Poseidon at Ephesos (Athen. 10. 425c), the *poloi* in Laconia (on whom see Bodson 1978: 159–60; for the Leukippidai: Hesych. s.v. † *polia*; for Demeter and Kore: IG V.1 594); the Peleiai or Peleiades at Dodona (cf. Paus. 10. 12. 10; Hesych. s.v. *peleiai*; cf. also Soph., *Trach.* 172; Schol. Soph., *Trach.* 172; Easterling 1982: 99; Bodson 1978: 101–14).
31. On the relationship between Delphi and the oracle of Trophonios cf. Parke and Wormell 1956: 368.
32. On this passage cf. Radermacher 1931: 169–73; Amandry 1950: 61 ff.; Latte, *RE* xviii. 832; Feyel 1946: 5–22; Jacoby, *F Gr. H* III

B Suppl. 559–60; cf. also the commentary on the passage in Allen, Halliday, and Sikes 1936: *ad loc.*

33. On *apaneuthe* of v. 556 cf. Amandry 1950: 62 n. 4.
34. Amandry 1950: 63 n. 1.
35. *RE* xviii, col. 832.
36. *F Gr. H* III B Suppl. 560.
37. Amandry 1950: 63 n. 1.
38. As cit. n. 36.
39. Amandry 1950: 61.
40. Cf. Latte, *RE* xviii, col. 832. Radermacher 1931: 169–73; Feyel 1946: 6; Jacoby (as cit. n. 32).
41. The problems concerning the name of the bee-women in the hymn (on which cf. Feyel 1946: 7–8, and Amandry 1950: 62, 64 n. 2) and the relationship between this passage of the hymn and Philochoros, *F Gr. H* 328F 195 and Apollod., *Bibl.* 3. 10. 2 (on which relationship cf. Jacoby (as cit. n. 32) do not affect the argument that the hymn concerns an oracle involving the behaviour-pattern of bees or women-bees in the myth—reflecting the cult phenomenon of bee-divination (cf. esp. Jacoby, p. 560).
42. Cf. for Delphi Diod. Sic. 16. 26; Paus. 10. 5. 7.
43. There were of course also other modes of divination practised at Delphi (cf. Amandry 1950: 57–65). The Delphic oracle having become the most important Greek divination centre, it attracted to its divination methods that may not have originally belonged there.
44. It has been suggested that the mythological temple of wax and feathers had been transposed into architecture, reproduced as a real building. The excavation at Delos revealed some blocks of stone, belonging to the outer walls of two buildings, which were decorated on the outside with continuous hexagons forming a honeycomb pattern (Vallois 1944: 23; 1966: 62–3; Orlandos 1958: 257, fig. 225, 257–8; Bruneau and Ducat 1983: 152–3). The buildings from which the blocks came are the Letoon, built *c*.540 and the so-called 'Monument with the hexagons', built *c*.500 (cf. Bruneau and Ducat 1983: 153; Gallet de Santerre 1958: 297). Blocks of stone with a similar pattern have been reported from Thasos (cf. Bruneau and Ducat 1983: 153; Rumpf 1964: 8). The Delian finds led Rumpf to suggest (Rumpf 1964: 5–8) that the mythological bees' temple of Delphi had been copied in stone at the Delian sanctuary. But there are serious objections. The stone blocks decorated with the honeycomb pattern do not all come from one building, as Rumpf appears to imply, but from two different ones. The earliest of these has a firmly established function: not a temple, or an offering, to Apollo, but a temple, the temple of Leto, situated in a distinct temenos belonging to Apollo's mother (cf. Gallet de Santerre 1958: 144). The other, of more problematic identification

(Bruneau and Ducat 1983: 153), does not appear to have been a temple. It follows, first, that these blocks with the honeycomb pattern do not come from a building conceived and executed as the architectural transposition of the mythological Delphic temple; and secondly, that it is not the case that the honeycomb pattern was identified, in the architectural vocabulary of the period, with the translation of a mythical temple made of wax and feathers into stone—for otherwise their use in these buildings would have been incongruous. (On this wall decoration, against the notion that it is to be connected with bees and honeycombs see now Drerup 1986: 21 n. 83) It could in theory be argued that this type of outer-wall decoration, found in a sanctuary of Apollo, could have been inspired by the story of the second Delphic temple, in which case we would have a *terminus ante quem* c.540 for the creation of the myth. Though this possibility cannot be conclusively excluded, other possibilities appear less implausible: that the pattern did not imitate a honeycomb, but had a different source of inspiration (see Drerup 1986: 21 n. 83); alternatively, that the idea to create a pattern imitating a honeycomb was dictated by a purely decorative impulse; or that it may have been inspired by the Delphic Apollo's association with prophetic bees, probably housed in the Delphic sanctuary, without there having been a story about an Apolline temple constructed by bees; or that the cultic connotations of the pattern are to be associated with Artemis, in whose cult at Ephesos bee elements were prominent (cf. above, nn. 26, 28, 29. Cf. also Kraay and Hirmer 1966: 355, 356–7; Kraay 1976: 23, 256; cf. also pl. 53, nos. 901–4) —it should be remembered in this connection that the outlook of the Delian sanctuary in the relevant period is predominantly Aegean and Ionian. The Swiss excavations at the sanctuary of Apollo Daphnephoros at Eretria revealed part of the stone socle of a building which was contemporary with the so-called *daphnephoreion* and which appeared at first to be polygonal on the inside and rounded on the outside, with inner corners inside and a rounded outer surface, and to have what seemed to be an oval shape. This first impression gave rise to the tentative hypothesis put forward by the excavators that the building evoked a honeycomb on the inside and a beehive on the outside; and that therefore it may have been built as an architectural transposition of the second Delphic mythological temple (cf. Auberson and Schefold 1972: 120, 116, fig. 22). However, further investigation showed that while the west wall does have an inner corner, the east one appears to be rounded on the inside and the outside; and that beyond the west inner corner and the east rounded one both walls run straight for the whole of their preserved course. (Cf. the corrected plan in *Ant. K.* 17 (1974), 70, fig. 1. I am very grateful to Dr Auberson for discussing this building with me and providing information about it.)

The excavators therefore became rather sceptical about the beehive/honeycomb hypothesis. It is perhaps more likely that the plan of the building was simply apsidal. Unfortunately, no more information can ever be gained about this building, as the remaining part of it has not survived. In my opinion, the new circumstances indicate that this building is unlikely to have been built as an architectural transposition of the Delphic myth. Not only does the plan now appear much more likely to have been apsidal than anything else, but also the corrected plan of the surviving part shows the apparent similarity of the interior to a honeycomb to have disappeared. Moreover, as we have seen, the so-called *daphnephoreion* itself turned out not to have been a laurel temple connected with the myth under consideration.

45. Cf. vv. 76–9.
46. *Vit. Apollon.* 6. 11.
47. I understand the Keledones as being part-human: cf. *parthenia* in v. 80 (and cf. Lobel 1961: 46) and perhaps also *akeraton* in v. 81 (cf. ibid.); vv. 82–6 surely imply a human voice. Otherwise they appear to have had, wholly or partly, the form of wrynecks (cf. Snell 1964: 41, and Philostr., *Vit. Apollon.* 6. 11).
48. On Lamia cf. Vermeule 1977: 296–7.
49. Lobel 1961: 47.
50. Cf. West 1966: 166, comm. on v. 32.
51. I should note that polarization frequently takes place within the mythopoeic process; cf. also Brelich 1958: 277–8.
52. Thompson 1936, s.v. ἴυγξ.
53. On this general question, this type of rationalizing approach, as it pertains to our own interpretations of ancient texts see below Ch. IV.3, sect. 1, and *passim*.
54. Cf. Martin 1965: 155.
55. Cf. Martin 1965: 156.
56. On the meaning of *lainos oudos* in this passage (a 'layer of orthostats made of stone') cf. von Blumenthal 1928: 220–4.
57. The selection of these two as the contributors of the heroic element in the first place was determined by the fact that they were the architects of the heroic age *par excellence* in combination with the fact that they were associated with Apollo. Alternatively, it is possible that only one of these two features was earlier than, and contributed to the creation of, the mytheme 'Trophonios and Agamedes were involved in the building of a Delphic temple' and the other resulted from their association with the building of the Delphic temple.
58. It is interesting that, as we have seen, Strabo had ordered the Delphic temples according to a tripartite model expressing three different orders of 'reality': the mythical sphere represented by the temple of wax and feathers, the world of the heroic, legendary past, represented by the

temple of Trophonios and Agamedes, said to be the second temple, and the real world of humanity, represented by 'the present' temple constructed by the Amphiktyones.
59. Cf. also above, the discussion of the second temple.
60. *WD* 109–201. On this myth cf. West 1978: 172–7 with bibl.; Smith 1980: 145–63 with bibliography; Falkner 1989: 42–60.
61. If it had indeed been the case that the decoration of stone blocks with honeycomb patterns had been inspired by the story of the temple made of wax and feathers, then we would have had a *terminus ante quem* of *c.*540—for I am assuming that, as I argued above, the story of the second temple was invented in the framework of the creation of the whole myth. However, as we saw, it is extremely unlikely that this honeycomb decoration is to be associated with our myth.

IV.2

Myth as History: The Previous Owners of the Delphic Oracle*

Many Greek myths express important perceptions of the society that generated them and contain insights which are (or can be reinterpreted so as to become) significant for our own age; thus they can be said to be 'true' even today. But they are not 'true' narrative accounts of past events (though they present themselves in that guise) and they should not be taken at face value and assumed to contain descriptions of past realities—as they sometimes are. The myth I am discussing here (which claims that Apollo did not found the Delphic oracle but took it over from an earlier goddess) has often been assumed to contain true information about the oracle's early history. Moreover, this historical reading of the myth has functioned as an (implicit) perceptual filter shaping many scholars' interpretation of reality, that is, of the surviving information pertaining to the oracle's early history. My purpose is to show that the Previous Owners myth does not reflect cultic history but expresses certain important perceptions about the Delphic Apollo, the oracle, and the cosmos. First I will deconstruct the argument in favour of the historicity of the myth and show that it depends on a series of hidden, mutually supporting, a priori, and sometimes demonstrably wrong, assumptions and that it is fallacious. In the second part I will analyse the myth and show that, while it cannot be cultic history, it makes perfect sense as a myth, articulating perceptions also known to us from other sources.

A variety of deities are named as Previous Owners in the different variants, but all versions include Gaia or Themis, or both.[1] Many scholars[2] believe that this story reflects a memory of a time in which Gaia and/or Themis were the oracular divinities at Delphi, dispossessed by Apollo who did not evict them altogether but allowed them to maintain a cult of secondary importance in the Delphic sanctuary. As we shall see,[3] the only 'evidence' for the view that these goddesses had preceded Apollo as oracular deities at Delphi is the existence of the myth—which can only be considered to be 'evidence' if it is assumed that the most reasonable interpretation of such a myth is that it reflects historical reality. This is an unwarranted—and fallacious—a priori assumption which, I shall show, lies at the core of the orthodox discourse's hidden circularity; it is the product of an implicit, rationalizing, euhemeristic reading of myth, which, once explicitly set out, would be supported by few. For myths are not translations of events into mythological language, which scholars can translate back into history. The myths of resistance to Dionysos' cult, for example, are not, as some had imagined, reflections of a historical conflict; they articulate, and are articulated by, religious realities such as ritual tensions and symbolic oppositions.[4] Since myths are structured by, and express, the (religious, social, and intellectual) realities and mental representations of the societies that produced or recast them,[5] any echoes of cultic history that may have gone into the making of a particular myth are radically reshaped and adapted, by a process of *bricolage*, to fit the 'needs', the 'spaces', created by the mythological schemata structuring that myth, which express, and are shaped by, those realities and representations.[6]

Thus, the hypothesis that our myth is a reversible translation of history is invalidated. In any case, even if we cannot conclusively *prove* the fallaciousness of the assumption that the most reasonable interpretation of our myth is that it reflects historical reality, since that assumption is a priori and thus culturally determined (by a rationalizing mode of thought which privileges 'positivist' interpretations), and since it cannot be shown to be right, it must not be allowed to form the hidden centre of a discourse the validity of which depends on that assumption's validity. Given that alternative interpretations of the emergence and significance of the myth are possible—not to say more convincing—it is illegitimate to assume the myth's historicity and base the validity of the whole case on that. In fact, the myth's pattern of appearance offers a serious objection to the historical interpretation. For the two earliest accounts of the early history of the oracle, in the *Homeric Hymn to Apollo*[7] and Alcaeus' *Hymn to Apollo*,[8] contradict the Previous Owners myth and present Apollo as the founder and first owner of the Delphic oracle.[9] Thus, the presumption must be—especially since the two hymns originated in different religious environments, and the Pythian part of the *Homeric Hymn* reflected the Delphic priesthood's theology—that 'Apollo's foundation of the oracle' was the early cultic myth on the oracle's origins and that the Previous Owners story was invented at a later stage—unless some contrary evidence can be adduced, which, we shall see, it cannot. The data, when investigated in their own right, cannot support the historicity of the myth. They can only appear to support it when, in the context of attempts to validate that historicity, they are structured and questioned by means of conceptual schemata dependent on the very hypothesis that is being tested—a circular procedure leading to corrupted, and thus wrong conclusions.

To eliminate bias, these data must be investigated through a neutral methodology which excludes prior assumptions. As we saw in Chapter I, one strategy conducive to neutrality is to investigate each of the relevant grids of evidence (archaeological, cultic, mythological) separately and independently, to keep the deconstructive and the mythological analyses separate, and to compare the results of these independent investigations only at a later stage. This will prevent the common fallacy of combining elements from different grids, taken out of their proper context, to make up an apparently coherent case which is in fact radically flawed by hidden circularity. In addition, the proposed strategy allows cross-checks between grids, which can provide controls and, if appropriate, confirmations. A rigorous methodology also demands, we saw, that the data should be studied in the context of the wider nexuses to which each particular set belongs (e.g. Mycenaean figurines or divine succession myths); for only this context can help determine their meanings in the particular case that concerns us and so protect the investigation from a priori bias.

A fundamental plank of the case for our myth's historicity is the alleged Mycenaean cult of Gaia. The gist of my argument is that, though there may have been a Mycenaean shrine at Delphi, its possible existence is irrelevant to the myth's historicity. For it is only if we assume that the myth creates an a priori case for the existence of a Gaia cult—an assumption which our investigation purports to examine—that Gaia can be considered at all in connection with the Mycenaean cult; thus the relevance of the latter to the former rests on a circular argument. But even if we grant that special pleading as a working hypothesis, the notion that the supposed Mycenaean cult provides support for the myth's historicity has to rely on a further series of unwarranted assumptions—and in the end it proves untenable. There had probably been a Mycenaean shrine at Delphi, perhaps at Marmaria, at the later sanctuary of Athena Pronaia,[10] but not on the site of the temple of Apollo.[11] Since we know nothing about the deity or deities worshipped at this hypothetical Mycenaean shrine, the claim that it must have been an oracular shrine of Gaia is without foundation, wild. The female figurines (n. 10) may have come from a shrine, but they do not show that that shrine's divinity was female. For almost all Mycenaean figurines are female; we do not know whom they represent.[12] But even if we knew that the chief deity of the hypothetical Mycenaean shrine had been a goddess, we would still know nothing about her. There is certainly no reason for thinking she was Gaia; for, we know from the Linear B tablets, the Mycenaeans had a genuinely polytheistic religion, with a hierarchically articulated pantheon[13]—in which, incidentally, Gaia is not attested.

Thus the notion that the hypothetical Mycenaean cult at Delphi can support the view that Gaia's cult had preceded Apollo's is based on a circular argument; for Gaia can only be considered as a possibility at all if we begin with the assumption that the myth creates a presumption that Gaia's cult preceded Apollo's, and then look for evidence that can be made to support it. On that (hidden) assumption of historicity depends another, which in turn implicitly supports the first: the assumption that, since the myth tells us that Gaia preceded Apollo at Delphi, this must be presumed to be correct unless conclusively disproved. Given that only very rarely can anything be conclusively proved or disproved in early Greek religion, the fact that something as elusive as proving that a particular deity was not worshipped at a particular hypothetical Mycenaean shrine cannot be achieved has, obviously, no evidential value. And yet the orthodox discourse assumes implicitly that, failing conclusive proof against it, the view that Gaia preceded Apollo at Delphi stands.[14] Since, we saw, the assumption at the centre of this argument (the myth's presumption of historicity) is fallacious, and since, in fact, the myth's pattern of appearance suggests that it does not reflect historical reality, the whole case pertaining to the alleged Mycenaean cult of Gaia at Delphi and its relevance to our myth is clearly circular and rests on fallacies. The view that it is erroneous is strengthened by further arguments.

There is no cult activity at either Marmaria or the site of the temple of Apollo between the Mycenaean period and the late ninth century;[15] this absence of continuity argues strongly against the view that the hypothetical shrine of Mycenaean Delphi can be connected with the Previous Owners myth. For the only thing that could (conceivably) have survived through the centuries in those circumstances is the mere memory of an earlier cult. Thus, the cultic discontinuity invalidates another nexus of arguments for the historical interpretation of the myth, the notion (which, we shall see, is also discredited on other grounds) that various elements in the cult of the Delphic Apollo are hang-overs from Gaia's. For if all that had survived from the hypothetical Mycenaean cult had been the memory that it had existed, Apollo's cult could not have inherited any cultic elements from it. Moreover, in so far as it is possible to assess scarce and dumb data of this kind, the evidence cannot support the notion that Gaia was the mistress of a Mycenaean oracle. We do not know whether Mycenaean oracles existed, and if they did, what their diagnostic features would have been. However, what we can see is that, at Delphi, such Mycenaean elements as are capable of a religious interpretation are not of a type (or quantity) to suggest the presence of a cult-place in any way important or exceptional, anything other than an ordinary Mycenaean shrine. Given that the Pronaia deposit had been put together by seventh-century Greeks, who may, perhaps, be presumed to have selected the most impressive and unusual finds, and—to judge by the presence of the pottery—also a representative sample, I submit that this observation has more value than the usual argumentum *ex silentio*.

Now let us consider some more specific hypotheses connecting the hypothetical Mycenaean cult with the Previous Owners myth. Roux argues that, since Athena had been a Mycenaean goddess, there is no reason to think that it was not she who had been worshipped at Marmaria in Mycenaean times.[16] There are serious objections to this argument. First, *a-ta-na po-ti-ni-ja* does not mean, as Roux thinks, 'auguste Athena' but 'potnia (Mistress) of *atana* [probably a toponym]'.[17] Second,[18] it is illegitimate—especially since *a-ta-na po-ti-ni-ja* may suggest a geographically circumscribed deity—to conclude that Athena had been worshipped at Mycenaean Marmaria because many centuries later, and after a break in cult use, Athena's sanctuary was situated on the site where the Mycenaean shrine may have stood. Third, Athena is not a Previous Owner in the myth but, both in cult and myth, a collaborator and friend of Apollo.[19] Consequently, even if we assume that there had been a Mycenaean cult of Athena at Marmaria, and further that the memory of it had lingered through the Dark Ages despite the break, the myth of the Previous Owners would still not be reflecting that cult. Thus this would be an argument against interpreting the myth of the Previous Owners in terms of a relationship between the cult of Apollo and the supposed Mycenaean cult.

In Béquignon's view,[20] a Mycenaean Gaia shrine at Marmaria was replaced by Apollo's sanctuary. But even leaving aside all the objections to the historical interpretation, if (as this view presupposes) the memory of the cult had been preserved through the Dark Ages, the archaic sanctuary would have been dedicated to Gaia, not Athena. For Cassola[21] divine names are not important, they allude to a female chthonic deity whose heir was Athena. But, we saw, there is no evidence whatsoever that the Mycenaean cult involved a female deity, let alone that she was chthonic. Two interdependent (implicit) assumptions sustain Cassola's argument—and all variations of this hypothesis. First, that the most plausible interpretation of the Previous Owners myth is that it reflected cultic reality. Second,—implicitly supporting the first—an underlying evolutionary model which, though discredited as a serious account of the development of Greek religion, nevertheless still unconsciously informs many discourses: the model according to which Greek religion progressed from dark, chthonic (and female) deities to light and celestial ones[22]—derived from, and sustained through, the misinterpretation of classical Greek symbolic articulations (mistaken for reflections of past events) in this and other myths. These underlying assumptions make the historical interpretation of the Previous Owners myth seem eminently logical, for it conforms with the expectations which it helped form.

Now Poseidon: it has been claimed that, since he is a Mycenaean god and husband of Gaia, his cult at Delphi must go back to the Mycenaean period; and that this provides an additional argument for the early Gaia cult, and thus the historicity of our myth,[23] one version of which (Paus. 10. 5. 6) says that Gaia and Poseidon had shared the oracle before Apollo. However, Gaia's Mycenaean existence, we saw, is phantomatic, and we do not know whether Poseidon was worshipped in Mycenaean Delphi. Furthermore, the notion that Poseidon's name designates him as 'husband of the Earth' is very far from certain;[24] nor is there any mythological support for the notion that he was the Earth's husband.[25] In addition, Poseidon's consort in Mycenaean cult is *Po-si-da-e-ja* (PY Tn 316.4);[26] if the evidence of the Pylos tablets is to be used, as it is by Roux for Poseidon's importance in Mycenaean religion (see n. 23), it should not be used selectively, and Posidaeja must not be ignored in favour of a phantomatic union with Gaia (who is unattested in the Mycenaean period), a union whose claim to existence at any period is highly dubious. Thus, we are left, once again, with a myth which, we shall see, makes perfect sense in its own, mythological terms.

There is no evidence for a cult of Gaia and/or Themis at Delphi before the first half of the fifth century[27]—a period when its emergence should be seen as a response to the myth.[28] The case for an earlier cult of Gaia at Delphi runs as follows. We know from a fourth-century inscription and Plutarch's description that Gaia had a shrine south of the temple of Apollo.[29] After the temple's destruction in 548, its terrace was extended and a polygonal retaining wall built;[30] in the process, several buildings were destroyed. Because the later shrine of Gaia was in this region, it is assumed by some that the area belonged to Gaia before the rearrangement; on that view, the extension of the terrace of Apollo's temple encroached on Gaia's temenos and marked the god's final triumph.[31] However, the assumption that the spatial organization of the Delphic sanctuary did not change between the early sixth and the fourth centuries, a period during which drastic rearrangements of space indisputably took place, is extremely implausible and again depends on the a priori conviction that, given the myth's existence, Gaia's cult must be old. For it is illegitimate to assume, in the case of a continuously growing and developing sanctuary, that the fact that a deity was worshipped in one place in the fourth century entails that she had been worshipped in the same place in the early sixth, especially since we do not know whether or not she had been worshipped in that sanctuary at all in that early period—indeed this is what we are trying to find out. The earliest evidence for a Gaia cult probably belongs to the Kastalia area.[32]

Among the buildings buried under the new terrace is no. xxviii,[33] about the function of which we know nothing. Its southwest angle is built against a rock, and at the foot of the rock there is a small spring. Because of its association with the rock, and especially with the spring, it has been suggested that no. xxviii was a building with some religious function rather than a treasury. This is probably right. But there is no justification for calling it a 'temple of Gaia'. This identification depends entirely on two preconceived—and fallacious—assumptions: first, that there must have been an early cult of Gaia because the myth says so; and second, that springs are associated with Gaia because in the context of certain (simplistic) modern perceptions of Apollo (which ignore his complexity and ambivalence and the development of his divine personality) the Apollo–springs association appears illogical, while Gaia–springs seems 'natural'.[34] Thus the data are forced into perverse explanatory patterns and linked by circular arguments, to produce interpretations which only appear convincing when viewed through the perceptual filters of the culturally determined expectations which generated them.

The following facts show that the Gaia interpretation of building xxviii rests on fallacious foundations and is highly implausible. First, springs and water are connected with Apollo in his oracular function also in other important oracles, Didyma, Claros, and Ptoion.[35] Second, at Delphi, in the period that concerns us, c.600, there were two fountains associated with the temple of Apollo: fountain 24 and a spring behind the opisthodomos.[36] It is thus perverse to assume (on no evidence) that spring 16[37] had a different significance and association, and decide that it belonged to Gaia, and then identify building xxviii as the temple of Gaia *because* it is associated with this spring. Third, no. xxviii's entrance is at its north side, that is, it opens up towards the temple of Apollo. It thus related spatially to the temple, which suggests that it was associated with the cult of Apollo and not with a different, rival, cult.

Moreover, even if—despite what the evidence suggests—there had been a cult of Gaia earlier than the fifth century, and earlier than the myth, this would not be evidence for the view that Gaia preceded Apollo as mistress of the oracle. For, since Delphi was an established Apolline oracle in the eighth century (see e.g. *Od.* 8. 79–81), soon after the beginning of cult activity in the sanctuary, there is no place for Gaia as mistress of the oracle from the late ninth century onwards. Consequently, since Gaia did not have an oracular cult at Delphi before that date, even if her cult had begun before the myth's creation, it would not be evidence for the myth's historicity. Myth and cult interact, myths using existing cultic and theological material to weave their tales through *bricolage*. If a Gaia cult had preceded the myth, this would only entail that the chronological order of myth and cult, the two articulations of symbolic reality, would be the reverse of the one I envisage here;

it would not be evidence for the material existence of this symbolic reality, that is, for the myth's historicity.

The third part of the case in favour of Gaia's ownership of the oracle consists in the claim that some cultic elements—the chasm and pneuma, the laurel, the omphalos, and the altar of Poseidon, Gaia's alleged husband—are incompatible with Apollo's personality and thus a legacy from Gaia's chthonic oracle.[38] Some scholars claim that the Pythia's sex and the inspirational element in the divination also make better sense as a legacy from a chthonic goddess.[39] These arguments are wrong. First, the long gap in the cult-use of the relevant sites and in archaeologically detectable cult activities precludes any continuity in oracular or other cult practices of the kind they presuppose. Second, the notion of divine personality on which the above theory is based is fallacious. For it ignores the (empirically demonstrable) complexity and ambivalence of divine personalities and the fact that they develop in the course of time and are defined through their relationships with the other deities of the pantheon to which they belong, and with the worshipping group and its (changing) needs.[40]

Thus, the notion that the elements under consideration are 'un-Apolline' is simply a culturally determined judgement, the result of the fact that we have been looking at Apollo's personality and the oracle's early history through a series of distorting mirrors: partly through the perceptual filter of the classical Delphic Apollo's persona, which had developed in response to and interaction with the needs which the god had been called upon to fulfil in the Greek world and is not a good guide to the god's early profile; and partly through the filters created by our own constructs about his early history, which are based on culturally determined assumptions about, for example, what constitutes a logical connection between divine functions.[41]

The study of these elements' cultic history shows that they are not a legacy from Gaia's cult. Poseidon's marriage to Gaia, we saw, is almost certainly a mirage. The laurel is closely and widely associated with Apollo from an early date, and not simply as a result of Delphic influence; in some cults this important aspect of the god's persona is crystallized in his epithet Daphnephoros, Apollo defined as the carrier of the laurel (connected with the laurel from Tempe which had a central part in Delphic myth and ritual).[42]

The chasm with the vapours is a Hellenistic invention, though some, probably small, symbolic, opening of the ground with a stomion is perhaps suggested by Aeschylus, *Cho.* 806–7.[43] Such a small (artificial) opening in the earth would relate the temple's space (which belongs to the human world and to culture) with the inside of the earth with its 'other worldly' symbolic connotations, and thus help put the prophesying Pythia in symbolic contact with the 'other world', situate her between this and the 'other' world, in an appropriate symbolic position for receiving prophetic inspiration from the god. In the classical period at least the opening was not a vehicle of prophecy, nor was it connected with the myth of the discovery of the prophetic chasm, presented as the source of inspiration. For there are no classical references to such a role, and no sign representing, or signalling the presence of, the opening of the ground in the representation of the prophesying Themis (sitting on a tripod and holding a laurel-branch) on the cup Berlin 2538 (*ARV* 1269.5; *Para.* 471; *Add.* 177). More importantly, the notion that the 'chasm' was the source of inspiration presupposes the localization of the consultation at one unmovable, spot. Recent research has led Amandry to doubt the established view that the fourth-century temple had been built over the repaired foundations of its predecessor, and to think that it may have been moved to the north of the earlier temple;[44] this would imply that the opening in the earth—assuming that it existed at that time—was not a particular, special, prophetic chasm located at a particular spot in the adyton; and this fits my interpretation that this opening had simply a symbolic meaning— which was later reinterpreted.

As for the Pythia, Apollo had a female seer also at Didyma, and he was associated with inspired divination also at other oracles; the (well-established) relationship between ecstatic prophetess and god appears to have Near Eastern antecedents.[45] Thus there can be no support for the view that the Pythia's sex and the inspirational element of her prophecy are incompatible with Apollo and must be Gaia's legacy.

The omphalos[46] resembles closely in both shape and associations a particular type of oval stone (an actual example has recently been found) represented on some Minoan glyptic scenes, in which *an oval stone as a cultic object*, decorated with *fillets*, is associated with *eagle-type birds* and a *young male god* characterized by the *bow*. These scenes, together with some others, depict parts of a particular ritual which I examine elsewhere.[47] In my view, the young god involved in this ritual (after undergoing syncretism and changes) contributed significantly to the Cretan component of the historical Apollo's personality. The omphalos, I believe, is one of the elements which Apollo's Cretan component contributed to the Delphic Apollo's persona; the Cretan component entered the Delphic cult (perhaps together with the title Delphinios), probably in the late eighth century, when there were contacts between Crete and Delphi,[48] and the growing Delphic cult and its god were developing in response to the needs they were fulfilling with increasing success, and crystallizing into the main lines of the shape they were to have from then on. The stone's meanings in the Minoan ritual have similarities with and may be the ultimate origin (after reinterpretation and adaptation to fit a different cult nexus) of some of the Delphic omphalos' meanings and associations: the eagles in one of its myths and its funerary connotations—for that Minoan ritual involves death and renewal; it is also connected with hunting, and according to Burkert the omphalos pertains to the hunting ritual horizon, the category of ritual restoration.[49] Be that as it may, as Nilsson noted,[50] Apollo is the god most closely associated with cults involving stones in Greek religion; thus in any case the stone is anything but un-Apolline, and the notion that it is a legacy from Gaia is wrong.[51]

Now the mythological analysis. The myth's earliest known variants belong to the fifth century. In Aeschylus, *Eum.* 1–8 the transfer of the oracle's ownership from Gaia to Themis to Phoebe to Apollo is friendly. In Pindar fr. 55 it is a violent event: Apollo seized the oracle by force, hence Gaia wanted him cast into Tartaros. In Euripides, *Or.* 163–5 the Delphic tripod is referred to as Themis' tripod. (See the cup, c.440, with Themis sitting on the tripod: *ARV* 1269.5; *Para.* 471; *Add.* 177.) In Euripides' *Iphigeneia in Tauris*, 1242–82, Apollo takes over the oracle from Themis by violence and faces Gaia's hostility. In Pindar fr. 55 we are only given the bare structure of the myth. No other figure, apart from Gaia and Apollo, seems to be involved.[52] At this time, the Delphic Apollo is, above all, the (celestial, male) god who establishes order, a law-giver, guide, and purifier. Gaia[53] is a primordial female deity, involved with death, deceitful and threatening, dangerous, representing a stage in cosmic history in which vengeance and not regulated civilized law obtained. She gave birth to various creatures pestering gods and men. She is also a positive nurturing figure, but when contrasted to Apollo, as in this succession-by-conflict schema, she drifts towards the negative pole. The theme 'Apollo replaces another deity as master of the oracle', common to all variants of our myth, is a version of the mythological schema 'divine succession', which is shaped by and articulates social, religious, and intellectual realities and collective representations.[54] In the most potent of the established divine succession schemata, the Hesiodic *Theogony*, as in our myth, a god of the younger generation replaces an older deity. Like the primordial goddesses in the *Theogony*, Gaia is integrated into the new order in a subordinate position. Thus, the Pindaric myth is a sovereignty myth[55] in which the establishment of order is preceded by disorder and followed by the integration of the primordial powers in the new order. Gaia's revenge, also found in the *Theogony*, depends on the fact that she represents a cosmic era in which vengeance, and not regulated civilized law, obtained. The Gaia–Apollo relationship has several meanings in this myth.[56] First, through the defeat of the female primordial goddess by Apollo the law-giver and establisher of order, the triumph of law and order and the Delphic oracle's contribution to it are articulated. Second, this relationship expresses the two deities' complementarity. Gaia's chthonic (including her prophetic) powers are harnessed in the service of Apollo; this is the meaning of the mytheme, and the corresponding cultic reality, 'Gaia's cult continues in a subordinate place at Delphi'. The Gaia–Apollo relationship articulates also certain perceptions pertaining to prophecy which we shall discuss below.

This myth is structured by, and expresses, the perception that at Delphi the chthonic, dangerous, and disorderly aspects of the cosmos have been defeated by, and subordinated to, the celestial guide and law-giver. Apollo's oracle has tamed the darker side of the cosmos—both at the theological (Gaia's defeat) and at the human level: it gives men divine guidance through which they can cope with that dark side of the cosmos. A comparable perception is expressed in the motif 'killing the baneful dragon' in 'Apollo's foundation of the oracle' in the *Homeric Hymn to Apollo*.[57] The motif 'god or hero kills a chthonic monster' is connected with a foundation also in other myths.[58] It represents the establishment of order and the elimination of disorder, evil, and danger to humanity, symbolized by a chthonic monster, a representation of raw nature at its most frightening and savage. Thus the dragon-killing in the *Homeric Hymn* expresses in symbolic terms the significance of the oracle's foundation: Apollo founded it in order to guide mankind, to give laws, and establish order. Consequently, the mythological representation

'Apollo defeats the chthonic and integrates some of its aspects in his cult',[59] contained in the Previous Owners myth, appears in connection with Apollo's oracle already in the *Homeric Hymn*. Moreover, in that hymn, through the dragoness' association with Typhoeus, the last challenger to Zeus's power, the disorder and chaos preceding the oracle's foundation which she represented are symbolically equated with the conditions preceding, and opposed to, the establishment of Zeus' rule. Thus Apollo's killing of the dragon and founding of the Delphic oracle are represented as corresponding symbolically to the establishment of Zeus' reign. The dragon-killing is also a 'replay' or that struggle and victory which ensured that Zeus' order will be served by the oracle.

The Previous Owners myth contains the same symbolic equivalence between Apollo's oracle and Zeus' rule. This equation is earlier than the Homeric hymn. For the mytheme 'Zeus set up the *sema* of his assumption of sovereignty at Delphi' (see appendix at end of this chapter) established a direct association between Delphi and Zeus' triumph over the old order; this was underpinned and strengthened by, and perhaps elaborated under the impetus of, Delphi's central role in promoting order in the Greek world, with Zeus as its ultimate guarantor. It is probably in the context of this elaboration that the 'dragon-killing' motif of the foundation legends was adapted so as to connect the monster with Zeus' enemies. Because it was a monster, it was connected with another monster among Zeus' enemies, Typhoeus; because it was associated with raw nature and, like all challengers to Zeus' rule and their allies, thought of in terms of the earlier gods, it was partly modelled on Gaia, presented as a savage transformation of Gaia: a dangerous death-bringing female monster and (like Gaia) a kourotrophos—of the plague Typhoeus (*Hom. H. Ap.* 353–5). In the Previous Owners myth the earlier order is represented by the older goddesses themselves, so the 'dragon-killing' motif was reinterpreted. The dragon—modelled on the motif 'serpent/dragon as guardian of a spring/sanctuary'[60]—became the guardian of Gaia's oracle, thus making explicit the symbolic equivalence 'Apollo kills the dragon' ≡ 'Apollo takes over the oracle from Gaia by force'; for the violent take-over is focused on the killing of the oracle's guardian dragon.[61] While in the *Homeric Hymn* Apollo creates order out of chaos, in the Gaia myth he establishes a higher type of order, which supersedes that of the primordial goddess. Its symbolic equivalence with the order of Zeus' reign articulates the view that the Delphic oracle has a central role in establishing that order among men.

The fact that the myth 'Gaia as a Previous Owner' contains formal elaborations of motifs and notions which appear in a simpler (and wilder) form in the *Homeric Hymn*'s dragon-killing, and is itself a more elaborate, acculturated, version of that myth, offers support for the presumption, enunciated earlier on, that the Previous Owners myth was later than 'Apollo's foundation of the oracle'.

In Euripides' *IT* 1234–83 Apollo took over Themis' oracle after killing the dragon who guarded it; to avenge her daughter, Gaia sent prophetic night dreams which made Apollo's oracle redundant; Zeus, whose help Apollo sought, removed the night dreams' truthfulness and restored men's confidence in Apollo's prophecies. The revenge and the Apollo–Gaia conflict are also found in Pindar; in *IT* the oracle's owner is Themis, who, though a primordial goddess and Gaia's daughter, is associated with Zeus' order[62] and with Apollo—in myth (*Hom. H. Ap.* 123–5) and personality. Themis, then, was a symbolically mediating figure between Apollo and Gaia. In one variant the oracle passes from Gaia to Themis to Apollo.[63] Its transfer from Gaia to Themis is a transfer from a primordial and often savage goddess to one associated with order and justice; that from Themis to Apollo a transfer to the male (and thus symbolically superior) law-giving and civilizing god of the new order. When contrasted to Apollo, Themis drifts towards her primordial female, older goddess aspect;[64] thus Apollo's ownership is symbolically correlative with the establishment of Zeus' rule.

Given the symbolic correlation between Apollo's Delphi and Zeus' rule (seen already in *Hom. H. Ap.*), Gaia's possession of Delphi after Zeus became sovereign was symbolically unsatisfactory (at that point Apollo had not been born). Thus, when the oracle acquired a pre-Apolline past, the myth created a 'space' for an intermediate figure, defined by the traits (*a*) 'older goddess somehow associated with Gaia' (for the structuring schema was 'Apollo replaces an older goddess', and its established form involved Gaia) and (*b*) figure associated with values pertaining to Zeus' order. This space corresponds to Themis' persona, and, in my view, it is in this context that she became a Previous Owner of the Delphic oracle.[65] This variant stresses the oracle's close association with Apollo and Zeus, and its high claims to justice and order, and thus also its important role in establishing them. In some ways, 'Themis' ownership' can be seen as an elaboration of the formulation in Alcaeus' hymn *propheteu[s] onta diken kai themin*, which describes Apollo's mission to Delphi and expresses the same perceptions of the role of the Delphic Apollo and his oracle.

Given the model of a violent take-over leading to a higher order in Hesiod's *Theogony*, the violent transfer schema was one potential articulator of Apollo's take-over of Themis' oracle (cf. Apollod. 1. 4. 1). But the pull was towards the friendly transfer, with the conflict gravitating towards Themis' mother Gaia. Themis and Apollo were positively related. The myth's structure creates a contrast between them at the same time as it brings out their similarities; but the value of the Apollo–Themis relationship in this myth is also determined through their relationship as a pair to the pair Gaia–Apollo which is their alternative. When related to the Gaia–Apollo pair, the relationship between Themis and Apollo drifts towards the friendly pole, with Gaia–Apollo occupying the hostile one, as in *IT*.

In the *IT* version another set of relationships also comes into play: the pair Gaia–Themis is implicitly compared with, and presented as inferior to, the pair Zeus–Apollo. Zeus is the sovereign, thus his offspring, Apollo, wins. This is one of the myth's meanings. Gaia was a guarantor of the old order, but she is subordinate to Zeus, the guarantor of the new, higher, order and of Apollo's prophecies. The (intertwined) representations 'male is superior to female' and 'the father–son relationship is superior to the mother–daughter one' structure, and are articulated in, this myth. To understand fully the myth's meanings we must consider its dramatic context. It is part of a song praising Apollo at a crucial moment in the action, thus presaging a happy ending, since it suggests that Orestes' doubts were mistaken, and Apollo's guidance was right (see especially v. 1254). Within the song, the Previous Owners myth foreshadows that ending most potently. For it says that Apollo's prophecy is guaranteed by Zeus, which is equivalent to saying that Apollo's prophecy to Orestes was right, that they will be saved. The violent take-over of the oracle in the myth, which led to the establishment of a superior cult, foreshadows—and thus symbolically characterizes, and will in its turn be characterized by—the end of the play: the violent take-over of an especially holy statue and the establishment of a new, superior, civilized cult of Artemis Tauropolos, presented as an acculturated version of the Tauric cult.[66]

Prophecy is an important theme in *IT*, as in the Previous Owners myth. It is mysterious and in some ways frightening (as well as order-creating and helpful); it is also uncertain and vulnerable to misinterpretation. In *IT* these negative characteristics gravitate to Gaia's prophecy, which is defeated in the myth and also proved fallacious within the play—for Iphigeneia misunderstood her prophetic dream (which only told part of the truth); it is also limited, and offers no guidance.[67] In the myth the prophetic dreams sent by Gaia are negatively characterized: they are born of malice, they come unbidden (and are thus not controllable), and they are associated, through language and content, with darkness and night. Thus, in both myth and play, the dark side of prophecy drifts to Gaia, and this allows Apollo's prophecy to emerge as wholly positive. Prophecy's dark side has been articulated, but, because it was attributed to the defeated and superseded Gaia, it has not contaminated Apollo's oracle; on the contrary, that oracle has contributed to the dark prophecy's defeat, and is thus presented as its opposite, strengthened by the other's failure.

This variant, then, was also shaped by, and expressed, a belief in progress—in the cosmos, and in prophecy, the instrument of communication between men and gods. It reaffirms the Delphic oracle's reliability as guide, and emphasizes the association with Zeus and his order, which supersede the darker and more dangerous aspects of the cosmos, as of prophecy. It is a tale of reassurance, faith in progress in the divine order and in the possibility of divine guidance for humanity—through the Delphic oracle. In the play also the reliability of the Delphic Apollo's prophecy—after it had been repeatedly questioned (vv. 78–103, 573–5, 711–15, 723)—is proved; it offered guidance, salvation, and happiness beyond Orestes' expectations and led to the foundation of a new cult beneficial for all time. This focal dramatic strand of the play is condensed, and foreshadowed, in the Previous Owners myth in vv. 1234–83.

According to Aeschylus, *Eum.* 1–8, Gaia gave the Delphic oracle to Themis, succeeded with her consent by Phoebe, who gave it to her grandson Apollo on his birth. That this friendly transfer foreshadows the play's conclusion has been noted by others, as has the passage's relationship with Hesiod's *Theogony*.[68] Since in the early fifth century the established schema for the replacement of a primordial deity by a younger god was the violent transfer of the *Theogony*'s succession myth—through which Apollo replaced Gaia—the friendly transfer variant was perhaps created, in the context of the play's needs and aims, by Aeschylus. This would explain why there is, uniquely in his version, an extra mediating figure, Phoebe, whose close kinship with Apollo allows a friendly power-transfer from an older goddess to a younger god, through the schema 'gift on a special occasion' (compare e.g. Diod. 5. 2. 3). Phoebe is also a representation (in this play where male–female family relationships are an important issue) of a positive relationship between Apollo and the maternal side of his family—perhaps a symbolic counterweight to Orestes' matricide and Apollo's role in it and in its aftermath. The Aeschylean myth's meanings are a more ethical, 'civilized' version of the violent variants, ascribing a higher ethical tone to the oracle (and its god)—again represented as instrumental in establishing order, and symbolically homologous to Zeus' reign of justice.

One Ephoros fragment (*F. Gr. H* 70 F 31*b*) tells us that Apollo and Themis founded the oracle together, to guide and civilize humanity, another (F 150) that Apollo obtained Delphi from Poseidon in exchange for Tainaron. The relationship between the two is unclear (cf. Jacoby, *F Gr. H* IIC, 49). They could be harmonized if Apollo had obtained Delphi as a region (with or without a sanctuary) from Poseidon, and then founded the oracle with Themis. This joint foundation is a transformation of the mytheme 'Apollo succeeds Themis', stressing the two deities' similarity and complementarity. In one story (Paus. 10. 5. 6) Poseidon had owned the oracle jointly with Gaia, who gave her share to Themis, who gave it to Apollo, to whom Poseidon ceded his share in return for Kalaureia. In both versions Apollo obtains Delphi from Poseidon through gift-exchange. Since it characterized Zeus' rule in the *Theogony*,[69] gift-exchange was the most fitting mode of succession in changes of ownership between 'younger gods', especially when, as here, it is differentiated from ownership changes involving symbolically charged generational differences.

Pausanias (10. 24. 4) explains the presence of Poseidon's altar in the temple through his Previous Ownership of the oracle, thus showing that one function of the myth was to explain Poseidon's role in Delphic cult[70] and articulate his relationship to Apollo. The presence of certain significant physical elements and phenomena which belonged to Poseidon's sphere (springs, rocks, and earthquakes) may also have been seen as tokens of that god's claim on the locality. Apollo and Poseidon are antithetical: Apollo belongs to the symbolic pole of culture, Poseidon to that of wild nature;[71] in the Delphic oracle—the myth says and the cult shows—Poseidon and his values are subordinate to Apollo and the Apolline. Poseidon and Gaia are semantically related; their relationship to each other is comparable to that between Apollo and Themis. As a pair co-operating at Delphi, they are opposed to (and the myth of their partnership may have been inspired by) the pair co-operating in the cult of the present: Apollo and Athena, both symbolically opposed to Poseidon[72]—and Gaia. Thus, these variants represent the Delphic oracle as a civilizing centre, in which the 'wilder' deities—and what they represented—were subordinated to Apollo the law-giver and civilizer.

Clearly, the Previous Owners myth, once established, became the vehicle for articulating relationships between Apollo and the other Delphic deities, especially those symbolically antithetical to the order and civilization represented by Apollo; thus, different variants of the Previous Owners myth, expressing different variations of the meaning 'from savage to civilized', were created, by filling the 'wild Previous Owner' slot with different deities.[73]

The mytheme 'Gaia herself prophesied at her oracle' (Paus. 10. 5. 6), and the representation of Themis prophesying on the tripod, connect the Pythia with these two goddesses, ascribe this divination rite to them. This is correlative with, and so articulates and explains, a tension between on the one hand the prophetic ritual's order-creating function and Apollo the civilizing god of order, and on the other a divination rite involving disorder (the Pythia's ecstatic state),[74] a mysterious access to the divine will, a temporary and partial blurring of the limits between mankind and the gods. Like Gaia, the Pythia is an ambivalent female figure who oversteps the normal limits; this, the myth implies, is because she is a legacy from Gaia, but now she operates under the control of Apollo the god of order, who has tamed the previously disordered (and fearsome) divination rite.

Thus, all variants of the Previous Owners myth are shaped by and express positive representations of the Delphic oracle and its god and of the role and nature of prophecy, as well as perceptions pertaining to the ritual and to relationships between deities—and through them also to the Greek conception of the cosmos. The Previous Owners myth, then, which does not fit the facts of, and therefore cannot be explained as, cultic history, makes perfect sense as a myth, expresses and is structured by significant Greek collective representations. In this sense, this myth is 'true'.

APPENDIX

The Omphalos: Some Further Remarks

An important transformation of the Minoan ritual nexus 'oval stone, eagle-hawk, and young god' in Delphic cult is the nexus 'omphalos, eagles, and Zeus'[75] in the story that the omphalos marks the centre of the world, which was determined by Zeus, who released two eagles, one from the East and one from the West, who met at Delphi (cf. Pindar fr. 54). Here the god connected with the omphalos is Zeus; it is therefore interesting that the Minoan god involved in that ritual nexus had contributed—or rather, his later transformations did—to the creation of Zeus' (especially the young Zeus') persona[76] as well as Apollo's. Thus the fact that the Minoan god connected with the stone contributed to the creation of both Apollo and the young Zeus is reflected in the omphalos' association in the Delphic cult of the historical period with both Apollo (the sanctuary's presiding deity in whose adyton the omphalos stood) and Zeus— through the myth of Zeus' eagles.[77]

Zeus is also associated with another sacred stone at Delphi, which, in my view, is another transformation of the Minoan god's stone: the stone swallowed by Kronos which Zeus set up at Delphi as a *sema* (Hes., *Th.* 498–500) when he became the world's sovereign.[78] In my view, this mytheme arose in connection with the stone which (on my hypothesis) entered the Delphic cult as part of Apollo's Cretan component, through the interaction between four elements. First, the Minoan stone's association with the god who had contributed to the young Zeus' persona—which included the myths surrounding his birth and upbringing in Crete;[79] for this brought that stone within the orbit of the mythological nexus of Zeus' birth and its sequel. Indeed, in my view, the motif 'stone swallowed by Kronos instead of Zeus', the second element that went into the making of the mytheme we are considering, was probably itself a mythological transformation of the ritual association between the stone and the Minoan god who contributed to the creation of the young Zeus' persona; for in both cases (in the Minoan ritual and in the Greek myth) there is a symbolic equivalence between the god's symbolic death and a stone. The third element is the fact that Apollo prophesied at Delphi under Zeus' supreme authority, which entailed an association between Delphi and the sovereign god.

Finally, Delphi's identity as a major Panhellenic sanctuary created the symbolic space in which Zeus' victory could be connected with Delphi, made Delphi a plausible setting for the *sema* of Zeus' victory.

All interpretations of the omphalos can be made sense of if we understand it to be one transformation of the Minoan stone (the mythicoritual nexus of which was reinterpreted so as to fit the Delphic cultic context), with Zeus' *sema* being another such transformation. The centre of the world interpretation and the myth of Zeus' eagles can be seen as an elaboration (in interaction with the (reworked) Minoan stone's associations with eagles) of the mytheme 'Zeus set up the *sema* marking his sovereignty at Delphi', which gave a cosmic dimension to the notion of a sanctuary as in some sense a centre of the world[80]—an enlargement underpinned at another level by Delphi's central place in archaic Greece. In any case, in this (centre–eagles) story the omphalos is also a *sema* of Zeus, also connected with his sovereignty of the world—which in the myth he is mapping. The two stones, then, are semantically very close, and this supports the view that they are related transformations of one earlier cult object. The omphalos' funerary interpretations[81] resulted from the interaction between the Minoan stone's funerary connections[82] and the funerary 'spaces' of Delphic myth and cult, which involved Dionysos and the Python. On this view, the Minoan stone gave rise to different cult objects, associated with different mythemes and rituals, through the interaction between on the one hand the mythemes and rituralemes associated with that stone when it entered Delphic cult, and on the other the 'spaces' in Delphic cult and myth, as they were developing in response to the needs which the oracle and its god fulfilled in archaic Greece. Through fission and conflation these transformations were apparently distributed between two physical objects: the omphalos in the adyton and Zeus' *sema*.

NOTES TO CHAPTER IV.2

* I am very grateful to Prof. C. Rolley for discussing this paper with me at great length. The late Prof. H. W. Parke was kind enough to discuss Gaia with me, despite our disagreement.

1. The myth: Aesch., *Eum.* 1–8; Pind. fr. 55; Eur., *IT* 1234–83; Eur., *Or.* 163–5; Ephorus, *F Gr. H* 70 F 31B, F 150; Aristonoos, *Paean to Apollo* (Colin 1909–13: no. 191/iii; Paus. 10.5. 6–7, 10.24. 4; Diod.

16. 26; Plut., *Pyth. orac.* 402C–E; Schol. Eur., *Or.* 164; Photius, *Lex.*, s.v. *themisteuein*; Hypoth. Pind., *Pyth.* a; Apollod. 1.4. 1; Menander, *Rhet. Gr.*, ed. Spengel iii. 441–2; Theopompus, *F Gr. H* 115 F 80; Orph., *H.* 79; Hygin., *Fab.* cxl; Lucan 5. 79–81; Ovid, *Met.* 1. 320–1, 4. 643. Cf. also Plut., *Def. orac.* 414A–B.

2. See e.g. Parke and Wormell 1956: 6–13; Gallet de Santerre 1958: 150–1; Delcourt 1955: 28–32; Martin and Metzger 1976: 15, 28–33; Béquignon 1949: 62–8; Roux 1976: 21–34; Herrmann 1959: 100–16; Herrmann 1982: 54–66; Dietrich 1978: 5; West 1985: 174–5. Sceptical/against: Nilsson 1967: 171–2; Rolley 1977: 137–8; Fontenrose 1978: 1, 4; Amandry 1950: 214.

3. And Fontenrose 1978: 1 noted.

4. Discussion and bibliography: Burkert 1983: 177–8.

5. See e.g. Detienne 1979: 14–16; Loraux 1980: 108–10.

6. See e.g. for Delphi Ch. IV.1 above; Graf 1985*a*: 106–7.

7. The bibliography is vast; most recently Janko 1982: 99–132; Thalmann 1984: 64–73.

8. On which: Page 1955: 246–50.

9. Apollo is the first owner also in Paus. 10.5. 7–8 which gives two versions of Apollo's 'foundation of the oracle': (*a*) it was founded for Apollo by Hyperboreans, (*b*) shepherds discovered it—an alternative to Diod. 16. 26 (chasm taken to be Gaia's oracle). The goat element is probably earlier than the Previous Owners myth. (Cf. below n. 47 and cf. also Apollo's pastoral function.) Paus. 10.5. 7 shows that the goats were not perceived as inextricably bound with Gaia's ownership.

10. A deposit of Mycenaean objects (pottery, a few objects of metal, stone, and glass paste, and about 175 female terracotta figurines and one animal figurine) was found in the archaic sanctuary of Athena Pronaia (Demangel 1926: 5–36); this is a 7th-cent. deposit, probably buried during the construction of the temple (cf. Lerat 1957: 708–10), and made up of the 'holy' Mycenaean objects found by the locals while building and ploughing (cf. Rolley 1983: 113).

11. See Amandry 1950: 205–7; Rolley 1977; 136–7; see also Martin and Metzger 1976: 30–1. On the finds: Lerat 1938: 187–207. The presence of rhyta does not entail a cult-place. Rhyta appear in domestic, funerary, and cultic contexts; on the Mainland most come from graves, a few from domestic contexts; in Minoan Crete large groups of rhyta are found in repositories of cult implements, but in Mycenaean shrines rhyta are rare (Koehl 1981: 179–88). Themelis 1983: 248–50 claims to have identified some Mycenaean capitals, which he assumes to have come from a Mycenaean colonnaded room with a cultic function. The argument relies on unwarranted, mutually supporting assumptions. Even if the objects are (*a*) capitals (which is doubtful) and (*b*) Mycenaean, they cannot support Themelis' claims.

12. So when Herrmann (1959: 100; 1982: 54) states that Gaia's 'Idole' were found, he is completely misrepresenting and distorting the facts. Perhaps the archaic Greeks assumed that these female figurines pertained to a female deity, and so deposited them in Athena's sanctuary; but this says nothing about their Mycenaean significance. On Mycenaean figurines and their function most recently: French 1981: 173–8; 1985: 209–80.
13. Brelich 1968: 924–7. So vague notions such as 'the Great goddess of old Aegean religion' who lived on in Delphic tradition under the names of Gaia and Themis (Herrmann 1959: 100) are entirely out of place.
14. This e.g. is the underlying implication in Roux 1976: 26.
15. See Rolley 1977: 135–8, 142–3; 1983: 109–14.
16. Roux 1976: 23. He takes Gaia to have been worshipped in the later sanctuary of Apollo.
17. On *a-ta-na po-ti-ni-ja*: Gérard-Rousseau 1968: 44–5; Chadwick 1976: 88–9; Burkert 1985: 44, 364 n. 17, 139, 403 n. 3.
18. See also Rolley 1977: 136.
19. As Roux 1976: 25 admits.
20. Béquignon 1949: 66–7. He thinks Mycenaean Gaia also had a small shrine at the site of the later Apollo sanctuary.
21. Cassola 1975: 89. Cf. also Herrmann 1959: 100.
22. See e.g. Gallet de Santerre 1958: 136, 150.
23. See Roux 1976: 25, 29–30.
24. Burkert 1985: 136; against the etymological argument also Chadwick 1976: 86–7.
25. Burkert 1985: 136–8.
26. Cf. Gérard-Rousseau 1968: 184–5; Chadwick 1976: 94–5.
27. The date of the statue bases; on the latter: de la Coste-Messelière and Flacelière 1930: 283–95; Amandry 1950: 208 n. 3.
28. Martin and Metzger 1976: 30 and 33 acknowledge that there is no archaeological evidence to support the priority of Gaia's oracle.
29. Plut., *Pyth. orac.* 402C–D; Bourguet 1932: 25, col. III/A, 3–4; on Gaia's sanctuary: Bourguet 1932: 129 n. 1; Pouilloux 1960: 96; Courby 1927: 183–4.
30. Concise history of the site: de la Coste-Messelière 1969: 730–58. On this point: cf. also Amandry 1981: 677–9.
31. Cf. e.g. Courby 1927: 201; de la Coste-Messelière 1936: 69–72; *contra*: Amandry 1950: 210 n. 2.
32. Some have argued that the bases had been moved there from a different location (cf. short discussion with bibliography: Amandry 1950: 208 n. 3).
33. On this building: de la Coste-Messelière 1969: 734.
34. Cf. Martin and Metzger 1976: 14–15, 28.

35. On Didyma, Claros, and Ptoion see: Parke 1985: 1–111 (Didyma), 112–70 (Claros); Martin and Metzger 1976: 35, 43–53, 53–60; Burkert 1985: 115; Fehr 1971/2: 14–59; Gruben 1963: 78–177; Touloupa 1973: 117–123. At Didyma a laurel + spring combination as at Delphi. The hypothesis (see e.g. Martin and Metzger 1976: 44) that these Apolline oracles' associations with springs are a legacy of earlier Gaia cults replaced by Apollo, for which there is no evidence whatsoever, is another example of the fallacy just discussed.
36. de la Coste-Messelière 1969: 736.
37. On which: de la Coste-Messelière 1969: 736–7.
38. Roux 1976: 25, 28–33, 116; Delcourt 1955: 31, 32, 144; Parke and Wormell 1956: 6–7; Herrmann 1959: 100–16; Harrison 1899: 205–51; Dietrich 1974: 308–9. For Martin and Metzger 1976: 14–15, 28, the 'natural elements, water, tree, animals, chasm', were originally attached to Gaia. On springs see above; on animals nn. 9 and 47; and cf. Apollo's connection with wolves and deer.
39. Parke and Wormell 1965: 10, 12–13; Herrmann 1959: 101 n. 303.
40. See Ch. III above; and cf. Vernant 1974: 105–10; Detienne and Vernant 1974: 176.
41. On this: Gernet and Boulanger 1932: 221–31; see also n. 40 above.
42. See Ch. IV.1 above.
43. If it refers to Delphi, and not, as the scholium (on 806) claims, to Hades. On the chasm: Amandry 1950: 214 ff.; Dodds 1951: 73–4, 91–2 n. 66; Martin and Metzger 1976: 34–8; Fontenrose 1978: 197–203; Parke and Wormell 1956: 19–24; Roux 1976: 110–17; Burkert 1983: 122–3; Price 1985: 139–40.
44. Amandry 1981: 687–9; see also Bommelaer 1983: 193. Against the view (which implies that prophesying is tied up with one spot) that the sekos was rebuilt first, because of the special needs imposed by the cult: Bommelaer 1983: 192–215.
45. See Burkert 1985: 115, 116–17. Dietrich 1978: 5 speaks of contamination between chthonic and Apolline oracles; but this is a simple assumption, based, moreover, on an a priori—and mistaken—construct: it depends on the existence of Bronze Age chthonic oracles, which itself depends on the historical interpretation of the Previous Owners myth and similar legends.
46. On the omphalos: Herrmann 1959; Nilsson 1967: 204 and n. 6; Richards-Mantzoulinou 1979: 72–92; and esp. Burkert 1983: 126–7. A list of representations of omphaloi: Blech 1982: 442.
47. In: *Reading Dumb Images: A Study in Minoan Iconography and Religion* (in preparation). Actual stone found: Renfrew 1985: 102, pl. 7. Scenes: stone + bird (eagle/hawk: not naturalistic, a conflation combining the characteristics of both birds): Sellopoulo ring, *BSA* 69 (1974), pl. 37; Kalyvia ring, *Corpus der minoischen und mykenischen*

Siegel (CMS), II.3, no. 114, in which the stone appears to be decorated with fillets; fillets also on the object in a fresco fragment which may, as Evans suggested, be an oval stone (Evans 1921–35, vol. II.2, 839, fig. 555, and 840). Stone (with pithos and plant) and young male god with bow: ring AM 1919.56, Sourvinou-Inwood 1971a, 60–9, pl. I; the ring's authenticity is now accepted: see e.g. Pini 1981: 147. In the forthcoming book I argue, on the basis of autopsy, microscopic examination of a cast, and the study of many parallels, that the object in the god's other hand is a wild goat's horn. In my view, Apollo's association with goats—cf. Delos *keraton* (Callim., *H Ap.*, vv. 60–4), and the goats in Delphic myth—originated in Minoan Crete, but this is not the place to discuss this question. The Minoan god is also closely associated with a tree in the ritual involving the stone (cf. Sellopoulo and Kalyvia rings)—not a laurel, but a fig-tree (the laurel pertains to the Dorian–North-West Greek component of Apollo. (On Apollo's components see Burkert 1985: 144–5.) These remarks are based on the conclusions of my study of the Minoan ritual, itself based on internal Minoan evidence alone, to the complete exclusion of historical Greek data.

48. Rolley 1983: 110–11; 1977: 145 ff. On Apollo Delphinios: Graf 1979a: 1–22.
49. Burkert 1983: 126–7. For more on the omphalos see appendix above, at end of this chapter.
50. Nilsson 1967: 204 and see p. 202.
51. For Herrmann 1982: *passim*, the tripod originated in the Mycenaean figurines' high-backed three-legged throne/chair, whose occupant he identifies as the Mother Goddess worshipped at Mycenaean Delphi, in myth Gaia–Themis, whom he associates with the Pythia sitting on the tripod. Apart from the implausibility of the identification of the tripod with the high-backed Mycenaean 'throne', Herrmann's reliance on the circular 'Mycenaean Gaia at Delphi' hypothesis invalidates his case. Amandry's suggestion, *REG* 97 (1984), pp. xx–xxi (I owe this reference to Prof. C. Rolley), and esp. Amandry 1986: 167–84, that the Pythia's prophetic tripod (which, he says, had not been seen by the ancient writers and artists who spoke of or represented it) may have been not a proper tripod but something related to the three-legged Mycenaean throne (survival of a tradition or preservation of a relic) is, in my view, wrong. 1) Though the Pythia was probably not in view when prophesying, we cannot know that the part of the adyton in which her tripod stood was not visible at other times. 2) There is no reason to suppose that the description of the instruments of divination would be kept secret, since the consultation procedure was spoken of freely. 3) The Delphian priesthood certainly did know what the prophetic tripod looked like, and it is highly implausible that they

would have allowed its misrepresentation e.g. on coins (e.g. Delphic Amphictyony coinage: Kraay 1976: 122, pl. 22, no. 414: Apollo, omphalos, and laurel, and with them, and thus part of the cult (which identifies it as the prophetic tripod) a normal tripod). (On the Delphic tripod: Burkert 1983: 121–5; Parke and Wormell 1956: 24–6; Roux 1976: 119–23; Willemsen 1955: 85–104.)

52. If Themis was an owner of the oracle in Pi., *P.* 11. 9–10, which is unlikely (the case against: Vos 1956: 62–3 with bibl.), the two versions could be harmonized if in fr. 55 Gaia was, as in Eur., *IT*, avenging Themis. The Gaia–Apollo conflict also in Theopompos, *F Gr. H* 115 F 80.
53. On Gaia: Arthur 1982: 64, 65, 66, 70–1, 76; Nilsson 1967: 456–61; Farnell 1907: 1–28, 307–11; Deubner 1969: 26–7.
54. See Vernant 1976: 23.
55. On Zeus' conquest of sovereignty: Detienne and Vernant 1974: 61–124.
56. Cf. Vernant 1976: 25–6 on Hermes–Hestia.
57. See Burkert 1983: 121; Thalmann 1984: 72; Fontenrose 1959: 13–22, 77–93.
58. Trumpf 1958: 129–57; Vian 1963: 94–113.
59. The monster's rotting corpse gave Delphi the name Pytho and Apollo the epithet Pythian (*Hom. H. Ap.* 372–4) (compare Eur., *Ion*, 989–1119).
60. Bodson 1978: 70 and n. 89 (In the *Hom. H. Ap.* (v. 300) the dragoness was associated with a spring.) In Eur., *IT* the monster is male and Gaia's son.
61. In Menander Rhetor and in Hypoth. Pind., *Pyth.* a, the dragon does not guard the oracle; in the latter it usurps it and in the former it devastates the countryside and keeps pilgrims away. (Cf. Plut., *Def. orac.* 414A–B.)
62. Hes., *Th.* 901–2. On Themis: Vos 1956: 39–78; Hamdorf 1964: 50–1, 108–10; Burkert 1985: 185–6; Harrison 1977: 156–60; Lloyd-Jones 1971: 166–7 n. 23; Nilsson 1967: 171–2; Pötscher 1960: 31–5. On the primordial goddesses' integration in Zeus' order: Detienne and Vernant 1974: 102; Arthur 1982: 65. In my view, Ge-Themis is a later syncretism, Themis was not identified with Gaia in 5th-cent. religion; Aesch., *PV* vv. 211–13 is surely a theological statement similar to Heraclitus' (B15 Diels/Kranz) Hades–Dionysos identification. Perhaps it was inspired—given the mantic context—by our myth, under the impulse of the dramatic context: Themis is Prometheus' mother in *PV* (18 and 874). Her identification with Gaia may depend on Prometheus' ambiguous generational affiliation (Detienne and Vernant 1974: 81–2; affiliated to the Titans in vv. 206–20, while as a Titans' son he should be of Zeus' generation) which it helps to blur.

63. Schol. Eur., *Or.* 164; cf. Paus. 10. 5. 6. Cf. also Aristonoos' *Paean* iii; Photius, *Lex.*, s.v. *themisteuein*.
64. A comparable drift in Kronos' relationships with Ouranos and Zeus: Detienne and Vernant 1974: 101.
65. The connection of *themistes* (On *themistes*: Lloyd-Jones 1971: 6–7, 84) and *themisteuo* (on *themisteuo*: Vos 1956: 20–1) with prophecy enhanced her appropriateness as owner. (Cf. Diod. 5. 67. 4.) But the original meaning of *themistes* was not, as has been claimed, 'oracular pronouncements' (see Vos 1956: 17–22; Themis not oracular before the 5th-cent.: ibid. 62–5; Hamdorf 1964: 51). In Delphic cult Gaia was more important than Themis.
66. In the play the transition from savage to civilized is effected through a movement from a barbarian land to Attica, in the myth through a movement in time and divine generations.
67. In strict logic, since Zeus removed the prophetic dreams' truthfulness, Iphigeneia's dream would be different from those sent by Gaia in 1262 ff. But in symbolic logic they are the same; thus Iphigeneia believes in, and acts on, her dream.
68. Zeitlin 1978: 163–4; Finley 1955: 277; Robertson 1941; 69–70. Cf. also Vidal-Naquet in Vernant and Vidal-Naquet 1972: 154–5.
69. Arthur 1982: 64.
70. Daux 1968: 540–9; Pouilloux 1960: 92–8.
71. Cf. Burkert 1983: 134; Parker 1987: 199.
72. Cf. Burkert 1985: 139.
73. Cf. Plut., *Pyth. orac.* 402C–D: the Muses Gaia's *paredroi* at the oracle. (On the Delphic Muses: Kritzas 1981: 195–209; their dark side: ibid. 209 n. 93.) In Hypoth. Pind., *Pyth.* a, the oracle was owned by Nyx, then Themis, then Apollo, with a separate line of succession for the tripod: first Dionysos prophesied on it, then Python took it over and was killed by Apollo—probably reflecting the tradition that the tripod held the remains of Dionysos or Python (cf. Burkert 1983: 123–5, also on the Apollo–Dionysos relationship at Delphi), reshaped through the Previous Owners schema which was a vehicle for articulating Apollo's relationships with other Delphic deities. Gaia's replacement by Nyx confirms that it was the slot 'primordial, antithetical to Apollo goddess' that was important. Nyx is more negative than Gaia, so the contrast was greater.
74. See also Burkert 1983: 130.
75. Another such transformation may underlie Apollo's close association with a particular type of hawk, the kirkos. (On Apollo and birds: Bodson 1978: 94–8, and Ch. IV.1 above with bibliography.)
76. The Minoan young god is already syncretized as Dictaean Zeus (*di-kata-jo di-we*) in the Linear B tablets of Knossos (KN Fp 1.2). I discuss this syncretism elsewhere (cf. n. 47). For *di-we* see Chadwick,

Killen, and Olivier 1971: 182; Olivier, Godart, Seydel, and Sourvinou 1973: s.v. *di-we* (p. 50).
77. A third god whose persona contained transformed elements of the young Minoan god is the god who became the dying Dionysos, and whom we may call, for convenience's sake, Dionysos/Zagreus. Thus it cannot be excluded that (the dying) Dionysos' association with the Delphic omphalos which is said to be his grave in Tatian, *Adv. Graec.*, may be another transformation of the association between the stone and the gods to whose persona the (transformations of the) Minoan god had contributed, especially since we have seen that the funerary connections of the omphalos correspond to similar connotations of the stone in the Minoan ritual.
78. Unworked wool was placed on it (Paus. 10. 24. 6), as on the omphalos. Frickenhaus 1910: 271–2 saw this stone as the omphalos' 'Vorbild'.
79. On which Willetts 1962: 199–220.
80. See on this Burkert 1983: 127, who notes that the image of the navel expressed anthropomorphically the concept 'centre of the world'.
81. Python's or Dionysos' tomb (references in Parke and Wormell 1956: 14 n. 17).
82. See text above.

IV.3
'Myth' and History: On Herodotos 3.48 and 3.50-53

1. INTRODUCTION AND METHODOLOGY

It is generally accepted that the Herodotean story about the relationship between Periander and his son Lykophron and Lykophron's death and its aftermath told in Herodotus 3.48 and 50-3 contains substantial elements of folk-tale, is more legendary saga than history.[1] However, in my view, the implications of the 'mythological' nature of the narrative for the historicity of the events recounted have not been correctly drawn. Attempts have been made to recover the historical basis of this story,[2] disentangle the historical facts from the 'dramatic embroidery',[3] which are inevitably based on a priori, culturally determined, assumptions. A common—if implicit—central assumption is that a legend of this kind results from the translation of historical events into the 'legendary idiom', and the addition of 'dramatic embroidery'; and that therefore it is possible simply to separate out what seems likely to be the product of such additions, and retranslate back to the historical events reflected in the saga; and then, allowing for political bias, we have access to what actually happened. In my view, these assumptions are incorrect. For—as I hope to show—the way in which the mythological mentality operates on, and manipulates, historical 'raw material' is much more drastic and radical. But in any case, even if these assumptions are not fallacious, since they cannot be proved to be correct, they must not form the basis of an investigation the validity of which depends on their correctness. We must aim at a neutral methodology which does not rely on a priori assumptions and does not deploy culturally determined notions of 'likelihood', 'dramatic embroidery', and the like;[4] a methodology which makes no assumptions about the nature of the narrative and its relationship to historical events, but attempts to reconstruct that narrative's conceptual idiom, the conventions, aims, and modes of thought that shaped it,[5] and to read it through perceptual filters constituted by the ancient assumptions (knowledge, beliefs, and ways of organizing the world and of structuring narratives) which helped create it. Consequently,[6] the first step in the process of ascribing meanings to the narrative and to its constituent elements should be a series of comparisons with other ancient narratives, the meanings and values of which are clearer to us. This is what I do in the second part of this paper.

I will be concentrating on one important category of assumptions which has been largely neglected in the study of texts like Herodotus 3. 48–53, despite its fundamental significance: the 'schemata' which structure ancient texts, and which function as matrices shaping the elements that make up the story. I use the term 'schemata' to denote particular models of organizing experience which structure myths, collective representations, and 'mythological' narratives, and are themselves structured by and express certain beliefs and representations, the perceptions and ideology of their society.[7] An example of such a schema, which, as we shall see, structures significantly the Herodotean narrative, is the one I call 'father–son hostility', which determines that stories involving hostility between father and son always end in some form of (greater or lesser) disaster, expressing and reflecting (among other things) the fundamental Greek representation that a son owes obedience and respect to his father. Since it is the products of the mythological imagination that are structured by such schemata—rather than actual historical events—when (as, I will try to show, is the case with Herodotus 3. 48 and 50–3) a narrative can be convincingly shown to be structured by such 'mythological' schemata, we must conclude that it is a 'mythological' text, articulated by the mythological mentality—even if it presents itself (like Hdt. 3. 48 and 50–3) as the narration of past historical events. This entails that any historical 'raw material' that may have gone into its making has been radically recast; for the mythological mentality, operating through a process of *bricolage*, structures narratives by means of the schemata through which it operates, and articulates 'new' stories using any available historical material as building blocks, to be recast and reshaped to fit the 'spacings' required by the schema.[8]

2. ANALYSES: SON AGAINST FATHER; THE WOLF-MINDED EPHEBE, PUNISHMENT, INITIATION, AND DEATH

Herodotus 3. 48 and 50–3 tells the following story. Periander had two sons from his wife Melissa (whom he had killed), an 18-year-old whose name is not mentioned and a 17-year-old called Lykophron. The two youths went to stay with their maternal grandfather, Prokles tyrant of Epidaurus, who, as they were leaving, asked them ἆρα ἴστε . . . , ὃς ὑμέων τὴν μητέρα ἀπέκτεινε; The elder boy did not understand Prokles' 'riddle',[9] but Lykophron took it to mean that his father had killed his mother, and when he returned to Corinth he refused to speak to his father, who responded by throwing him out of the house. Having reconstructed what had happened by questioning his elder son, Periander instructed the friends with whom Lykophron was staying not to have him in their homes, so that the boy was driven from friend's house to friend's house, until eventually Periander issued a proclamation that whoever received Lykophron in his house or spoke to him made himself liable to a fine to Apollo. As a result, no one would speak to him or receive him, and Lykophron ἐν τῇσι στοῇσι ἐκαλινδέετο.[10] On the fourth day Periander saw him starving and unwashed and tried to persuade him to abandon his defiance and return home, saying that although he was the son of Corinth's basileus he had chosen a vagrant life by opposing his father, and urging him to remember (3. 52. 5), ὁκοῖόν τι ἐς τοὺς τοκέας καὶ ἐς τοὺς κρέσσονας τεθυμῶσθαι. Lykophron did not respond positively and Periander sent him away to Corcyra. But eventually he needed Lykophron as his heir—the elder son not being very bright—and sent for him, but Lykophron did not deign to answer. Periander then sent his daughter, Lykophron's sister, to persuade him. She urged Lykophron (following Periander's instructions) not to allow the tyrannida to pass to others and to return home to take up his rightful place as his father's successor, remarking that many τὰ μητρώια διζήμενοι τὰ πατρώια ἀπέβαλον.[11] Lykophron replied that he would never return to Corinth as long as his father was alive. Periander then offered to go to Corcyra while Lykophron would succeed him and rule Corinth; this Lykophron accepted. But the Corcyraeans, who did not want Periander in their territory, killed Lykophron. Periander punished them by selecting 300 boys from the most important Corcyraean families and sending them to the Lydian king Alyattes to become eunuchs. But the boys were saved: when their ship put in at Samos the Samians urged the boys to take sanctuary at the shrine of Artemis, refused to allow the Corinthians to drag them away from the shrine, and finally saved them from threatening starvation by instituting a new festival involving χοροὺς παρθένων τε καὶ ἠιθέων carrying offerings of cakes, so that the Corcyraean boys might snatch them and thus obtain food. This ritual was kept up until the Corinthians guarding the boys gave up and left, and then the Samians took the boys back to Corcyra.[12]

The first indication that this story[13] does not reflect historical reality is provided by the incompatibility of the chronological indications.[14] If we compress its early part—which will be considered later—the pattern of the story involves a ruler avenging the killing of his son by punishing the community at fault through selecting some of its young people to be taken (by ship) to another country, where they will suffer a horrible fate; the young people are in fact saved and a festival involving adolescents is instituted in connection with their salvation.[15] The pattern of this mythicohistorical/ritual nexus is the same as that of the Minotaur myth, according to which Minos punished the Athenians for the death of Androgeos by imposing the tribute from which Theseus freed Athens—with the Pyanopsia and the Oschophoria, involving adolescents, being associated with Theseus' victorious return.[16]

The similarity between the Lykophron–Periander story and the Minotaur myth is not, it will become clear, accidental; it reflects a similarity in an important semantic dimension of the two stories. (This—empirically demonstrable fact—is to be expected, if it is correct that myths are structured by schemata and 'messages' reflecting important facets of the society's beliefs, realities, and representations.) In the Theseus story the journey of the seven boys and seven girls reflects the structure of, and is the mythical model for, rites connected with the initiatory ritual sphere.[17]

Starting at the end of the Periander–Lykophron nexus, we note that the festival instituted by the Samians is—as has been recognized[18]—an adolescence rite connected with the initiatory ritual sphere. I will now try to show that the Corcyraean boys' story also pertains to the initiatory sphere, that both its general pattern and the motifs that make it up find precise parallels in Greek rites of an initiatory character. First, the pattern, the schema, of the story, corresponds to that of initiation rituals. The correspondence pertains (a) to the relationship of the story's general pattern to the general pattern of initiation rituals:[19] the boys are taken outside the community to be submitted to hardship and danger, and then brought back as a happy ending; and (*b*) it pertains to the particular form which this pattern takes in this story: it corresponds to the journey of Theseus and the Athenian youths and maidens to Crete and back, which, we saw, reflected an initiatory pattern: both groups were taken by sea (at the order of a vengeful ruler) to another country, an island where, after danger and hardship, the threat was overcome and the young people returned to their homeland.[20] The spatial aspects of the journey of Theseus and the other Athenian youths and maidens to Crete and back, and the significance of 'home', 'sea', and '(different) islands' have been analysed by Calame,[21] who argued that Athens, the young people's home community, represented the pole 'civilization' in an articulation of space along the axis civilization/non-civilization, with the sea standing at the 'non-civilization' pole and Crete and Delos being near that of civilization. Theseus' *dokimasia* takes place in a space belonging to the non-civilization pole:[22] the Labyrinth.[23] In the Corcyraean boys' story, Lydia, the truly foreign, eastern, 'other' country corresponds to the position of the Labyrinth in the Theseus story. This is where the boys will be castrated. In 'actual fact', in the reality of the narrative, the projected punishment in Lydia is replaced by a 'trial' at Samos, in a sacred spot belonging to a deity, in a place belonging to civilization. The permanent loss of manhood is replaced by a temporary period of danger and segregation, at the end of which a festival involving youths and virgins is instituted, and the formerly segregated youths return home. The close correspondences between the two stories are clear. The divergences are correlative with the differences in the narrative idiom in which the two are articulated: in one case myth, in the other 'historical narrative' purporting to describe events of the relatively recent past with no supernatural element, no encounter with 'the other world'. The story of the Corcyraean boys has close similarities also with another initiatory nexus, the rite of the Locrian maidens;[24] this also involved travelling (by sea) to a different community, there to be punished in expiation for a crime (in the case of the Locrian maidens a crime committed by the ruler of their community in the remote past). There are, we shall see, further and more specific similarities between the Locrian maidens and the Corcyraean youths.

The second element of the Corcyraean boys' story which belongs to the ritual sphere of initiation is their segregation in the shrine (placed in the 'middle phase' of the story), on a (Greek) island abroad: segregation is a common element in Greek initiation rites and in myths connected with, and refracting, such rites[25]—always in that 'middle phase'. It is found, for example, in the case of the Locrian maidens, of the Brauronian *arktoi*, and the youths and maidens at the Corinthian Akraia.[26] Brelich[27] takes the ritual segregation to symbolize ritual death. The Locrian maidens and the Corcyraean boys share one particular version of that segregation:[28] they both took sanctuary at a shrine outside which they were in grave danger.[29] The fact that the shrine in the Corcyraean boys' story is a shrine of Artemis is significant, since Artemis presides over initiatory rites in many Greek communities.[30] It is interesting to note that in other versions of our story, which do not follow this pattern of segregation in the shrine,[31] the deity involved is—not unexpectedly—Hera, the most important Samian deity, who presides over Samos' most important sanctuary. This detail, and, more generally, the clustering of so many different initiatory motifs—at the same place in the structure of the story as they (or their mythological refractions) occupy in the mythicoritual nexuses of initiations—confirm that the initiatory schema that we have identified in the Herodotean story is indeed significant, and not the result of a mirage and the imposition of modern readings. The third initiatory element is the boys' snatching of food offered to the deity. Stealing and trickery motifs are associated with the period of segregation in boys' initiatory institutions in Greece[32] as elsewhere.[33] The fact that both in the Corcyraean boys' story and in the known initiatory institutions this motif is associated with the period of segregation shows that the similarity is not superficial, and it is not a case of a motif torn from its context and reused. This, then, helps confirm that the Corcyraean boys' story is cast in the shape of an initiatory institution, that, like the myths refracting initiatory rites, it is structured by the schema of adolescent initiations. This interpretation is strengthened by the more specific close correspondence between the particular form of the motif in the Corcyraean boys' story and a ritual motif found in another Greek initiatory complex: the Spartan boys stealing *tyrous* from the altar of Artemis Orthia.[34] The story of the Corcyraean boys involves a complex interplay of motifs, in which the schema 'someone seeks sanctuary in a shrine, his enemies starve him out'[35] has been interwoven with the initiatory ritual schema, and in particular with the versions including the motif 'initiands steal/snatch food', to shape a story which purports to relate events of the not too distant past and cast it into a mythical pattern, articulate it through mythicoreligious schemata pertaining to the ritual sphere of adolescent initiation.

The fourth element pertaining to this initiatory sphere and confirming the above analysis is the castration, the fact that the boys were threatened with loss of their manhood. The Athenian tribute to Minos, the Locrian maidens, and other stories refracting initiation rituals involve threatened loss of life which is interpreted as symbolic death, passage into a new status. The threatened loss of manhood here pertains to the sphere of the symmetrical inversion and to a known mode of dramatization of the transition between childhood and adulthood in initiatory rituals;[36] it corresponds to the ritual motifs of transvestism found in some initiatory rites,[37] and of passive homosexuality, sometimes imposed on the initiands in the Greek world and elsewhere.[38] In these rites, as in the motif 'threatened loss of manhood', the initiands become in some way ritually and/or symbolically non-men[39] before obtaining full manhood.[40] Finally, the association of the Corcyraean story with the χοροὺς παρθένων . . . καὶ ἠιθέων which, we saw, appears to refer to an initiatory rite, adds further confirmation to the above analysis and interpretation: the story which, on this view, is structured by an initiatory ritual schema serves as an aition for an initiatory rite.

We can summarize the Periander–Lykophron story as follows. 1. The son turns against the father for the sake of the mother (at the instigation of the mother's father). 2. The father's response is to throw him out of his house and eventually also (symbolically and socially) out of the community. The youth wanders, rolling about in porticoes, unwashed and starving. 3. The father offers to reinstate the son if he will lay aside his opposition. The son refuses and is banished. 4. The father sends for the son to succeed him, the son refuses. 5. The father sends his daughter (the son's sister) to intercede. 6. The outcome is an agreement by which the son will replace the father, succeed to him as ruler, and the father will replace the son: it is he who will now live outside the community.

Myths are polysemic and operate at different levels. They are shaped by different aspects of the collective representations of the society which generated them—as well as by the mode and conventions of the idiom in which their particular formulations are expressed, and by their relationship to the rest of the artefact (text or image) of which these formulations are part. Thus each version of a myth, besides the basic 'messages' structuring, and expressed through, the common schema articulating all versions of that myth, also has other meanings, special to that particular formulation; and all myths express 'messages' of different kinds at different levels. In studying myths of patricide, matricide, and other parent–child hostility themes, I tried to show how the individual narratives telling these myths were structured by schemata such as patricide, which reflect, are structured by, and thus also express as 'messages', the realities, beliefs, and ideologies of the society which produced them.[41] I tried to show that two important messages shaping, and expressed in, patricide myths are the following. First, the message 'if the father abandons and does not care for his son, disaster will follow', which corresponds to the social reality that, in Athens at least, a child was totally and absolutely dependent on his father's goodwill. The second message is of universal validity—though the way it is expressed is culturally determined, shaped by ancient Greek society and its collective representations: 'for society to go on, sons must destroy (replace) their fathers'. I will return to this below, for patricide is the extreme, polarized form of 'father–son conflict', one variant of which is the theme/schema 'father–son hostility', foregrounded in the Periander–Lykophron story.[42] We shall see below how different (themselves polysemic) schemata interact to structure, and so help ascribe meanings to, this narrative.

The explicit theme of the Periander–Lykophron story, stressed by explicit statements in the narrative,[43] is father–son hostility. This theme is also stressed through statements presenting the notion of inheritance and the father's substance as crucial to the whole situation;[44] for the notion of inheritance, including disordered inheritance, and of replacing the father, is central to 'father–son conflict' themes. Moreover, Periander's daughter's statement to Lykophron (3. 53. 4) that many lost their father's substance out of concern for the *metroia* once again stresses the importance of the notion 'inheritance' in this narrative, and also identifies a locus for opposition to the father, of potentially conflicting loyalties: the mother and the mother's *oikos* which function as a potential locus of conflict also in other father–son hostility stories and in real life.[45]

The father–son hostility theme in Herodotus 3. 50–3 is structured through an established schema reflecting Greek realities and idealities; for this narrative shares a common structure—and several motifs—with other father–son hostility stories, and they all relate in systematic ways to the (related but different) schema/theme 'patricide'. The three stories structured by the same (elementary) schema as that of Periander and Lykophron are the myth of Theseus and Hippolytos,[46] of Phoenix,[47] and of Tenes.[48] The schema structuring all four is as follows: (i) the hostility begins—or is falsely and unjustly considered by the father to begin—with the son and (ii) it is centred on the father's wife. (iii) The father retaliates with an act of hostility. (iv) The son goes into exile; the father does not die, he is only harmed through the loss of his son/heir—which (at least in the case of Lykophron, Hippolytos, and Tenes) the narrative tells us he comes to regret. The variations within this elementary schema are as follows: (i) Lykophron and Phoenix were truly guilty of initiating the hostility against the father and both (ii) did so for the sake of their mother; Phoenix's act of hostility took the form of seduction of the father's concubine (i.e. the father's 'surrogate wife'), so that the motif 'hostility centred on the father's wife' is duplicated here, and the story of Phoenix pertains both to the 'for the sake of the mother' variant to which belongs the Lykophron story, and to the 'real or fictitious seduction of father's wife' variant which is found (in the 'false accusation' version) in the stories of Hippolytos and Tenes. (iii) Lykophron, Tenes, and Hippolytos are banished by their father, while Phoenix chose to exile himself. In the final part of the story we can distinguish two significantly different variants. In the first (Phoenix and Tenes), the son lived on in the foreign land in which he made his home. In the second (Lykophron and Hippolytos), the son dies as a consequence of the conflict and of his father's actions.[49]

The outcome of the conflict—whether the son lives on in a foreign land or dies—does not depend on whether or not the son had been truly guilty of originating the hostility. For Lykophron belongs to the first category[50] and Hippolytos to the second. This (though perhaps surprising to those reading the stories through modern assumptions) is entirely consistent with the mentality and collective representations reflected in this schema, as will become clear below. The individual motifs that make up this narrative are also paralleled in other 'father–son hostility' stories, reflecting the fact (and so demonstrating) that these motifs are meaningful representations in the stories structured by the 'father–son hostility' schema.[51]

In order to protect (as far as possible) the attempt to recover this schema's meanings from the intrusion of our own (culturally determined) assumptions, we will first consider how the 'father–son hostility' schema relates to patricide. First, in both (as in all 'father–son conflict' schemata) some form of disaster follows; this is correlative with the fact that the notion 'son turning against the father' conflicts with the established Greek ideology which decreed that sons owe total loyalty to their fathers, and also conflicts with, and threatens, social order, the social realities in which father and son are interdependent.[52] The 'father–son hostility' schema differs from patricide in the following ways. I. In the former the father does not die; in one of the variants the son dies, and in all the son permanently abandons the home community. II. In patricide the hostility begins with the father;[53] in 'father–son hostility' the hostility begins—or is falsely and injustly considered by the father to begin—with the son. III. In patricide the conflict arises directly out of the relationship between father and son, in 'father–son hostility' it centres on the father's wife who is sometimes also the son's mother (and begins at the instigation of either the mother herself (Phoenix) or of her father (Lykophron)). IV. In patricide the son replaces the father, in 'father–son hostility' the son does not. This last element, combined with the explicit and foregrounded concern with inheritance and the replacement of the father by the son, shows that, like patricide, the 'father–son hostility' schema is concerned with this socially important notion, that for society to go on, sons must replace fathers; but, unlike patricide, the 'father–son hostility' stories involved a failed replacement. This failure is one fundamental common element in all variants of the 'father–son hostility' schema; the other is the role of the father's wife at the centre of the conflict.

In these circumstances, and given that the outcome of the conflict does not depend on whether the son had been truly guilty of originating the hostility (all go into exile and one of each guilty/innocent son lives on in exile and one of each dies), we conclude that, according to the mentality structuring this schema, the succession is disturbed, there is disorder, and the son does not replace the father, when father and son do not privilege their relationship with each other absolutely and above everything else—the emphasis here being primarily on the danger that the son's absolute loyalty to the father may be subverted. This is correlative with the social realities and the collective representations associated with them, with the importance in both of the interdependence between father and son.[54] Patricide stresses the danger that the father may not care for his son; 'father–son hostility' stories deal with potential loci of subversion of this relationship with special emphasis on the danger of the son's loyalty being subverted from the father: one potential locus of subversion, pertaining to the loyalty of the son, is the mother and the mother's *oikos*, which may be subversively privileged by the son, or at least not totally subordinated to an exclusive loyalty to the father.[55] The second potential locus of subversion pertains to both father and son: it is the former's second wife/surrogate wife. As far as the father is concerned, the danger is that he may privilege the interests (in the widest sense, including the children) of his second wife over those of his elder son(s). As far as the son is concerned, the danger is that he will attempt to replace the father sexually by seducing the father's wife.

All these stories, I must stress again, are polysemic, and articulated through the interaction of more than one schema. Thus the Theseus–Hippolytos–Phaedra story is also articulated through the schema 'indirect stepmatricide', which interacts with the 'father–son hostility'. 'Indirect stepmatricide' is a version of 'indirect matricide' with the stepmother replacing the mother;[56] in both variants the notion 'wife's disloyalty to her husband's *oikos*' is an important element—reflecting the fundamental importance of the wife's loyalty and fear of that disloyalty.[57] The motifs making up the narratives (as well as the narratives and the schemata structuring them) are polysemic.[58] Thus, for example, apart from all its other meanings (some of which will be discussed below) the motif 'father casts son outside the community' expressed and refracted representations pertaining to a social fact: the father's real-life power (at least in Athenian society) not to admit his son into his family or to remove an adult son from his house, a move which put the son's citizenship in jeopardy.[59]

I said earlier that the 'father–son hostility' schema expresses the failure of the son to replace the father, a disordered succession; Hippolytos, one of the instances of this schema, is known to be the 'failed ephebe' *par excellence*.[60] It is therefore significant that, in my view, the second schema structuring—and thus helping create the meanings of—the Lykophron–Periander narrative is the initiatory schema. I base this reading of the Herodotean narrative on the following arguments. First, the basic schema of this story corresponds to the basic pattern of Greek initiation rituals. A boy is cast out of the community and goes 'outside', to a foreign land;[61] he is about to return as a full member of his community (to replace his father as ruler) when he is killed. This pattern is the same as that of the initiatory schema, transformed to indicate failure—though not, we shall see, only and always failure. Second, Lykophron's 'outsideness' takes two forms: first social exclusion, placed under the guarantee of Apollo, and thus given a ritual overtone; and then spatial exclusion, when he is banished from the community and sent to a foreign land. During the phase of social exclusion he wanders about without food and unwashed, in a state of abnormality which reminds us of the Spartan youths during the *krypteia*.[62]

My third argument relates to Lykophron's age: when the narrative begins he is 17 years old, an age pertaining to the transition between boyhood and adulthood/full citizen status, and to the institutions associated with this transition.[63] Fourthly, there is the boy's name: he is called Lykophron, wolf-minded. The symbolic connections between the figure of the wolf and the Greek male initiand, the wild adolescent, are too well established to need further discussion here.[64] Thus Lykophron has a name-paradigm for an adolescent initiand, a name appropriate to a mythical prototype of ephebic initiation.[65] Another motif found in this story which also belongs to the initiatory nexus is the 'riddle'; it is the solving of the riddle that sets Lykophron apart from his brother, that triggers off the schemata 'hostility to the father'/'initiation' which structure the story. Answering riddles and generally questions requiring judgement is part of some Greek initiatory institutions.[66] In myth it can be a test to be passed by candidates for kingship.[67] Another argument in favour of the view that the Lykophron story is also structured by, and relates to, the schema reflecting adolescent initiation rituals is provided by the context of the story: as we have seen, the whole nexus of which the Lykophron narrative is part pertains to the ritual sphere of initiation. Moreover, the comparison of the nexus 'Lykophron—Corcyraean youths—Samian ritual' to 'Androgeos—Athenian tribute—Theseus—Athenian festivals associated with Theseus' return' shows that they are similar. In both an initiatory exploit takes place, and an initiatory ritual is founded, as a result of the attempt—by the failed ephebe's father—to extract revenge for the death of a failed ephebe.[68]

The initiatory pattern structures also the other versions of 'parent–son conflict', and this provides further confirmation for our conclusions. First, patricide. We saw that one of the important 'messages' of the patricide schema is 'for society to go on, sons must destroy (replace) their father'; this, it will become clear, is symbolically equivalent to 'successful initiation'. The 'successful replacement' aspect of patricide myths corresponds above all to, and is expressed by, the heroization of the patricides after death.[69] Related to this is the fact that one of the schemata structuring patricide myths is that of initiation: the son leaves the community and goes outside;[70] then he returns and kills his father.[71] That the parent–son conflict schemata should be structured by, and express, the schema of puberty rituals is not surprising. For puberty rituals effect and signal the transition from boyhood to manhood, and one area which is affected by this transformation, and in which the transformation effects itself, is parent–son relationships. It is also an area that lends itself to the generation of symbols to represent social transformation, and the transformation of family relationships, through the manipulation (such as through reversal or polarization) of the 'normal' family relationships. Thus the separation from, and the killing and replacement of, the father express symbolically what the ritual effects, the transformation of the boy into a man. Because in myth the ideal ephebe is usually—like Theseus—the son of a king who will become king, the initiatory schemata structuring such legends relate to both initiation into manhood and succession to kingship; the two processes are condensed into one (complex and polysemic) pattern which functions at two levels. This is one of the matrices determining the symbolism in which 'killing the father' is symbolically equivalent with 'succeeding, replacing, the father' and becomes an image for the successful completion of initiation. Thus the mythological representation 'killing the father' represents the notion 'replacing the father', which in social reality takes the form of 'replacing the father as a citizen', in the sense of first acceding to citizenship and then, when the father is no longer active, or after he has died, replacing him in the strict sense.[72]

Let us see how the 'father–son hostility' schema, which structures the Lykophron story, compares with 'patricide' with respect to their relationship to the initiation pattern. The 'father–son hostility' schema appears in two basic versions when considered from this point of view. In the first (Phoenix, Hippolytos, Tenes) the son has replaced or is deemed to have (improperly) replaced the father sexually, and is driven out not to return, while the father stays where he is. In the Lykophron—Periander story the notion of replacement is expressed in terms of an exchange, by which the father will be replaced by the son at the centre, and the son will be replaced by the father at the 'outsider', so that the son moves from the margins to centrality and the father moves from centrality to marginality. The first corresponds (in historical reality) to the young Athenians' accession to citizenship, the second to the notion of reduced centrality coming with old age, which precedes death. Since acceding to full citizenship (the first stage in the replacement of the father in social reality) entails successful completion of one's 'initiation', there is a symbolic equivalence between the successful completion of initiation and the notion 'replacing the father' expressed symbolically through the motif 'killing the father'. In this narrative, instead of the son killing/replacing the father, son and father agree to change places; at one level, then, this has the same meaning as 'killing the father': the son will replace the father as ruler, moving from marginality to the centre, while the father moves from the centre to marginality. But this does not come about, because the son is killed. This schema, then, articulates a case of failed initiation; and the death of the failed ephebe is made the aetiology for the initiatory rite: the failed initiation causes the creation of an initiatory institution; so at one level the failed initiation does not wholly fail.

However, Hippolytos and Lykophron are not always and only failed ephebes; there were variants of their stories in which their death is followed by some sort of reintegration into another status. In these stories the death of Lykophron and Hippolytos is to be read as the symbolic death of the initiand before the assumption of the new status.[73] Indeed, it is surely a perception of this kind, (i.e. the articulation of the story in this alternative way) that underlies the 'sequels' to the two stories found in some variants. With regard to Hippolytos, one variant (as old as the *Naupactica*) says that Asklepios restored him to life.[74] Significantly, Hippolytos leaves his home community, is not reconciled with his father, and does not in fact replace him.[75] He becomes a full adult member of another community. So this story expresses a notion like that of Phoenix and Tenes: successful adulthood somewhere else—disordered succession. This is one possible mode of failure to replace the father (as a social unit in his own community—in myth, as a king). In my view, some versions of the Lykophron story present the nephew of Periander,[76] who is said to have succeeded him after Lykophron's death, as a double of Lykophron; for he is said to come from Corcyra to succeed to the tyranny,[77] thus making the journey from marginality to centrality which Lykophron had been about to make, to replace his father, when he was killed (and which is correlative with the reintegration of the initiand in his new status). In this reading, the old Lykophron dies to be replaced by the new, adult Lykophron, presented here in the guise of another character, his cousin. The latter is soon overthrown, so in another way he does not successfully replace his symbolic father either. Lykophron, wolf-minded, is the name of a role, that of the marginal ephebe; it is thus conceivable that Lykophron's name 'predicts', is correlative with, the failure of the ephebeia: a young male is not supposed to remain an ephebe, wolf-minded, for ever; he must return to society and become a full citizen. But a youth called Lykophron is 'permanently' characterized as wolf-minded, and this is correlative with his failing to end his ephebeia successfully in the narrative. It is interesting to note in this connection that the form of the 'sequel' to his story, the variant in which—if I am right—Lykophron's death is symbolic and the 'failed ephebe' survives and becomes reintegrated, does not involve (as with Hippolytos) the resurrected Lykophron, but someone with a different name who takes over and slips into Lykophron's place:[78] Lykophron dies and the fully adult Psammetichos succeeds the father/father-figure Periander.

Another version of the father–son conflict is 'indirect patricide' in which some action of the son's causes the death of the father (guilty of an act of hostility against the son at an earlier stage). One example is that of Theseus,[79] the ephebe *par excellence*, who at the end of the Minotaur expedition caused his father's death[80] through his oversight about the sails, and replaced him as king of Athens.

Mythical matricide is also connected with puberty rites. In the ancient collective representations every ephebe had to overcome and 'defeat' the mother, to 'kill her' symbolically, in order to become fully male and join the world of men. The 'message' of matricide corresponding to patricide's 'sons must replace fathers', the representation symbolized in the killing of the mother, is: 'sons must leave behind them the world of childhood and their mothers, in order to become fully male and join the world of men'.[81]

We have seen that Lykophron's exclusion and marginality corresponds to the ritual marginality of the initiand. The modality of this exclusion and marginality is partly articulated by another schema, pertaining to ritual behaviour and sometimes found in 'parent–child conflict' myths, the ritual exclusion of the polluted killer. It is the first stage of the marginality and exclusion, brought about by Periander's proclamation, that involves Lykophron in a state which is the same as that of the polluted killer. The following motifs (and their relationships within the schema which they constitute) belong to this sphere of exclusion resulting from pollution. No one is allowed to speak to him or take him into their house, but (before the proclamation) he is driven from house to house.[82] Eventually, he is sent into exile.[83] Apollo is the 'guarantor' of Lykophron's exclusion.[84] In this context, it seems to me that *ekalindeeto* should also be associated with the pollution nexus and with the use of the form *kylindeomai* in pollution contexts.[85] At the more explicit level of the narrative Lykophron's exclusion is presented as social in character, imposed by the ruler as a punishment[86], for reasons pertaining to the fundamental social relationship between father and son—as well as that between ruler and subject.[87] There is, then, a symbolic equivalence between (*a*) the marginality (presented as social) of a son who has rebelled against, and been repudiated by, his father (corresponding to the overt character of the story as a politico-historical narrative purporting to recount real events pertaining to the life of a ruler of the not too distant past); (*b*) the marginality of the son banished by his father in the myths and schemata of 'father–son hostility' according to which (*a*) is cast; (*c*) the marginality of the polluted killer in the mode of which the marginality of (*a*) is cast; (*d*) the marginality of the initiand which is correlative with and is expressed through (*a*) (cast through (*b*)) in the narrative.

Though various forms of marginality are ascribed to the son in the other 'father–son hostility' myths, that marginality is not articulated through the schema of pollution behaviour, as it is in the Lykophron–Periander story. In function and definition Lykophron's being declared 'to be treated as though polluted' by his father is homologous with the curses uttered by Hippolytos' and Phoenix's fathers against them; both are a form of punishment of the son consisting of a pronouncement by the father which—for one reason or another, and in one sphere or another—had great power.[88] In terms of form, the son's marginality articulated as pollution (because there it is due to pollution) is also found in the myths of patricide and matricide.[89] Patricide involves killing the father, other forms of 'father–son conflict' involve a symbolic form of killing the father—as does initiation with which they are correlative. As we saw, both versions of this mythological image represent (among other things) the social notion of replacing the father, and are correlative with the ritual notion of successful initiation and its mythological refractions. The fact that the polluted murderer schema shapes the mode of the son's marginality in this 'father–son hostility' story suggests that the patricide nexus 'son cast out, son kills father, son polluted' is homologous to the 'father–son hostility' nexus 'son against father, son cast out, father declares him to be like a polluted killer'. To put it differently, the mode of Lykophron's marginality, the fact that it is articulated by the polluted killer schema, suggests that the 'father–son hostility' is thought of as in some sense 'like' patricide. Lykophron is treated as though he had killed Periander. It is possible that one of the 'messages' structuring and expressed in this particular version of 'father–son hostility' is that 'turning against the father' and 'killing the father' are in some ways symbolically equivalent. This would fit with what we have seen above concerning the prescribed behaviour towards the father. That 'turning against the father' is symbolically closely related to 'killing the father' is further suggested by the fact that a 'bracketed' (or censored) version of patricide is also found in other 'father–son hostility' stories. Thus, Phoenix had wanted to kill his father (after the latter had cursed him) but in the end did not.[90] Tenes committed an act of aggression against his father and his ship which became proverbial.[91]

The Lykophron–Periander story as narrated by Herodotus contains also another motif, that of the oracle, which is also found—in different forms—in myths of patricide and in stories (such as those involving the killing of the mother's brothers or the mother's father) which are—among other things—transformations of patricide[92]—that is represent both the relationships with which they are explicitly concerned and also a censored version of the relationship to, and killing of, the father.[93] An oracle had foretold[94] that Periander would not be succeeded by his sons.[95] The whole Lykophron–Periander story is, in a way, crystallized in this formulation which is presented as part of an oracular response;[96] for the notion 'failed replacement of the father by the son' was, in the Greek collective representations, articulated through the schema 'father—son hostility', correlative with 'failed ephebeia', as it is in the Lykophron—Periander story. And this brings me, once again, to the question of historical reference.

3. HISTORY AND MYTH: CONCLUSIONS

These analyses have, I hope, shown that Herodotus 3. 48 and 50–3 is articulated through the same schemata and in the same idiom (involving mythological modes of articulation such as polarizations and symbolic correspondences) as mythological narratives. Since it is the products of the mythological imagination that are structured by such schemata (rather than actual historical events), we conclude that this narrative is a 'mythological' text, the product of mythopoeic creation. The fact that the narrative was structured and articulated through the schemata recovered above entails that any historical 'raw material' that may have gone into its making has been radically recast and reshaped to fit the spacing required by the schemata.[97] The results of the analyses conducted in the preceding section, then, show that it is fallacious to assume (*a*) that Herodotus 3. 48 and 50–3 is basically a 'translation' of historical events into legendary idiom with some dramatic embroidery added, and (*b*) that it is possible to recover the underlying historical events by retranslating the narrative, removing the dramatic embroidery, and purging the political bias of the sources. These analyses cannot prove that the elements considered historical by the exponents of the above approach do not reflect historical fact; but they show that the case for considering them historical rests on flimsy arguments and culturally determined assumptions. First, by demonstrating that, since the narrative is articulated in the mythological idiom, the assumption that it can be considered the product of a translation of historical facts into 'folk-tale idiom' and the addition of some extra elements of 'embroidery' (according to this view easily detectable and separate) is incorrect; and that therefore any study which relies on that assumption is fallacious and likely to corrupt the evidence and lead to incorrect conclusions—even leaving aside the dangers of culture determination in the assessment and 'translation' of the narrative elements. For our analyses showed that elements which may appear historical in origin to us are an integral part of narrative nexuses which are versions of the mythological schemata structuring the text; thus any historical material these elements may have contained has been reshaped and manipulated to fit the role required of them in these narrative nexuses. The second way in which the above analyses deconstruct the 'translation' approach is by offering alternative scenarios which, given the narrative's idiom, appear much more plausible. They show how the alleged historical elements can be more plausibly explained as created out of more generic historical stuff (such as the historical relationship between Corinth and Corcyra) used as raw material by the mythological imagination to fill the 'spacings' created by its mythopoeic activities in which stories are articulated through mythological schemata.

I will now consider *exempli gratia* a nexus of narrative elements generally believed to reflect historical fact:[98] 'Lykophron banished in Corcyra which was subject to Periander; Periander about to retire to Corcyra and hand the tyranny of Corinth over to Lykophron; Corcyraeans kill Lykophron; Periander's vengeance and the projected punishment of the 300 boys'. These narrative elements have been interpreted[99] as meaning that in historical reality Periander had subjugated Corcyra and that Lykophron had ruled for a time in that island.[100] The story of the Corcyraean boys has also often been taken to reflect historical reality.[101] But the analyses conducted in the preceding section have shown us that although these elements may appear (to our culturally determined judgement) unlikely to have been invented, and therefore likely to have been part of the factual core, they are in fact an integral part of narrative nexuses which are structured by, and can be shown to be versions of, established mythological schemata, and thus can most plausibly be accounted for as products of the mythopoeic process. First, as we saw, the element 'banishment in Corcyra' is an integral part of the schemata structuring the narrative ('father–son hostility', 'initiation', 'exclusion of the polluted murderer'), all of which involve the element 'hero cast out of the community'. Moreover, the particular version found here, consisting of banishment in a foreign land, is an established 'casting out' model in the mythological versions of all three schemata. Given these established models, the historical relationship between Corinth and Corcyra (close, but antagonistic, with Corinth hierarchically superior) made Corcyra an obvious selection for filling the spacing 'the outside community' in the narrative. The element 'Periander retiring to Corcyra' is, we saw, an integral and significant part of the 'initiatory' schema according to which son and father change places symbolically.

Let us now consider the notion of Lykophron's governorship of Corcyra and the account of the succession to Periander given by Nicolaos of Damascus. According to this author, Lykophron died in connection with his setting up a tyranny among the *perioikoi*, while the son of Periander who was assassinated in Corcyra just as Periander was going to hand over the tyranny of Corinth to him was called Nicolaos.[102] Periander was succeeded by his nephew (sometimes called Kypselos[103] and sometimes Psammetichos[104]) who came from Corcyra to succeed to the tyranny.[105] The consideration of this account shows that it is basically a variant of Herodotus's story, in a slightly different idiom, but expressing the same meanings: father–son hostility, failed ephebeia and death, failed succession. The notion of Lykophron's *tyrannida . . . para tois perioikois* and that of his exile in Corcyra are homologous; both are articulated by the same three categories: opposition to Periander, marginality, failure of succession. The notion 'tyranny among the *perioikoi*' is not known from anywhere else in Greek history, and it is difficult to see how such a category could have existed. It is, therefore, a narrative construct[106] which in mythological mentality and its symbolic classification expresses the notion of marginality—a marginality correlative with the marginality of the mythological Lykophron in the Herodotean story.[107] The fact that the son who is in revolt against his father is in both versions called Lykophron shows that this homology is real and significant, and that the account of Nicolaos of Damascus partially follows the schemata structuring the Herodotean story (despite the differences in the idiom). Given that Lykophron is the name of a role, that of the marginal ephebe, correlative with the son who is in conflict with his father in the father–son hostility schema, this coincidence of names confirms that these stories are indeed structured by significant mythological schemata.[108] We saw that, when Lykophron's story is taken together with that of Psammetichos' and considered as a myth, Lykophron's death acquires the function and meaning of the symbolic death of initiation: Lykophron dies and the fully adult Psammetichos succeeds Periander, but not for long; so, ultimately, Psammetichos' story also gives us a version of the notion 'failure of succession' which structures the Herodotean story. I will return to this. Thus the figure of

Psammetichos appears to be the result of mythopoeic duplication. A different duplication of the Herodotean Lykophron in Nicolaos of Damascus' account results from the attempt to make sense of the nexus 'Lykophron—hostility to the father—son banished—son replaces father' in historical terms. The nexus was divided into two: on the one hand the rebel son, to whom pertain the name Lykophron, the hostility to the father (translated in the historical idiom), and the marginality; and on the other the son to whom pertains the replacement of the father as ruler. Both die without replacing the father. This mentality, which perceived a mythological nexus through a rationalizing filter, failing to understand that, in the logic of myth, it is the son who is cast out who must replace the father, is not very different from the mentality behind modern arguments such as Berve's (see n. 108). The difference is that the ancient operator[109] had to some extent shared the assumptions and perceptual filters of the mythological mentality which had articulated the earlier versions. Thus, the rationalization of the story[110] is partly structured by the mythological schemata which had shaped the earlier versions.[111]

The fact that, as we saw, the story of the punishment of the Corcyraean boys is structured entirely on the basis of mythico-ritual schemata suggests that it is a mythological invention. With regard to its origin and circumstances of invention, we note that the element 'death of the failed ephebe in the hands of the foreign community' in the Lykophron—Periander story would have triggered off the established schema 'punishment of the guilty community through a group of adolescents who went on an "initiatory expedition" abroad, which was the aition of initiatory rites', and this provided the model for the invention of the Corcyraean boys' story and its structuring as an initiatory 'myth'. What, then, of the form of the punishment, 'castration in Lydia'? Does its specific and bizarre nature not suggest that it reflects historical fact? Far from it. Let us attempt to reconstruct (through a consideration of both the Herodotean narrative and the relevant Greek assumptions) the parameters which determined the creation of the motif 'castration in Lydia as punishment', or rather—to put it neutrally—which determined the form of the Corcyraean boys' punishment in the story. They are the following: (i) the fact that the events are represented as historical reality; possibly (ii) the Greek model of punishment of an enemy community 'selling its males in slavery';[112] (iii) the model suggested by the 'Athenian tribute to Minos' version of the schema: that the youths have to embark on an 'initiatory expedition', go somewhere else to be punished. The combination of (i) and (iii), possibly under the influence of (ii), suggested that the threatened punishment should be something other than death—which could have been inflicted on the spot by Periander himself. The combination of (i) and of another factor resulting from the context, (iv) that the form of the threatened punishment should be significant in terms of initiatory symbolism, and also of (v) the knowledge of the practice of castration associated in Greek representations with the kingdoms of the East[113]—together with (vi) the (greater or lesser) foregrounding of Lydia in the Greek collective representations of the archaic period[114]—can be seen to determine the selection of the form of punishment as 'castration in Lydia'.

Since we have recovered the 'mythopoeic structure' of the Herodotean narrative, let us attempt to ask what it is saying, what are the messages structuring, and expressed in, that narrative. Our analyses revealed the Periander–Lykophron story to be a transformation of an initiation schema (that of failed initiation) into a narrative purporting to recount historical events concerning the succession to a tyranny, with the help of the schema 'father–son hostility'. Both schemata, and the Herodotean narrative which they structure, express the notion 'the son does not replace the father'.[115] The central theme, then, structuring and expressed in the schemata 'father–son hostility' and 'initiation', and also the narrative constituted by the interaction of these schemata, is 'failed succession'. The deployment of the schema 'polluted killer' to articulate the mode of the exclusion of Lykophron confirms the centrality of this meaning: it is because the exclusion will never be reversed and the son will not succeed the father that his exclusion and marginality is articulated in terms of the polluted murderer's exclusion which cannot be reversed.

In these circumstances, would we be justified in concluding that the element 'Periander was not succeeded by his son' reflects a historical fact? Possibly, but not necessarily. For another version of 'failed succession' is 'disordered and discontinued succession'—not in the sense that Periander was not succeeded by his son, but in the sense that the tyranny did not last long after Periander's death, that it was overthrown in the third generation. That this may be the underlying notion is perhaps suggested by the versions in which the nephew who succeeds Periander is presented as a 'double' of

Lykophron. For Periander's alleged nephew Psammetichos is himself associated with another version of 'failed succession'. Thus it is possible that the extant versions of the stories concerning the succession to Periander contain two refractions of the historical fact 'failed succession', tell us the same story twice in different languages: once in the language of historical narrative, in the form 'overthrow and killing of Psammetichos shortly after he succeeded Periander';[116] and once, in the Herodotean version, in mythological language, through the schema 'failed ephebeia', itself articulated through the schema 'father–son hostility', resulting in the ephebe's death—a mythological narrative refracting the historical event 'failed succession in the sense that the son/successor did not last long and the dynasty came to an end'.[117]

The story of the alleged oracle to Kypselos, according to which he and his sons but not his sons' sons would be kings of Corinth,[118] confirms, I believe, the view that in this context the concept 'failed succession' can also apply to the succession by Psammetichos. For the oracle equates 'succession' with 'successful succession'; shows that it is not only the failure to succeed the father that is presented as failed succession, but also the failure to succeed him successfully: the oracle makes clear that the failed succession it purports to be predicting[119] is failed in the sense that it does not last long, that Psammetichos was quickly overthrown.[120] For Kypselos was not succeeded by his son's son only in the sense that the dynasty came to an end in the third generation. That is, the (allegedly) oracular formulation 'but not your sons' sons' is equivalent to 'the dynasty will come to an end in the third generation'. It is this same notion 'the dynasty came to an end in the third generation', that of Periander's son, that is articulated through the mythological schemata (perhaps under the influence of the oracular formulation or at the same time as its creation) in the Herodotean narrative, which expresses the concept 'failed succession' in the Greek mythological idiom.[121]

The exploration of the notion 'tyranny' through the casting of the story into traditional schemata is one of the main structuring themes in the Herodotean narrative pertaining to the Kypselids. For example, one question which is asked and answered in the Kypselid story is 'How do tyrannies start, where do tyrants come from?'; the answer given is that such a transition involved both continuity and change: Kypselos is, through his birth, both central and marginal.[122]

The transition is sanctioned by the oracle who prophesied it and who defined the new situation and set its limits: two generations, ending in the third. The latter element was subsequently doubly articulated in the story: Periander was not succeeded by his son, and his successor did not last long.[123]

In these circumstances, we can see that the narrative about Periander and Lykophron, expressing and structured by the 'message' 'the father is not succeeded by his son', articulates the notion 'interruption of the dynasty', the end of the Kypselids. And this, in my view, is the only element which we can be certain was truly historical: after Periander's death the Corinthian tyranny did not last long. I would not like to commit myself as to the date and authorship of the narrative in which the Kypselid saga was first articulated. But I would not exclude that Herodotus, operating through the same mythological patterns as those which had structured any earlier forms of that narrative which may have existed, played some part in the creation of the version which he sets out in the *Historiai*.[124]

NOTES TO CHAPTER IV.3

* I am very grateful to Prof. W. G. Forrest who read, and commented on, a draft of this essay.
1. The same is true of the earlier Kypselid story told in Hdt. 5. 92. Cf. e.g. Will 1955: 457–9; Aly 1969: 93–5; Murray 1980: 142–3; Andrewes 1956: 52. On the Kypselids, the sources, and their attitudes: Andrewes 1956: 45–53, cf. pp. 108–9; Will 1955: 441–570 (on Periander 502–16); Berve 1967: i. 14–27, ii. 521–31 (on Periander i. 19–25, ii. 525–30); Forrest 1966: 109–22; Murray 1980: 140–7; Salmon 1984: 186–230 (on Periander 197–205). Cf. the brief discussion of the study of folk tales and folk-tale motifs in Herodotean scholarship: L. Huber, 'Nachtwort', in Aly 1969: 317–25 *passim*.
2. Cf. e.g. How and Wells 1936: i. *ad* 50–3. On this question see sect. 3.
3. How and Wells 1936: i. *ad* 50–3.
4. For example, our own notions about what facts can plausibly be reconstructed as the basis of a narrative: e.g. the notion that a given narrative element must reflect a certain historical event, based on the (implicit or explicit) assumption that the most likely explanation for the presence of this element is that it is a transformation into the folk-tale idiom of a historical event which can be reconstructed. Such modern constructs can appear logically satisfactory to contemporary scholars precisely because they are implicitly structured by modern assumptions, reflecting our own modes

of thought. Hence e.g. the privileging of political bias and the propagandistic aspirations of powerful families in the study of oral tradition, because these concepts are nearer our own modes of thought.
5. Whether e.g. it is the historical idiom, determined by the desire to achieve maximum historical accuracy, or the mythological mentality. In my view, the account of Herodotus' approach to history in Hunter 1982: 97–115 underestimates the 'mythological' modes of thought and articulation in Herodotus' narrative, the modes which we associate with myth-telling. In fact, it is now being realized that they inform much non-mythological writing, so that the distinction mythological–non-mythological (historical, scientific, etc.) writing and thought is being deconstructed. I am here concentrating on one set of parameters among those which shaped the Herodotean narrative. Other sets, such as those pertaining to Herodotus' values and the didactic/paradigmatic aims of his stories (which determine his selection of stories and their placing in the narrative) have been discussed by others (see e.g. Heni 1977: 69 and esp. 81 with regard to the Periander–Lykophron story; see also Will 1955: 457–8).
6. Cf. Ch. I above.
7. Cf. Ch. I above. Some schemata are closely related to rituals, reflect the rituals' structure, and express notions similar to those associated with the rituals. Similar motifs may be found in many different cultures (see e.g. Lykophron compared with Hamlet: Aly 1969: 94). What is important, and specific to particular cultures, is the way in which motifs relate to each other to create meanings, the way they are structured. For it is this structuring that reflects the society's mentality. It is clear from all that has been said throughout this volume that motifs and schemata are polysemic.
8. On the concept of *bricolage* see: the writings of Lévi-Strauss, esp. 1966: 16–36; see also Douglas 1975: 170; Derrida discusses critically Lévi-Strauss's concept of *bricolage* and points out that all discourse is *bricoleur*. Derrida 1967: 418–21.
9. A kind of riddle, a question requiring judgement, articulated in a form which can be considered a version of Mastronarde's 'apodeictic questions' (1979: 8–9, a classificatory system developed for tragedy, but also applicable here), a category of 'rhetorical questions'. On 'rhetorical questions' in Herodotus see also Lang 1984: 37–51; Lang p. 81 (see p. 80) classifies Procles' question as an appeal, an utterance aiming at provoking action—in this case making the boys hostile to their father. But such classifications are always problematic and partial, and can only serve as a generic guide; indeed, the dichotomy between rhetorical and 'true', information-seeking, 'literal' questions has been deconstructed by contemporary literary criticism (see De Man 1979; 9–12; Culler 1983: 246–7). All that matters for our purposes is that

Procles' question was a kind of riddle, a question requiring judgement. (On the form of the questions and speeches in Hdt. 3. 50–3, and their formal place in the narrative see Hohti 1976: 28, 96, 107–8.)
10. 3. 52. 2.
11. 3. 53. 3–4.
12. 3. 48. A different version ascribed the boys' salvation to the Cnidians (Plut., *Mor.* 860B). As Salmon (who, however, takes the Corcyraean boys' story to be historical fact) noted (1984: 225 n. 151), Plutarch does not mean that the affair took place at Knidos, but that it was the Knidians who rescued the boys at Samos. Diog. Laert. 1. 94 gives a summary of the Periander–Lykophron story including the Corcyraean boys episode.
13. On Hdt. 3. 48–53 see How and Wells 1936: comm. *ad loc.*; Sayce 1883: *ad loc.*; on Hdt. 3. 48 see Davies 1988: 368–9; Aly 1969: 92–3; Andrewes 1956: 52.
14. On this: How and Wells 1936: *ad* 48–53; Sayce 1883: 251–2 n. 7; Tölle-Kastenbein 1976: 17; Davies 1988: 368. This incompatibility shows that the story is not a precise reflection of historical events; the analyses that follow will, I hope, demonstrate that it is mythological in nature—whether or not it contained recycled historical material.
15. That is, a festival's aition connects it with those young people's salvation.
16. It is irrelevant to the pattern whether and how many times the Minotaur had actually killed previously sent tributes; what matters is the eventual salvation. On Pyanopsia: Deubner 1969: 198–201; Parke 1977: 75–7; Simon 1983: 76–7. On Oschophoria: Deubner 1969: 142–7; Parke 1977: 77–81; Simon 1983: 89–92. On Oschophoria and ephebes: Vidal-Naquet 1983: 164–8. On Pyanopsia and Oschophoria and their relationship to the Theseus legend: Calame 1983b: *passim* and esp. pp. 11–12 (= Calame 1986: 163–76 *passim*, esp. pp. 172–4).
17. The myth of Theseus' and the boys' and girls' journey to Crete, and the defeat of the Minotaur structured as an initiatory myth and the model for Athenian 'initiatory' rites: Jeanmaire 1939: 256–7 and 312–75 *passim*; Calame 1983b: 10–14 (= Calame 1986: 171–6). I include in the category 'rites of an initiatory type' (like Brelich 1969) rites ranging from 'proper' (albeit integrated in divine cults) initiations, such as the *arkteia*, to rites (pertaining to divine cults) which have absorbed and reinterpreted features of initiation ceremonies, and in which the initiatory element is reduced and may reflect one phase of the initiation ritual only (cf. Brelich 1969: 364–5).'
18. Calame 1977: 185–6, 190. In addition it may be noted that the use of the term ἠίθεοι coupled with παρθένοι confirms that the ceremony in question pertains to initiatory type/origin rites (cf. Jeanmaire 1939: 333–4 on ἠίθεοι and esp. on the fact that the term seems to pertain especially to youths participating in rites of adolescence).

19. On which cf. esp. Brelich 1969: *passim*.
20. We know nothing of any Corcyraean festivals associated with the boys' salvation. (Plut., *Mor.* 860B, who (see n. 12) attributes the rescue to the Cnidians, mentions that the latter enjoyed *timai* and *ateleiai* and *psephismata* at Corcyra.)
21. See n. 16.
22. Leaving aside the sea, which pertains to his leap into the sea and is polysemic, with the dominant significance elsewhere.
23. Calame 1983b: 9 (= Calame 1986: 169–70): the Labyrinth is non-civilization within civilization.
24. On Locrian maidens: Graf 1978: 61–79 with bibl. Graf has shown that the textual evidence concerning the tribute of the Locrian maidens retracts an initiatory rite.
25. Brelich 1969: *passim*: see Index s.v. 'segregazione'.
26. On which Brelich 1969: 355–65. On the bears of Brauron see bibl. in Ch. II.3 n. 20; I discuss them in Sourvinou-Inwood 1988a. For segregation in the particular version found here ('abroad', in another community) see (besides the Locrian maidens) Brelich 1969: 30 and 67 n. 51 and pp. 422 and 425–6.
27. Brelich 1969: 359.
28. In both cases the segregation is located in a foreign land.
29. The Locrian maidens from the local inhabitants, the Corcyraean boys from the Corinthian guards.
30. See Burkert 1985: 407 n. 14; Lloyd-Jones 1983: 96–7, 100–1; Graf 1979a: 33–41 (see esp. 41); Vernant 1987: 280–90 and *passim*.
31. Cf. Diog. Laert. 1. 95.
32. On stealing by Spartan boys: see Brelich 1969: 119–20 where the connection between this stealing and hunger is also discussed. On the general trickery of the ephebe: Vidal-Naquet 1983: 162–3, 173–4, 204; see also pp. 27 and 30–1. On ritual stealing in general: Burkert 1985: 410 n. 26.
33. Brelich 1969: 35–6 and 85 n. 101.
34. Xen., *Rep. Lac.* 2. 9. On this passage and its content see Brelich 1969: 134–5 with bibl. *Harpazo* is used for the Spartan ritual, as for the Corcyraean boys by Herodotus. The similarity between the Spartan rite and our story was noted by Rose 1941: 1–5; see also Vernant 1987: 289 n. 58. The boys stealing the cheeses were whipped by another, specially appointed, group. On the ritual stealing of the *tyroi* from Artemis' altar see Nilsson 1906: 193–4; Burkert 1985: 152, 407 n. 32, 262, and 449 n. 18; Vernant 289–91; Graf 1985a: 88; Parker 1989: 148, 167 n. 33. Though we know the date given as a *terminus post quem* by the *aition*, which is attested in a very much later source, we do not know at what date the ephebes' flagellation at the altar was associated with a *pompe Lydon*, a procession of Lydians, which

Plutarch tells us (Plut. *Arist.* 17. 8) follows after the ephebes' beating at the altar, a *mimema* of an event which allegedly took place at the battle of Plataea: while Pausanias was sacrificing a little to the side of the battle line, some Lydians fell upon him and disrupted the sacrifice, *harpazein kai diarriptein ta peri ten thysian*; Pausanias and his entourage resisted: as they had no weapons they hit the Lydians *rabdois kai mastixi*. The meaning of this ritualeme is not unproblematic, and it is beyond my scope to consider it here. What is interesting for us is that in it 'Lydian' is a symbolic counter deployed in, and helping to articulate, a male initiation rite. On the analysis proposed here 'Lydian' also functions as a symbolic counter in the Herodotean story, which on my analysis is a 'mythological' narrative structured through 'initiatory' schemata—though in this story the symbolic counter 'Lydian' operates in a different modality from that in which it operates in the Spartan rite (on this rite see Nilsson 1906: 194–5; Graf 1985*b*: 86–9; 1986: 43–4).

35. Cf. e.g. Thuc. 1. 134–1–3.
36. On this see Vidal-Naquet 1983: 163–4, 168–9.
37. See ibid. 163–4; Brelich 1969: 72 n. 60, 164 n. 156; Jeanmaire 1939: 351–5.
38. Brelich 1969: 35 and 84 n. 100, pp. 120–1; Bremmer 1980: 279–98; 1989: 3–7; cf. also pp. 10–11.
39. In the Herodotean story which presents itself as the recounting of historical events of the not too distant past, the loss of manhood is symbolic: they are destined to become permanently non-men, but are saved.
40. The element 'boys from the best families' is itself a selection found in initiatory rites in which only a small number of boys and/or girls take part as representatives of their generation. (Cf. e.g. Brelich 1969: 358, 360, with reference to the Corinthian festival Akraia; initiation belonging to the élite in Greek rituals: Bremmer 1978: 18.) '300 boys from the best families' may be the result of a conflation between the concepts 'boys from the best families' and 'a whole year-crop' (or 'initiand crop' if the dominant model shaping this story's creation involved a non-yearly initiatory institution); this selection was perhaps determined by narrative considerations, the desire to indicate that the punishment was severe. That the number 300 may have had—at least in some places—some significance in connection with initiatory type practices is perhaps suggested by the Spartan 300 *hippees*, young élite warriors closely connected with the *agoge* (see Brelich 1969: 121–2, 155; Vidal-Naquet 1983: 201, 215).
41. Sourvinou-Inwood 1979: 14–15 and 65–6 nn. 62–78. In order to avoid, as far as possible, the intrusion of culturally determined assumptions, the analyses of the myths and recovery of the 'message'

structuring them must be conducted separately from the attempt to reconstruct the beliefs and collective representations (as revealed in law, in customs, statements, etc.) which will be compared to the recovered 'message'. If the results of the two separate studies converge the validity of the recovery is confirmed.

42. I call 'father–son hostility' this variant (described below) of the theme 'father–son conflict' (of which patricide is a more polarized version). The theme 'father–son conflict' is itself one variant of 'parent–son conflict' (on which see my discussion in Sourvinou-Inwood 1979).
43. See e.g. Periander's statements in 3. 52.
44. Periander to Lykophron in 3. 52. 3–4 and Periander's daughter to Lykophron in 3. 53. 4.
45. The formulation in Hdt. 3. 53. 4 stresses that this is indeed the case, for it is presented in the guise of a *gnome*. On *gnomai* in Hdt. 3. 50–3 see Will 1955: 452; Heni 1977: Index s.v. 'Gnomen'.
46. See my remarks in Sourvinou-Inwood 1979: 12. On the myth and its sources see Barrett 1964: 6–45.
47. *Il.* 9. 447–80. The emphasis here is on the son replacing the father as a sexual male, rather than as a figure of authority and power.
48. See Apollod., *Epit.* 3. 24–6; Paus. 10.14. 1–4; Konon, *F Gr. H* 26 F 1 (xxviii); Tzetzes, Schol. Lycophr., *Alex.* 232; Diod. 5. 83; Steph. Byz. s.v. Tenedos; Plut., *Quaest. Graec.* 297E–F; Suda, s.v. *Tenedios anthropos*; see Frazer 1921: ii. 193 n. 2. See also Luppe 1984: 31–2; and my remarks in Sourvinou-Inwood 1979: 66 n. 70.
49. Theseus causes Hippolytos' death directly, through a curse, Periander causes Lykophron's death indirectly, through his decision.
50. As far as the relationship between father and son is concerned, Periander's manslaughter of Melissa cannot, in Greek mentality, be considered an act of hostility against Lykophron.
51. Individual motifs of the Lykophron story not already mentioned which also appear (in the same or a variant form) in other 'father–son hostility', and in general 'father–son conflict', stories are: (*a*) the father attempts to conciliate (Lykophron, Tenes, Althaimenes, Katreus); (*b*) the sister plays a role (Lykophron, Tenes (his father cast him and his sister away together); in patricide: Althaimenes kills his sister who accompanied him). I should mention that there are also other variants of the 'father–son hostility' schema, structuring other stories, which have greater or lesser similarities to the variant discussed here. In one version of the myth of the Phineids—on which see Bouvier and Moreau 1983: 5–19; Preller–Robert 1921: 817–21; Pearson 1917: 262–7 (*Tympanistai*) and 311–20 (*Phineus* A and B); *Tr. GF* iv. 458–61 (*Tympanistai*); 484–9 (*Phineus* A and B); Kamerbeek 1978: 16; Frazer 1921: *ad* Apollod. 3. 15. 3—the two youths were blinded by their father, acting under the influence of the calumnies of their stepmother,

who had accused them of sexual advances or interference (cf. Schol. S., *Ant.* 981; Apollod. 3. 15. 3, 1. 9. 21; see Bouvier and Moreau 1983: 8–10, 11–12). This version has certain similarities with the stories of Hippolytos and Tenes. The motif of the slandering stepmother turning the father against his sons is also found in one version of the Oedipous myth in which Oedipous threw out Jocasta and married Astymedousa who slandered her stepsons and led to Oedipous' cursing them. (Eustath. *ad Il.* 4. 376, van der Valk 1971: 767, ll. 24–7; see Bouvier and Moreau 1983: 16–18.) Another variant schema, which, like the one discussed here, involves the death of the son, structures the story of Korythos. According to Konon, *F Gr. H* 26 F 1 xxiii, Paris killed Korythos, his son by Oinone, out of jealousy over Helen; Korythos had been sent to Helen by Oinone precisely because he was more beautiful than his father and she was hoping to arouse Paris' jealousy and make trouble for Helen.

52. On the son's dependence on the father see my remarks (1979: 15). The son's duty of total loyalty, obedience, and support to the father is correlative with the father's dependence on the son, and hedged with many safeguards and obligations. On father–son relationships: Dover 1974: 273–5; Harrison 1968: 70–8. Horror of patricide: Parker 1983: 124. On patricide and matricide see also Burkert 1983: 74–5.
53. See my discussion (1979: 14–15).
54. This remained unchanged in Greece from the societies whose realities and representations were reflected in the Homeric epics to the classical polis.
55. See Periander's daughter's comment. Lykophron's turning against his father because the latter caused his mother's death is in some ways the mirror-image of Orestes' and Alcmaeon's murder of their mothers who had killed their fathers. But while privileging the father in this way is ultimately 'acceptable' (see my remarks, 1979: 8–12, on matricide), privileging the mother is not. (Besides, for the father to kill the mother is an entirely different proposition from the fundamentally subversive motif of a woman turning against her husband, see ibid.) Thus Orestes who actually killed his mother eventually returned to normal life, while Lykophron who did not kill his father died.
56. See my remarks, 1979: 12. Stepmatricide also expresses, of course, meanings and representations pertaining to the stepmother herself.
57. See my discussion, 1979: 9–11.
58. Thus the motif of the seduction, real or fictitious, of the father's second wife/surrogate wife expresses a variety of representations (which interact with the other elements of the narrative to create meanings): 1) Son protecting mother's/dead mother's interests 2) Son replacing father in the social sense 3) Son replacing father sexually—a motif including the father–son rivalry: between the father/older generation

with power and authority and son/younger generation with sexual potency and attraction. (This may include a censored representation of some form of Oedipous complex if such a thing exists outside a particular culturally determined horizon.) 4) The woman is a locus/cause of evil, disorder, and menace 5) A father's second marriage threatens to damage the interests of the son(s) of the first marriage.

59. See on this power of the father my brief discussion with bibl., 1979: 15 and 66 nn. 70–4.
60. See e.g. Vidal-Naquet 1983: 172–3; Tyrrell 1984: 85.
61. We saw that a foreign land can play the role of space 'outside the community' in initiation. As we shall see, the location of the 'place outside the community' (here the place of banishment in the context of the father–son hostility schema) in Corcyra depends on the interaction between narrative needs and historical circumstances. In patricide myths also the 'outside' is situated in a 'foreign' (Greek) land: for Oedipous it is Corinth that is 'outside' and Thebes the home community; for Katreus Crete the home community and Rhodes the outside.
62. On which Jeanmaire 1939: 550–4; Brelich 1969: 155–7; Vidal-Naquet 1983: 201–4.
63. In Athens a 17 year old is at the stage after the *koureion* sacrifice and before the ephebic oath and his inscription in the deme register at 18 (see Labarbe 1953: 358–94; Vidal-Naquet 1983: 155; Brelich 1969: 216–17, (my views on Brelich's treatment of the ephebeia are to be found in *JHS* 91 (1971), 174)).
64. See Jeanmaire 1939: 550–88 and esp. p. 554; Vidal-Naquet 1983: 201; Burkert 1983: 84–93, esp. pp. 89–90. Jameson 1980: 231–3 notes that in Athens Apollo Lykeios represents the completion of the initiatory process, since he is the god associated with the adult males. On the wider question of the perceptions, and symbolic values, of the wolf in Greek thought see Buxton 1987: 60–79.
65. The name Lykophron, of course, existed outside, and prior to, this story (e.g. *Il.* 15. 430–2: it is interesting that this Lykophron had left his homeland because he had killed a man), but its importance here lies in the fact that it is correlative with the many other elements characterizing Lykophron as an adolescent initiand. That names are used significantly in this story to express social roles is also suggested by the fact that Periander's wife (about whom Hdt. 5. 92 tells the story of Periander's necrophilia and her apparition) is called Melissa; bee is a Greek image for the ideal (from a male viewpoint) woman/wife—Detienne 1981: 101; Lefkowitz 1981: 36, 73; Parker 1983: 83. The role-significance of this name is made even clearer in Diog. Laert. 1. 94 who tells us that Periander's wife's name was Lyside but Periander called her Melissa. (There are, of course, significant names

also in the earlier part of the Kypselid saga, such as Labda and Eetion (see on these Oost 1972: 17; Vernant 1982: 19–38). On Melissa and her sons see also Berve 1967: 1. 526.
66. See e.g. Plut., *Lyc.* 18. 2; see Brelich 1969: 124.
67. See e.g. the Oedipous myth; esp. the version in Paus. 9. 26. 3–4 (see also Vernant 1982: 24). Periander's elder son fulfils an 'acting' role required by the plot, that of informer to Periander about what Procles had said. But at another level he expresses another version of failure to succeed one's father (in a variant in which the latter holds high hereditary office)—through lack of required natural abilities. Will 1955: 458 n. 5 failed to understand the elder son's significance because he tried to make sense of him in rationalizing terms.
68. Androgeos as a failed ephebe: Jeanmaire 1939: 340–3. Initiatory rituals connected with the killing of the mythical prototypes of the segregated initiands: the Akraia festival at Corinth (on which Brelich 1969: 355–65 *passim*, esp. p. 365). More generally on rites instituted in honour of prematurely dead young heroes: Jeanmaire 1939: 341–2.
69. See my articulation of the myths with references, 1979: 14. The motif is, of course, polysemic.
70. In two out of the three cases of patricide not as an adult: Telegonos is born elsewhere and Oedipous was taken elsewhere as a baby after he had been exposed. Variations of this kind result from the interaction with other schemata and/or other demands of the particular narratives.
71. This happens in the case of Oedipous and Telegonos. In Katreus' story (Katreus left home as a young man) his father goes to meet him outside the home community (he lands on Rhodes where Katreus has settled and it is there that he is killed). The initiation pattern interacts with other schemata in 'father–son conflict' stories, and this is correlative with the fact that the replacement/non-replacement of the father by the son is only one of the representations structuring (and thus are also messages expressed by) these stories. The particular demands of the patricide schema made such stories transformations—rather than precise reflections—of the initiatory schema. The two variants of the son's departure from, and return to, the home community are: either the son is a youth when he leaves the community, or he returns and kills his father.
72. The psychological dimension in patricide and generally themes of son against father conflict do not concern me here.
73. On this symbolic death see Burkert 1983: 91; Brelich 1969: 33–4, 36, 359, 376, 386.
74. Kinkel fr. 11; Apollod. 3. 10. 3. On this variant see also Frazer 1921: ii. 17 n. 4.

75. See Frazer 1921: ii. 17 n. 4.
76. Sometimes said to be called Kypselos, sometimes Psammetichos. On the significance of the name Psammetichos: Berve 1967: i. 21, ii. 527; Salmon 1984: 225–7.
77. See Nicolaos of Damascus, *F Gr. H* 90 F 60; see also F 59: the son of Periander killed by the Corcyraeans is called Nicolaos while Lykophron is said to have set himself up *tyrannida . . . para tois perioikois.* On this see below.
78. The fact that resurrection was not possible in this everyday-world variant of the schema is an important factor in the 'selection' of this form.
79. I have discussed this elsewhere (1979: 21–2, 28).
80. See Sourvinou-Inwood 1979: 21, 22–8, on Aigeus' acts of hostility against Theseus. The Theseus legend can be said to be exploring all possibilities pertaining to the ephebeia; through the figure of Androgeos (cf. Apollod. 3. 15. 7. Paus. 1. 17. 10; Jeanmaire 1939: 340–3) the failed ephebe (see n. 68), whose death was (according to some versions) caused by Aigeus and in whose honour the initiatory institution around the Minotaur was instituted: failed ephebeia [according to Propert. 2. 1. 64 Androgeos too had been restored to life by Asklepios—so that in this version (the date of which is not known), as with Hippolytos, his death is the symbolic death of initiation. In connection with this it should be noted that in Apollod. 2. 5. 9 Androgeos has had sons; also, of course, that Androgeos' brother Glaukos was also brought back from the dead]—redeemed by Theseus' ephebeia *par excellence*, which in some respects duplicated successfully Androgeos' (see e.g. the version according to which Androgeos was killed by the Marathonian bull, the fate Medea and Aigeus had in some versions intended for Theseus— but Theseus won). Androgeos can be read as a double of Theseus (Jeanmaire 1939: 340). Aigeus as Androgeos' 'surrogate father' causes his death; Minos as Theseus' 'surrogate father' and Aigeus, his 'real' father, try to cause Theseus' death and fail; Theseus causes his father's death and succeeds him at the end of his final initiatory exploit. So failure and success in the ephebeia and in the replacement of the father by the son are both explored in the Theseus legend—as befits the myth of the Athenian ephebe *par excellence* in whose person ephebic representations were crystallized. Of course, the dominant pattern is that of success: the son succeeds in his exploits and is not killed, he 'kills' and replaces the father.
81. On the relationship between matricide and the initiatory pattern in the Oresteia see Zeitlin 1978: 149–84. (On Orestes' connection with puberty rites *passim* and 176–7 n. 13.)
82. As Stein 1883: *ad* iii. 52 and Parker 1983: 125 and n. 83 note, Oedipous' proclamation against Laios' killer in Soph., *OT* 223–41,

is very similar to that of Periander against Lykophron. This similarity has not hitherto been placed, as I think it should, in the wider context of 'father–son conflict', the basic schema to which both Oedipous' and Lykophron's stories belong. Vernant 1982, who compared the Kypselid with the Labdacid saga, was concerned with general themes and with the whole sagas, and thus did not place the significant similarities between Periander and Lykophron on the one hand and Laios and Oedipous on the other within that particular context that ascribes them meaning. On the proclamation against a killer (deemed to be polluted) see Parker 1983: 125–6.

83. See *OT* 228–9. On polluted murderers going into exile: Parker 1983: 118.

84. For in this context of pollution behaviour, this must be connected with Apollo's function as the god concerned with pollution *par excellence*; Apollo is probably polysemic here, pertaining to the different levels of the story's signification and structuring schemata, especially since he also presides over boys' initiations (see Burkert 1975: 1–21; 1985: 144–5). On fines to gods and the ways they function in Greece: Latte 1920: 48–61 and *passim*.

85. See *Il.* 22. 414, *kylindomenos kata kopron* (see also *Il.* 24. 165): mourning behaviour of males which (I hope to have shown (see Sourvinou-Inwood 1983: 37–9)) pertains the 'taking on' of pollution by the grieving relatives in the first part of the death ritual, immediately following the death.

86. Parker 1983: 194 notes that in the Herodotean story Periander is using excommunication like that of Oedipous' proclamation as punishment.

87. This pollution motif in some way also takes up Lykophron's behaviour towards his father as towards the killer of kin. Parker (1983: 122–3 nn. 71 and 77) connects Lykophron's not speaking to his father (after the incident with Procles) with the behaviour pertaining to the murder of kin, according to which the murderer of a member of the family must not associate with its other members. Aly 1969: 95 takes Lykophron's not speaking to his father as showing the son behaving towards the father as to an excommunicated murderer, and considers Periander's excommunication of Lykophron as simply the answer to the latter's behaviour.

88. On the father's curses having great power: Parker 1983: 192–7.

89. Pollution motif in patricide: Parker 1983: 124; see also my remarks, 1979: 65 n. 67. On the pollution motif in matricide: see briefly with references, Sourvinou-Inwood 1979: 8–9 and 62–3 *passim*.

90. *Il.* 9. 458–61. These verses may or may not be Homeric (see on this Erbse 1971: 498 with bibl.); this is not important for our purposes, since, within certain parameters of fluctuation (and we are not here concerned with the motif in sufficient detail for that fluctuation to

affect the argument), the schema does not change over the period of time involved. The controversy is, in my view, itself significant, but it is beyond my scope to discuss it here.
91. See Paus. 10. 14. 3; and my remarks 1979: 66 n. 70.
92. Their relationship to patricide is comparable to the relationship between stepmatricide and matricide.
93. On a child's relationship to the mother's brother and father see Bremmer 1983: 173–86.
94. Hdt. 5. 92.
95. On the motif of the oracle in the patricide schemata see my articulation, 1979: 14–6; in 'killing of the mother's brothers' combined with 'averted matricide' (Telephos): ibid. 13; in 'killing the mother's father' (Akrisios): Apollod. 2. 4 and Frazer's comm. (Frazer 1921: i. 153 n. 3). In stories of patricide and the like, what is predicted is the successful—if violent—replacement of the father/father-figure by the son. In the story of Lykophron what is predicted is the failure of the son to replace the father. The motif of the inescapable oracle in the former case represents (among other things) the notion that it is inescapable that fathers die and are replaced by their sons.
96. See also below.
97. On the process of *bricolage* through which the mythological mentality operates see above n. 8.
98. Further examples: How and Wells 1936: *ad* 50–3, suggest that the historical facts in 3. 50–3 are the following: Periander killed his wife; he conquered his father-in-law Procles; he reduced Corcyra to subjection; (probably) Lykophron ruled for a time in Corcyra; Periander left no son to succeed him, his successor was the son of his brother. But our uncovering and articulation of the mythopoeic structure of the narrative leaves very little of the above that has not been shown to belong to that structure, and thus to be explicable as an 'invention' in the framework of the mythopoeic process of the creation of that narrative. (On Lykophron's rule see below.) For all we know, Periander may have conquered Epidaurus, but the notion that Procles was his father-in-law is dependent on the motif 'Periander's hostility to his father-in-law' which was demanded by this version of the 'father–son hostility' schema. It cannot be excluded that this was combined with a historical element 'conquest of Epidaurus', which, however, may equally well be wholly invented; it cannot even be excluded that the ruler of Epidaurus had indeed been truly Periander's father-in-law; but the point is that the presence of the element 'conquest of Epidaurus and defeat of his father-in-law Procles' in the narrative can be explained plausibly in another way than through the assumption that it was historical truth; since its invention can be motivated in the context of the articulation of the

'father–son hostility' schema, there is no reason for thinking it reflects a historical event. The same is true of Periander's alleged murder of his wife: her name, we saw, suggests a role designation, and the murder of Lykophron's mother can be motivated in the father–son hostility schema. So the story of her murder cannot be assumed to be historical fact. (Paus. 2. 28. 8 saw the grave of Melissa and of Procles at Epidaurus but this does not tell us anything about the date of the story that invented them.) Berve 1967: i. 20 considers it to be a historical fact that Lykophron, incited by Procles, set up his own short-lived *tyrannis* in a peripheral area. On Lykophron's rule see below. Mossé 1969: 35 thinks that 'de sombres histoires de famille' did mark the end of Periander's tyranny. Salmon 1984: 197–205 believes most of the legend to be historically based.

99. On the (explicit or implicit) basis of this 'translation of historical events into "novelistic" idiom' approach, in symbiosis with a culturally determined judgement as to what is the most plausible explanation of an element.

100. Usually this is formulated as 'a son of Periander was governor in Corcyra', to take account of Nicolaus of Damascus who gives the name of Periander's son in Corcyra as Nicolaos (see Jacoby, *F Gr. H* IIC, p. 249; Sayce 1883: 253 n. 4; see Aly 1969: 94). This view is taken by, among others: Will 1955: 521; White 1979: 191–2; with a qualification such as 'probably' about Lykophron's/Nicolaos' governorship: How and Wells 1936: *ad* 50–3; Aly 1969: 94; Salmon 1984: 222; Berve 1967: i. 21.

101. See e.g. Andrewes 1956: 52; Berve 1967: i. 21.

102. See Jacoby (n. 100), p. 249; Sayce 1883: 253 n. 4; Aly 1969: 94. On Nicolaus of Damascus see most recently Toher 1989: 159–72.

103. In *F Gr. H* 90 F 60; in Diog. Laert. 1. 94 Kypselos is the name of Lykophron's elder brother who is anonymous in Herodotus.

104. In Nicolaus of Damascus *F Gr. H* 90 F 59; Ar., *Pol.* 1315^{b}27; see n. 76 on the name Psammetichos.

105. *F Gr. H* 90 F 60. Taking the accounts of Herodotus and Nicolaus of Damascus together, some scholars have surmised that the governor of Corcyra was first a son of Periander, and then after he was assassinated, a nephew of Periander, Psammetichos (Will 1955: 521 and n. 5; Berve 1967: i. 21; Salmon 1984: 222).

106. The category is a narrative construct, even if we were to accept the existence of *perioikoi* in Corinth, which is itself problematic (on the question of the *perioikoi* in archaic Corinth: Oost 1972: 26 n. 66 with bibl.). All that we can be certain about is that at some stage they were assumed to have existed (and to have been characterized by marginality—which real *perioikoi* were) by someone.

107. The view that the tyranny among the *perioikoi* is a transformation of the Herodotean theme 'Lykophron against his father—Lykophron at Corcyra' is not an *ad hoc* hypothesis produced to fit the approach suggested here. Historians who accept the notion of Lykophron's governorship of Corcyra as historical have also suggested that it may underlie the *tyrannida . . . para tois perioikois*, that the latter may be a version of the governorship of Corcyra. (Will 1955: 522 n. 5, cont. from p. 521, suggests that perhaps we should understand *para tois apoikois* and think of the Corcyraeans, but considers the problem insoluble. Salmon 1984: 222 n. 139 thinks that the notion of Lykophron establishing a tyranny among the *perioikoi* may be a confused reference to Lykophron at Corcyra.) And this suggests a perception of the homology of the two narrative elements which transcends the rationalizing interpretations of each individual element.

108. And this is an important argument against Berve 1967: ii. 526, who takes the notion that Lykophron set up a tyranny seriously: according to the 'novellenartigen Erzählung' Periander would have banished Lykophron to Corcyra; but Berve argues that Periander would not have sent his refractory son to Corcyra which is an island especially suitable for rebelling against Corinth from, and thinks that Lykophron had set up a tyranny as Nicolaos of Damascus says, and that Herodotus' account is mixing up Lykophron with Nicolaos to whom Periander had transferred Corcyra. Berve's argument relies on the (implicit) presupposition that the story is a translation of historical events, and argues and evaluates the historicity of sources around that. Moreover, he has made the further assumptions that the following historical events underlie the stories: (i) that a son of Periander ruled Corcyra and (ii) that a son of Periander rebelled against Periander. Having thus centred the discourse through these a priori assumptions, he arranged the data around them and thus privileged Nicolaos of Damascus. The hypothesis, incidentally, that there is a mix-up of names is clearly incorrect, since there is, we saw, a significant consistency. As for Berve's arguments concerning Corcyra, they should be turned on their head: it is because Corcyra can satisfactorily fill the slot 'close but antagonistic relationship with Corinth'/ 'appropriate geographical position', that in the course of the creation of this 'myth' it came to fill the spacing for Lykophron's place of exile.

109. Ephoros or whoever. Who did this and when and how is not very important.

110. The rationalizing aspect of Ephoros hardly needs commenting on (see e.g. Will 1955: 462; Oost 1972: 14 n. 17). Jacoby postulated that the historical fact is primary. He suggests (in his comm. on Nicolaus of Damascus *F Gr. H* 90 F 59–60, *F Gr. H* IIC, pp. 249–50) that Lykophron's tyranny may have been the historical

fact underlying the notion of Lykophron's hostility to his father Periander. This is an assumption depending on culturally determined notions concerning the nature of the narrative and its reference to historical reality. Once the discourse is centred in this way, this interpretation sounds plausible; but the analyses in the preceding section have shown that this preliminary centring of the discourse is itself culturally determined and unwarranted. I have tried to show that the hostility to the father is a mythological motif (shaping the Periander–Lykophron story for reasons to be discussed below), which in the story of Lykophron's tyranny is translated into the historical idiom and presented as historical fact (still on the basis of 'mythological' modes of articulation: the marginal youth is a tyrant among the marginal *perioikoi*; this last element in itself shows that we are dealing with myth here, myth translated into history and read as such by some modern scholars.)

111. Some scholars think (see Oost 1972: 14 n. 17; see also Will 1955: 462) that Ephoros/Nicolaos of Damascus had also had a source other than the 'rationalized Herodotus'. Whether or not they did, what matters to us is that Nicolaos of Damascus' account partly conforms to the same mythological schemata as those structuring Herodotus' account. As for the divergences, we have found good reasons for thinking that Nicolaos of Damascus' variant does not reflect a source preserving different historical facts, but is simply the result of further mythologizing which operated on, and elaborated, earlier material under the impulse to rationalize, to make sense of things, e.g. to make sense in historical terms of the schema 'father–son hostility'—but operating partly through the mythological mentality and mythological schemata. Of course, all this cannot disprove the hypothesis that Periander's conquest of Corcyra and his son's governorship of the island and/or tyranny among the Corinthian *perioikoi* are historical facts; but it does remove all the arguments on which that hypothesis was based and shows the implausibility of their being historical fact rather than mythological constructs.
112. That slavery may have been a model was suggested by Will 1955: 552, who thinks the story may reflect historical reality.
113. See e.g. Hdt. 3. 77; see Tyrrell 1984: 86.
114. On the question of the relations between Corinth (and Periander in particular) and Lydia see Will 1955: 551–4; Salmon 1984: 225 with bibl. Of course, the creation of this particular motif would be even easier to understand if the symbolic counter 'Lydian' had already been established in the mythico-ritual nexuses of male initiations (cf. n. 34 above).
115. Of course, the story is polysemic. Another dominant 'message' is the danger of disloyalty to/rebellion against, the father (here in the particular version 'for the sake of the mother's *oikos*').

116. On the reign of Psammetichos and the fall of the Kypselid tyranny see: Berve 1967: i. 25, ii. 530; Salmon 1984: 229–30; Forrest 1966: 114–16.
117. If Periander was not succeeded by his son, the casting of the narrative through these schemata, having created its own momentum, would have led to the story of the successor (whether or not it contained historical elements) being cast into a mould fitting the schema, as a double of Lykophron (a successful one but short-lived, because overthrown).
118. Hdt. 5. 92e. 1; see Parke and Wormell 1956: ii. 5–6, no. 8. On the oracles given to Kypselos see: Forrest 1966: 111; Oost 1972: 17–18; Salmon 1984: 186–7.
119. The oracle here predicts failure in the succession, while oracles connected with patricide and 'patricidal' themes foretell successful—if violent—replacement of the father or father-figure by the son.
120. For this is the solution to the problem raised by Parke and Wormell 1956: i. 118–19 and n. 12 on p. 124 that, though the oracle was written after the event, it does not in fact correspond to the historical events except approximately, since, while Periander was not succeeded by his son, Kypselos *was* succeeded by his son's son, since Psammetichos was the son of Periander's brother.
121. The notion that tyrannies are short-lived is found in Ar., *Pol.*, 1315^{b}12–27. A related notion in Ar., *Pol.* 1312^{b}21–6 is that those who won tyrannies kept them, while almost all of those who inherited them lost them. This model does not apply to the Kypselids, who in 1315^{b}22–7 are said to be the second-longest lasting tyranny, with Kypselos' 30 years, Periander's 44, and Psammetichos' 3.
122. Both his parents are presented as both central and marginal. Labda is more central than marginal, but she belongs to the marginal sex: she is central because a Bacchiad, and marginal because being a woman with a physical defect she could not marry (and thus become integrated in society as a mature member) at the high level at which she had been born. Eetion was more marginal than central; he was marginal because non-Dorian, central because an aristocrat. (See Paus. 2. 4, 4 and 5. 18. 7–8.) Historians' views vary about the historical Kypselos' descent: e.g. Drews 1979: 261 (see p. 271) suggested that Kypselos was neither Bacchiad nor Corinthian; on Kypselos as a Bacchiad see Oost 1972: 23. Salmon 1984: 189 n. 9 accepts Kypselos' Bacchiad parentage on his mother's side as historical fact—which it could well be, for all we know—adding that Kypselos' parentage is not, strictly speaking, part of the legend and there was no reason to invent it. Both these statements are based on a culturally determined judgement which I consider incorrect: in my view, Kypselos' parentage is a 'mythological' way of characterizing him as 'both central

and marginal'; and of making a statement through him about the beginning of tyranny, answering the questions 'How do tyrannies come about? Where do they come from?' The answer given may or may not reflect historical fact; it certainly reflects a particular perception of historical events.

123. I agree with Vernant (1982: *passim*, see esp. p. 33) that 'tyranny, power gained and lost, the continuous or blocked sequence of generations, straight or indirect succession, . . . agreement or misunderstanding in the communication between fathers and sons' are some of the themes contained in the Herodotean narrative about the Kypselids. But (while Vernant is concerned with themes) my concern has been to recover particular structuring schemata and their interaction in a close reading of the narrative. As I tried to show, the schemata structuring the Lykophron–Periander story are found in many myths, relating to diverse figures; the closest parallel to the relationship between Periander and his son is that between Theseus and his son. The Kypselid story was modelled on stories pertaining to mythical kings because the only available model for the notion 'tyrant' was 'king'. (The term *basileus* is used for Periander, e.g. 3. 52. Oost 1972: 19 thinks that Kypselos' formal title was king. On the relationship between tyrant and king: Oost 19: 23–4.) The kings elsewhere associated with the schemata which were here applied to Periander were diverse in character, and included the positive model Theseus. Many important motifs and schemata associated with Periander are of the type which, though crystallized in a polarized form in the figures of mythical kings like Theseus, express the common experiences and problems of all men. Theseus crystallized not only exceptional, but also common experiences (in an exceptional form) in his own person—as e.g. through his being the ephebe *par excellence*. In my view, the emphasis in Hdt. 3. 48 and 50–3 is very much more on the tyrant/king as a paradigm (a polarized figure suitable to the idiom of myth) for all men, than as an abnormal being standing apart from society. For even the historical notion 'interruption of dynasty' is a particular version of 'interruption of father–son succession'. (Of course, the polarization does set tyrants and kings apart; but it is beyond my scope here to discuss this facet of the polarizing modality in Greek mythopea.)

124. All that can be said about the insoluble problems concerning Herodotus' sources and his contribution to the Periander–Lykophoron narrative is that at least some elements, such as the Samian rite (whether or not it had been associated with the Herodotean *aition* in this particular form) must have predated the Herodotean elaboration. Long 1987 explored the extent, the modalities, and the significance of repetition in Herodotus' short

stories. Leaving aside the repetition of words, which does occur, but would take too long to discuss, significant repetitions in our story are the following. First, on the analysis presented here, the story is structured by the 'initiatory schema' in a variety of ways—corresponding to the importance of this schema in the story as read here: as we saw, not only is the whole story structured by this schema, but also each of the two main parts that make it up (the narrative concerning the Corcyraean boys and the story of Lykophron). Second, in terms of stylistic articulation, in a pattern comparable to patterns discussed by Long 1987 (cf. e.g. Long 1987: 11), in each phase of the story of Lykophron we find a 'rhetorical' question (cf. n. 9) addressed to Lykophron (to Lykophron and his brother by Procles in 3. 50 by his father in 3. 52; and by his sister in 3. 53) which ushers in a new phase of the story. The first of these rhetorical questions was a kind of riddle (cf. above n. 9) and so it stood on its own, and no course of action was explicitly advocated by the questioner; but once Lykophron has solved the riddle he follows the course of action implicitly advocated by the questioner. In the second the rhetorical question is followed by an explicit advocacy of a particular course of action; Lykophron refuses to follow it. In the third also the rhetorical question is followed by an explicit advocacy of a particular course of action; Lykophron responds in a manner that leaves the way open to a possible compromise. Lykophron's actions in response to each of these questions has a progressively more sinister result, leading the first to alienation from his father, the second to his exile, and the third (which is on the surface the most conciliatory to his father's interests) to his death. This pattern of repetition punctuates Lykophron's trajectory from adolescent boy to failed ephebe and death.

BIBLIOGRAPHY

This bibliography includes, and consists of, all the books and articles cited in this volume except for sales catalogues, articles in lexica and encyclopedias, excavation reports in publications such as *Notizie degli scavi di antichità*, *Archaeological Reports* and the like, and also for reviews which are only cited once.

ARV J. D. Beazley, *Attic Red-figure Vase-painters*² (Oxford, 1963).
Para. *Paralipomena: Additions to Attic Black-figure Vase-Painters and to Attic Red-figure Vase-painters*² (Oxford, 1971).
Add. *Beazley Addenda: Additional References to ABV, ARV*² *and Paralipomena*, compiled by Lucilla Burn and Ruth Glynn (Oxford, 1982).
CB L. D. Caskey and J. D. Beazley, *Attic Vase Paintings in the Museum of Fine Arts, Boston* (Oxford, 1931–63).
Cité *La Cité des images: Religion et société en Grèce antique* (Mont-sur-Lausanne, 1984).
LIMC *Lexicon Iconographicum Mythologiae Classicae* (Zurich and Munich, 1981–).
Tr. GF iii *Tragicorum Graecorum Fragmenta*, ed. S. Radt, vol. iii. Aeschylus (Göttingen, 1985).
Tr. GF iv *Tragicorum Graecorum Fragmenta*, ed. S. Radt, vol. iv. Sophocles (Göttingen, 1977).

AGARD, W. R. 1966. Boreas at Athens, *CJ* 61 (1966), 241–6.
ALFIERI, N., ARIAS, P. E., and HIRMER, M. 1958. *Spina: Die neuenentdeckte Etruskerstadt und die griechischen Vasen ihrer Gräber* (Florence, 1958).
ALLEN, T. W., HALLIDAY, W. R., and SIKES, E. E., eds. 1936. *The Homeric Hymns*² (Oxford, 1936).
ALY, W. 1969. *Volksmärchen, Sage und Novelle bei Herodot und seinen Zeitgenossen*² (Göttingen, 1969).
AMANDRY, P. 1950. *La Mantique apollinienne à Delphes: Essai sur le fonctionnement de l'oracle* (Paris, 1950).
—— 1954. Notes de topographie et d'architecture delphiques IV: Le palmier de bronze de l'Eurymédon, *BCH* 78 (1954), 295–315.
—— 1981. Chronique delphique (1970–1981), *BCH* 105 (1981), 673–769.

AMANDRY, P. 1986. Sièges mycéniens tripodes et trépied pythique, in: ΦΙΛΙΑ ΕΠΗ εἰς Γεώργιον Ε. Μυλωνᾶν διά τά 60 ἔτη τοῦ ἀνασκαφικοῦ του ἔργου (Athens, 1986), A, 167–84.
ANDERSON, J. K. 1955. *Handbook to the Greek Vases in the Otago Museum* (Dunedin, 1955).
ANDREWES, A. 1956. *The Greek Tyrants* (London, 1956).
ANDRONIKOS, M. 1956. Lakonika anaglypha, *Peloponnisiaka*, 1 (1956), 253–314.
ANGERMEIER, H. E. 1936. *Das Alabastron: Ein Beitrag zur Lekythen-Forschung* (Diss. Giessen, 1936).
ARRIGONI, G. 1977. Atalanta e il cinghiale bianco, *Scripta Philologa*, i (Milan, 1977), 9–47.
ARTHUR, M. B. 1982. Cultural Strategies in Hesiod's Theogony: Law, Family, Society, *Arethusa*, 15 (1982), 63–82.
ASHMOLE, B. 1922. Locri Epizephyrii and the Ludovisi throne, *JHS* 42 (1922), 248–53.
—— 1936. *Late Archaic and Early Classical Greek Sculpture in Sicily and South Italy*, from *PBA* 20, separately printed (London, 1936).
AUBERSON, P. 1974. La Reconstitution du Daphnéphoreion d'Erétrie, *Ant K.* 17 (1974), 60–8.
—— and SCHEFOLD, K. 1972. *Führer durch Eretria* (Berne, 1972).
BAMBERGER, J. 1974. The Myth of Matriarchy: Why Men Rule in Primitive Society, in M. Z. Rozaldo and L. Lamphere, eds., *Woman, Culture and Society* (Stanford, 1974), 263–80.
BARRETT, W. S., ed. 1964. Euripides, *Hippolytos* (Oxford, 1964).
BARRON, J. P. 1980. Bacchylides, Theseus and a Woolly Cloak, *BICS* 27 (1980), 1–8.
BEAZLEY, J. D. 1929. Notes on the Vases in Castle Ashby, *PBSR* 11 (1929), 1–29.
BENTON, 1970. Nereids and Two Attic Pyxides, *JHS* 90 (1970), 193–4.
BÉQUIGNON, Y. 1949. De quelques usurpations d'Apollon en Grèce centrale d'après des recherches récentes, *RA* 1949. 1, 62–8.
BÉRARD, C. 1971. Architecture erétrienne et mythologie delphique: Le Daphnéphoreion, *Ant.K.* 14 (1971), 59–73.
—— 1974. *Anodoi: Essai sur l'imagerie des passages chthoniens* (Rome, 1974).
—— 1976. Axie taure, in: *Mélanges d'histoire ancienne et d'archéologie, offerts à Paul Collart* (Lausanne, 1976), 61–73.
—— 1980. Review of Sourvinou-Inwood 1979, *Gnomon* 52 (1980), 616–20.
—— 1983. Iconographie-Iconologie-Iconologique, *Études de Lettres*, 1983, 4, 5–37.
—— 1984. L'Ordre des femmes, *Cité*, pp. 85–103.
—— 1986. L'impossible femme athlète, *AION Archeologia e storia antica*, 8 (1986), 195–202.

—— 1988. La chasseresse traquée: Cynégétique et érotique, in Schmidt 1988: 280–4.
—— and DURAND, J.-L. 1984. Entrer en imagerie, Cité pp. 19–34.
BERVE, H. 1967. *Die Tyrannis bei den Griechen* (Munich, 1967).
BLECH, M. 1982. *Studien zum Kranz bei den Griechen* (Berlin and New York, 1982).
BLUMENTHAL, A. VON. 1928. Der Apollontempel des Trophonios und Agamedes in Delphi, *Philologus*, 83 (1928), 20–4.
BOARDMAN, J. 1958/9. Old Smyrna: The Attic Pottery, *BSA* 53–4 (1958/9), 152–81.
—— 1975. *Athenian Red Figure Vases: The Archaic Period. A Handbook* (London, 1975).
—— 1976. The Kleophrades Painter at Troy, *Ant.K.* 19 (1976), 3–18.
—— 1983. Atalanta, *The Art Institute of Chicago Centennial Lectures: Museum Studies*, 10 (Chicago, 1983), 3–19.
—— and LA ROCCA, E. 1978. *Eros in Greece* (London, 1978).
BODSON, L. 1978. *'IEPA ZΩIA: Contribution à l'étude de la place de l'animal dans la religion grecque ancienne* (Brussels, 1978).
BOMMELAER, J. F. 1983. La construction du temple classique de Delphes, *BCH* 107 (1983), 191–215.
BOON, J. A. 1982. *Other Tribes, Other Scribes: Symbolic Anthropology in the Comparative Study of Cultures, Histories, Religions and Texts* (Cambridge, 1982).
BORGEAUD, P. 1979. *Recherches sur le dieu Pan* (Rome and Geneva, 1979).
BOTHMER, D. VON. 1957. *Amazons in Greek Art* (Oxford, 1957).
BOULTER, C. G. and BENTZ, J. L. 1980. Fifth-Century Attic Red Figure at Corinth, *Hesperia*, 49 (1980), 295–308.
BOURGUET, E. 1932. *Fouilles de Delphes III.V. Epigraphie: Les Comptes du IVe siècle* (Paris, 1932).
BOUVIER, D., and MOREAU, P. 1983. Phinée ou le père aveugle et la marâtre aveuglante, *Revue belge de philologie et d'histoire*, 61 (1983), 5–19.
BOWIE, M. 1979. Jacques Lacan, in Sturrock 1979: 116–53.
BOWIE, A. M. 1987. Ritual Stereotype and Comic Reversal: Aristophanes' *Wasps*, *BICS* 34 (1987), 112–25.
BRELICH, A. 1958. *Gli eroi greci: Un problema storico-religioso* (Rome, 1958).
—— 1968. Religione micenea: Osservazioni metodologiche, *Atti e Memorie del primo Congresso Internazionale di Micenologia* (Rome, 1968), 924–7.
—— 1969. *Paides e parthenoi* (Rome, 1969).
BREMMER, J. 1978. Heroes, Rituals and the Trojan War, *SSR* 2 (1978), 5–38.
—— 1980. An Enigmatic Indo-European Rite: Paederasty, *Arethusa*, 13 (1980), 279–98.

BREMMER, J. 1983. The Importance of the Maternal Uncle and Grandfather in Archaic and Classical Greece and Early Byzantium, *ZPE* 50 (1983), 173–86.

—— ed. 1987. *Interpretations of Greek Mythology* (London and Sydney, 1987).

—— 1989. Greek Pederasty and Modern Homosexuality, in J. Bremmer, ed., *From Sappho to de Sade: Moments in the History of Sexuality* (London and New York, 1989).

BRENDEL, O. 1936. Symbolik der Kugel, *Röm. Mit.* 51 (1936), 1–95.

BROMMER, F. 1979. Theseus-Deutungen, *AA* 1979, 487–511.

—— 1982. *Theseus: Die Taten des griechischen Helden in der antiken Kunst und Literatur* (Darmstadt, 1982).

BRON, C., and LISSARAGUE, F. 1984. Le vase à voir, in *Cité*, pp. 7–17.

BRULÉ, P. 1987. *La Fille d'Athènes* (Paris, 1987).

BRUNEAU, P. 1985. Deliaca V, *BCH* (1985), 545–67.

—— and DUCAT, J. 1983. École française d'Athènes, *Guide de Délos*[3] by P. Bruneau and J. Ducat (Paris, 1983).

BRUNNSÅKER, S. 1971. *The Tyrant-Slayers of Kritios and Nesiotes: A Critical Study of the Sources and Restorations*[2] (Stockholm, 1971).

BUITRON, D. M. 1972. *Attic Vase-Painting in New England Collections* (Cambridge, Mass., 1972).

BURKERT, W. 1975. Apellai und Apollon, *Rh. Mus.* 118 (1975), 1–21.

—— 1983. *Homo Necans: The Anthropology of Ancient Greek Sacrificial Ritual and Myth*[2] (Berkeley, Los Angeles, and London, 1983).

—— 1985. *Greek Religion: Archaic and Classical*[2] (Oxford, 1985).

BUSCHOR, E. 1944. *Die Musen des Jenseits* (Munich, 1944).

BUXTON, R. G. A. 1980. Blindness and Limits: Sophokles and the Logic of Myth, *JHS* 100 (1980), 22–37.

—— 1987. Wolves and Werewolves in Greek Thought, in Bremmer 1987, 60–79.

BYVANCK, A. W. 1912. *Gids voor de Bezoekers van het Museum Meermanno-Westreenianum* (1912).

CALAME, C. 1977. *Les chœurs de jeunes filles en Grèce archaïque* (Rome, 1977).

—— 1983a. Introduzione: Eros inventore e organizzatore della società greca antica, in C. Calame, ed., *L'amore in Grecia* (Bari, 1983), pp. ix–xl.

—— 1983b. L'espace dans le mythe, l'espace dans le rite: Un exemple grec, *Degrés* 35/6 (1983), 1–15. (Publ. also in a somewhat modified form in Calame 1986, 163–76.)

—— 1986. *Le Récit en Grèce ancienne: Enonciations et représentations de poètes* (Paris, 1986).

CALLIPOLITIS-FEYTMANS, D. 1970. Déméter, Corè et les Moires sur des vases corinthiens, *BCH* 94 (1970), 45–65.

—— 1974. *Les Plats attiques à figures noires* (Paris, 1974).

CAMERON, A., and KUHRT, A., eds. 1983. *Images of Women in Antiquity* (London and Canberra, 1983).

CASSOLA, F. 1975. *Inni Omerici* (Verona, 1975).
CHADWICK, J. 1976. *The Mycenaean World* (Cambridge, 1976).
—— KILLEN, J. T., and OLIVIER, J.-P., ed. 1971. *The Knossos Tablets*[4] (Cambridge, 1971).
COHEN, B. 1978. *Attic Bilingual Vases* (New York, 1978).
COLE, S. G. 1984. The Social Function of Rituals of Maturation: The Koureion and the Arkteia, *ZPE* 55 (1984), 233–44.
COLIN, M. G. 1909–13. *Fouilles de Delphes III: Epigraphie*, ii (Paris, 1909–13).
COOK, A. B. 1895. The Bee in Greek Mythology, *JHS* 15 (1895), 1–24.
COSTE-MESSELIÈRE, P. DE LA. 1936. *Au Musée de Delphes* (Paris, 1936).
—— 1969. Topographie delphique, *BCH* 93 (1969), 730–58.
—— and FLACELIÈRE, R. 1930. Une statue de la Terre à Delphes, *BCH* 54 (1930), 283–95.
COULTON, J. J. 1988. Post Holes and Post Bases in Early Greek Architecture, *Mediterranean Archaeology*, 1 (1988), 58–65.
COURBY, M. F. 1927. *Fouilles de Delphes II: Topographie et architecture: La Terrace du temple* (Paris, 1927).
CULLER, J. 1981. *The Pursuit of Signs: Semiotics, Literature, Deconstruction* (London and Henley, 1981).
—— 1983. *On Deconstruction: Theory and Criticism after Structuralism* (London, Melbourne, and Henley, 1983).
DAUX, G. 1968. Le Poteidanion de Delphes, *BCH* 92 (1968), 540–9.
DAVIES, J. K. 1988. Religion and the State, *Cambridge Ancient History*[2], iv (Cambridge, 1988), 368–88.
DELCOURT, M. 1955. *L'Oracle de Delphes* (Paris, 1955).
DE MAN, P. 1979. *Allegories of Reading: Figural Language in Rousseau, Nietzsche, Rilke and Proust* (New Haven, Conn., and London, 1979).
DEMANGEL, R. 1926. *Fouilles de Delphes II: Topographie et architecture. Le Sanctuaire d'Athéna Pronaia* (Paris, 1926).
DEONNA, W. 1951a. L'Ex voto de Cypsélos à Delphes: Le Symbolisme du palmier et des grenouilles, *Rev. Hist. Rel.* 139 (1951), 162–207.
—— 1951b. L'Ex voto de Cypsélos à Delphes: Le Symbolisme du palmier et des grenouilles (Suite), *Rev. Hist. Rel.* 140 (1951), 5–58.
DERRIDA, J. 1967. *L'Écriture et la différence* (Paris, 1967).
—— 1972. *Positions* (Paris, 1972).
—— 1973. *Speech and Phenomena and Other Essays on Husserl's Theory of Signs* (Evanston, 1973).
—— 1974, 1976. *Of Grammatology* (Baltimore and London, 1974, 1976).
—— 1982. *Margins of Philosophy* (Chicago and Brighton, 1982).
DESPINIS, G. I. 1971. Συμβολὴ στὴ μελέτη τοῦ ἔργου τοῦ Ἀγορακρίτου (Athens, 1971).
DETIENNE, M. 1972. *Les Jardins d'Adonis* (Paris, 1972).
—— 1979. *Dionysos Slain* (Baltimore and London, 1979).

DETIENNE, M. 1981. The Myth of 'Honeyed Orpheus', in R. L. Gordon, ed., *Myth, Religion and Society* (Cambridge, 1981), 95–109.
—— and VERNANT, J.-P. 1974. *Les Ruses de l'intelligence* (Paris, 1974).
DEUBNER, L. 1925. Hochzeit und Opferkorb, *JdI* 40 (1925), 210–23.
—— 1969. *Attische Feste*^{3} (Vienna, 1969).
DEVAMBEZ, P. 1962. *Greek Painting* (London, 1962).
DIETRICH, B. C. 1974. *The Origins of Greek Religion* (Berlin, 1974).
—— 1978. Reflections on the Origins of the Oracular Apollo, *BICS* 25 (1978), 1–18.
DITTENBERGER, G. 1915–24. *Sylloge Inscriptionum Graecarum*^{3} (Leipzig, 1915–24).
DODDS, E. R. 1951. *The Greeks and the Irrational* (Berkeley, Los Angeles, and London, 1951).
DOUGLAS, M. 1975. *Implicit Meanings* (London and Henley, 1975).
—— 1982. Introduction to Grid/Group Analysis, in M. Douglas, ed., *Essays in the Sociology of Perception* (London and Henley, 1982), 1–8.
DOVER, K. J. 1974. *Greek Popular Morality in the Time of Plato and Aristotle* (Oxford, 1974).
—— 1978. *Greek Homosexuality* (London, 1978).
DOWDEN, 1989. *Death and the Maiden: Girls' Initiation Rites in Greek Mythology* (London and New York, 1989).
DRERUP, H. 1986. Das sogenannte Daphnephoreion, in K. Braun and A. Furtwängler, eds., *Studien zur klassischen Archäologie: Festschrift zum 60. Geburtstag von Friedrich Hiller* (Saarbrücken, 1986), 3–21.
DREWS, R. 1979. Die ersten Tyrannen in Griechenland, in Kinzl 1979: 256–80.
DUGAS, C. 1910. Fragment de bas-relief du musée du Louvre, *BCH* 34 (1910), 233–41.
DUNBABIN, T. J. 1948. *The Western Greeks* (Oxford, 1948).
DURAND, J. L., and LISSARAGUE, F. 1980. Un lieu d'image? L'espace du loutérion, *Hephaistos* 2 (1980), 89–106.
DURAND, J. L., and SCHNAPP, A. 1984. Boucherie sacrificielle et chasses initiatiques, in *Cité*, pp. 49–66.
EASTERLING, P. E., ed. 1982. Sophocles, *Trachiniae* (Cambridge, 1982).
ECO, U. 1976. *A Theory of Semiotics* (Bloomington, Ind., 1976).
—— 1981. *The Role of the Reader: Explorations in the Semiotics of Texts* (London, 1981).
ELDERKIN, G. W. 1962. *The First Three Temples at Delphi: Their Religious and Historical Significance* (Princeton, 1962).
ENGELMANN, R. 1903. Die Jo-Sage, *JdI* 18 (1903), 37–58.
ERBSE, H. 1971. *Scholia Graeca in Homeri Iliadem*, ii (Berlin, 1971).
EVANS, A. 1921–35. *The Palace of Minos* (London, 1921–35).
FALKNER, T. M. 1989. Slouching towards Boeotia: Age and Age-Grading in the Hesiodic Myth of the Five Races, *Class. Ant.* 8 (1989), 42–60.

FARNELL, L. R. 1896a. *The Cults of the Greek States*, i (Oxford, 1896).
—— 1896b. *The Cults of the Greek States*, ii (Oxford, 1896).
—— 1907. *The Cults of the Greek States*, iii (Oxford, 1907).
—— 1909. *The Cults of the Greek States*, v (Oxford, 1909).
FEHR, B. 1971/2. Zur Geschichte der Apollonheiligtums von Didyma, *Marb.Winckel.Progr.* (1971/2), 14–59.
FEYEL, M. 1946. Smenai: Étude sur le vers 552 de l'hymne homérique à Hermes, *RA* 25 (1946), 5–22.
FINLEY, J. H. 1955. *Pindar and Aeschylus* (Cambridge, Mass., 1955).
FOLEY, H. P. 1982. Marriage and Sacrifice in Euripides' Iphigeneia in Aulis, *Arethusa*, 15 (1982), 159–80.
FONTENROSE, J. 1959. *Python* (Berkeley, Los Angeles, and London, 1959).
—— 1978. *The Delphic Oracle* (Berkeley, Los Angeles, and London, 1978).
—— 1981. *Orion: The Myth of the Hunter and the Huntress* (Berkeley, Los Angeles, and London, 1981).
FORREST, W. G. 1966. *The Emergence of Greek Democracy* (London, 1966).
FRANCISCIS, A. DE. 1971. *Ricerche sulla topografia e i monumenti di Locri Epizefiri* (Naples, 1971).
—— 1972. *Stato e società in Locri Epizefiri: L'archivio dell'Olympieion locrese* (Naples, 1972).
FRAZER, J. G. 1921. *Apollodorus: The Library* (London, 1921).
FRENCH, E. B. 1981. Mycenaean Figures and Figurines, Their Typology and Function, in Hägg and Marinatos 1981: 173–8.
—— 1985. The Figures and Figurines, in Renfrew 1985: 209–80.
FRICKENHAUS, A. 1910. Heilige Stätten in Delphi, *Ath. Mit.* 35 (1910), 235–73.
FRIEDRICH, P. 1978. *The Meaning of Aphrodite* (Chicago and London, 1978).
FURTWÄNGLER, A. 1883–7. *La Collection Sabouroff* (Berlin, 1883–7).
GABRICI, E. 1927. Il santuario della Malophoros a Selinunte = *Monumenti Antichi*, 32 (1927).
GALLET DE SANTERRE, H. 1958. *Délos primitive et archaïque* (Paris, 1958).
GAUER, W. 1968. *Weihgeschenke aus den Perserkriegen* (Tübingen, 1968).
GÉRARD-ROUSSEAU, M. 1968. *Les Mentions religieuses dans les tablettes mycéniennes* (Rome, 1968).
GERNET, L. 1968. *Anthropologie de la Grèce antique* (Paris, 1968).
—— and BOULANGER, A. 1932. *Le Génie grec dans la religion* (Paris, 1932; repr. 1970).
GIANNELLI, G. 1963. *Culti e miti della Magna Grecia^2* (Florence, 1963).
GINOUVÈS, R. 1962. *Balaneutikè* (Paris, 1962).
GOMBRICH, E. H. 1977. *Art and Illusion: A Study in the Psychology of Pictorial Representation^5* (Oxford, 1977).
—— 1982. *The Image and the Eye: Further Studies in the Psychology of Pictorial Representation* (Oxford, 1982).

GRAF, F. 1978. Die Lokrischen Mädchen, *SSR* 2 (1978), 61–79.
—— 1979a. Apollon Delphinios, *Mus. Helv.* 36 (1979), 1–22.
—— 1979b. Das Götterbild aus dem Taurerland, *Antike Welt*, 4 (1979), 33–41.
—— 1985a. *Griechische Mythologie* (Munich and Zurich, 1985).
—— 1985b. *Nordionische Kulte* (Rome, 1985).
—— 1986. Boukoloi, *ZPE* 62 (1986), 43–4.
GRECO, G. 1985/6. Un cratere del Pittore di Talos da Serra di Vaglio, *RIA* 8–9 (1985/6: publ. 1987), 5–35.
GREGORY, R. L. 1966. *Eye and Brain: The Psychology of Seeing* (London, 1966).
GREIFENHAGEN, A. 1957. *Griechische Eroten* (Berlin, 1957).
GRENFELL, B. P., and HUNT, A. S., eds. 1922. *The Oxyrhynchus Papyri*, part xv (London, 1922).
GRUBEN, G. 1963. Das archaische Didymaion, *JdI* 78 (1963), 78–177.
GUARDUCCI, M. 1937–8. *L'istituzione della fratria nella Grecia antica e nelle colonie greche dell'Italia* (Rome, 1937–8).
GUIRAUD, P. 1975. *Semiology* (1st publ. as *La Semiologie* in 1971; Engl. transl. G. Gross: London, 1975).
HADZISTELIOU PRICE, T. 1978. *Kourotrophos: Cults and Representations of the Greek Nursing Deities* (Leiden, 1978).
HÄGG, R., ed. 1983. *The Greek Renaissance of the Eighth Century B.C.: Tradition and Innovation* (Stockholm, 1983).
—— and MARINATOS, N., eds. 1981. *Sanctuaries and Cults in the Aegean Bronze Age* (Stockholm, 1981).
HAMDORF, F. W. 1964. *Griechische Kultpersonifikationen der vorhellenistischen Zeit* (Mainz, 1964).
HARDER, R. E. 1988. Nausikaa und die Palme von Delos, *Gymnasium*, 95 (1988), 505–14.
HARRISON, A. R. W. 1968. *The Law of Athens: The Family and Property* (Oxford, 1968).
HARRISON, E. B. 1977. The Shoulder-Cord of Themis, in U. Höckmann and A. Krug, eds., *Festschrift für Frank Brommer* (Mainz, 1977), 155–61.
HARRISON, J. E. 1899. Delphika, *JHS* 19 (1899), 205–51.
HENI, R. 1977. *Die Gespräche bei Herodot* (Heidelberg diss., 1976; 1977).
HENRICHS, A. 1987. Three Approaches to Greek Mythography, in Bremmer 1987: 242–77.
HERRMANN, H.-V. 1959. *Omphalos* (Munster, 1959).
—— 1982. Zur Bedeutung des delphischen Dreifusses, *Boreas*, 5 (1982), 54–66.
HIGGINS, R. A. 1954–9. *Catalogue of the Terracottas in the Department of Greek and Roman Antiquities of the British Museum* (London, 1954–9).
HOFFMANN, H. 1963. A Clay Ball of Myrrhine, *BMus. Fine Arts*, 61 (1963), 20–2.

—— 1974. Hahnenkampf in Athen: Zur Ikonologie einer attischen Bildformel, *RA* 1974, 195–220.
—— 1977. *Sexual and Asexual Pursuit*, Royal Anthropological Institute Occasional Paper 34 (London, 1977).
—— 1980a. Review of Kaempf-Dimitriadou 1979, *Gnomon*, 52 (1980), 744–51.
—— 1980b. Knotenpunkte, *Hephaistos*, 2 (1980), 127–54.
HOHTI, P. 1976. *The Interrelation of Speech and Action in the Histories of Herodotus* (Helsinki, 1976).
HOLMBERG, E. J. 1947. Two White-Ground Vases in a Private Collection in Athens, *Act. Arch.* 18 (1947), 187–95.
HÖLSCHER, F. 1980. Der Raub der Kore im 5. Jahrhundert, in H. A. Cohn and E. Simon, eds., *Tainia: Roland Hampe zum 70. Geburtstag am 2. Dezember 1978 dargebracht* (Mainz, 1980), 173–9.
HOW, W. W., and WELLS, J. 1936. *A Commentary on Herodotus*, i (1912, repr. with corrections 1928, 1936).
HUNTER, V. 1982. *Past and Process in Herodotus and Thucydides* (Princeton, 1982).
IMMERWAHR, W. 1885. *De Atalanta* (Berlin, 1885).
JACOBSTAHL, P. 1927. *Ornamente griechischer Vasen* (Berlin, 1927).
JAHN, O. 1847. *Archäologische Beiträge* (Berlin, 1847).
JAMESON, M. 1980. Apollo Lykeios in Athens, *Archaiognosia*, 1 (1980), 213–36.
JANKO, R. 1982. *Homer, Hesiod and the Hymns: Diachronic Development in Epic Diction* (Cambridge, 1982).
JEANMAIRE, H. 1939. *Couroi et Courètes: Essai sur l'éducation spartiate et sur les rites d'adolescence dans l'antiquité hellénique* (Lille, 1939).
JENKINS, I. 1983. Is There Life after Marriage? A Study of the Abduction Motif in Vase Paintings of the Athenian Wedding Ceremony, *BICS* 30 (1983), 137–45.
JESSEN, H. B. 1955. Kirkos und Rabe in ihrer Beziehung zu Apollo und Hera, *AA* 1955. 281–309.
JUCKER, I. 1963. Frauenfest in Korinth, *Ant.K.* 6 (1963), 47–61.
JURRIAANS-HELLE, G. 1986. Apollo and the Deer on Attic Black-Figure Vases, in H. A. G. Brijder, A. A. Drukker, C. W. Neft, eds., *Enthousiasmos: Essays on Greek and Related Pottery presented to J. M. Hemelrijk* (Amsterdam, 1986), 111–20.
KAEMPF-DIMITRIADOU, S. 1979. *Die Liebe der Götter in der attischen Kunst des 5. Jhs. v.Chr.* (Berne, 1979).
KAHIL, L. GHALI, 1955. *Les Enlèvements et le retour d'Hélène* (Paris, 1955).
KAHIL, L. 1965. Autour de l'Artémis attique, *Ant.K.* 8 (1965), 20–32.
—— 1977. L'Artémis de Brauron: Rites et mystère, *Ant.K.* 20 (1977), 86–98.

KAHIL, L. 1981. Le 'cratérisque' d'Artémis et le Brauronion de l'Acropole, *Hesperia*, 50 (1981), 253–63.
—— 1983. Mythological Repertoire of Brauron, in W. G. Moon, ed., *Ancient Greek Art and Iconography* (Madison, Wis. 1983), 231–44.
KAMERBEEK, J. C. 1978. *The Plays of Sophocles: Commentaries, iii. The Antigone* (Leiden, 1978).
KAPLAN, A. 1970. Referential Meaning in the Arts, in M. Weitz, ed., *Problems in Aesthetics: An Introductory Book of Readings*^{2} (London, 1970), 270–91.
KELLER, O. 1920. *Die antike Tierwelt*, ii (Leipzig, 1920).
KEULS, E. C. 1982/3. The Hetaera and the Housewife: The Splitting of the Female Psyche in Greek Art, *Med. Rom.* 1982/3, 23–40.
—— 1983. Attic Vase-Painting and the Home Textile Industry, in W. G. Moon, ed., *Ancient Greek Art and Iconography* (Madison, Wis. 1983), 209–30.
KING, H. 1983. Bound to Bleed: Artemis and Greek Women, in Cameron and Kuhrt 1983: 109–27.
KINZL, K. H., ed. 1979. *Die ältere Tyrannis bis zu den Perserkriegen: Beiträge zur griechischen Tyrannis* (Darmstadt, 1979).
KOCH-KARNACK, G. 1983. *Knabenliebe und Tiergeschenke: Ihre Bedeutung im päderastischen Erziehungssystem Athens* (Berlin, 1983).
KOEHL, R. B. 1981. The Functions of Aegean Bronze Age Rhyta, in Hägg and Marinatos 1981: 179–88.
KONTOLEON, N. M. 1970. *Aspects de la Grèce préclassique* (Paris, 1970).
—— 1974 Ἡ στήλη τῆς Ἰκαρίας (Δευτερολογία), *Arch. Eph.* 1974, 13–25.
KRAAY, C. M. 1976. *Archaic and Classical Greek Coins* (London, 1976).
—— and HIRMER, M. 1966. *Greek Coins* (London, 1966).
KRAUSKOPF, I. 1977. Eine Attisch Schwarzfigurige Hydria in Heidelberg, *AA* 1977, 13–37.
KRIEGER, X. 1975. *Der Kampf zwischen Peleus und Thetis in der griechischen Vasenmalerei* (Munster, 1975).
KRISTEVA, J. 1981. *Le Langage, cet inconnu* (1st publ. under name of J. Joyaux in 1969; Paris, 1981).
KRITZAS, C. B. 1981. Muses delphiques à Argos, *BCH* supplément vi (1981), 195–209.
KRON, U. 1988. Zur Schale des Kodrosmalers in Basel: Eine Interpretatio Attica, in Schmidt 1988: 291–304.
KURTZ, D. C. 1975. *Athenian White Lekythoi* (Oxford, 1975).
LABARBE, J. 1953. L'Age correspondant au sacrifice du koureion et les données historiques du sixième discours d'Isée, *Bulletin de l'Académie Royale de Belgique*, Classe des Lettres V série vol. 39 (1953), 358–94.
LAMBRINUDAKIS, W. 1971. Μηροτραφής (Athens, 1971).
—— 1976. To anaglypho tes Ikarias. Zur Deutung des Ikariareliefs, *AAA* 9 (1976), 108–19.

LANG, M. L. 1984. *Herodotean Narrative and Discourse* (Cambridge, Mass., and London, 1984).
LATTE, K. 1920. *Heiliges Recht* (Tubingen, 1920).
LAURENCE, A. W. 1962. *Greek Architecture*² (Harmondsworth, 1962).
LEBESSI, A. 1985. Τὸ ἱερὸ τοῦ Ἑρμῆ καὶ τῆς Ἀφροδίτης στὴ Σύμη Βιάννου, I.1. Χάλκινα κρητικὰ τορεύματα (Athens, 1985).
—— and MUHLY, P. 1987. The Sanctuary of Hermes and Aphrodite at Syme, Crete, *National Geographic: Research*, 3 (1987), 102–13.
LEFKOWITZ, M. R. 1981. *Heroines and Hysterics* (London, 1981).
—— 1986. *Women in Greek Myth* (London, 1986).
LERAT, L. 1938. Fouilles de Delphes (1934–1935), *RA*, 1938, 187–207.
—— 1957. Delphes: Marmaria, in: Chronique des fouilles et découvertes archéologiques en Grèce en 1956, *BCH* 81 (1957), 708–10.
LE ROY, C. 1973. La naissance d'Apollon et les palmiers déliens, *Études déliennes*, *BCH* supplement i (1973), 263–86.
LESKY, A. 1966. *Gesammelte Schriften* (Berne and Munich, 1966).
LÉVI-STRAUSS, C. 1966. *The Savage Mind* (London, 1966).
LEZZI-HAFTER, A. 1976. *Der Schuwalow-Maler: Eine Kannenwerkstatt der Parthenonzeit* (Mainz, 1976).
LISSARAGUE, F. 1984. Autour du guerrier, in *Cité*, pp. 35–47.
LISSI, E. 1961. La collezione Scaglioni a Locri, *Atti Soc. MGrec.* 1961, 67–128.
LLOYD-JONES, H. 1971. *The Justice of Zeus* (Berkeley, Los Angeles, and London, 1971).
—— 1983. Artemis and Iphigeneia, *JHS* 103 (1983), 87–102.
LOBEL, E., ed. 1961. *The Oxyrhynchys Papyri*, part xxvi (London, 1961).
LONG, T. 1987. *Repetition and Variation in the Short Stories of Herodotus* (Frankfurt, 1987).
LORAUX, N. 1978. Sur la race des femmes et quelques-unes de ses tribus, *Arethusa*, 11 (1978), 43–87.
—— 1980. La Grèce hors d'elle, *L'Homme*, 20 (1980), 105–11.
LORRIMER, H. L. 1947. The Hoplite Phalanx with Special Reference to the Poems of Archilochus and Tyrtaeus, *BSA* 42 (1947), 76–138.
LUPPE, W. 1984. Zum Tennes-Mythos im 'Mythographus Homericus' P.Hamb. 199, *ZPE* 56 (1984), 31–2.
LYONS, J. 1977 *Semantics* (Cambridge, 1977).
MALKIN, I. 1987. *Religion and Colonization in Ancient Greece* (Leiden, 1987).
MARTIN, R. 1965. *Manuel d'architecture grecque*, i. *Matériaux et techniques* (Paris, 1965).
—— and METZGER, H. 1976. *La Religion grecque* (Vendome, 1976).
MASTRONARDE, D. J. 1979. *Contact and Discontinuity: Some Conventions of Speech and Action on the Greek Tragic Stage* (Berkeley, Los Angeles, and London, 1979).

MERTENS, J. R. 1977. *Attic White-Ground: Its Development on Shapes other than Lekythoi* (New York and London, 1977).
METZGER, H. 1951. *Les Représentations dans la céramique attique du IVe siècle* (Paris, 1951).
—— 1977. *Apollon spendon*. À propos d'une coupe attique à fonds blanc trouvée à Delphes, in *Études Delphiques*, BCH supplement iv (1977), 421–8.
—— 1985. Sur le valeur de l'attribut dans l'interprétation de certaines figures du monde éleusinien, in: *Eidolopoiia: Actes du Colloque sur les problèmes de l'image dans le monde méditerranéen classique* (Rome, 1985), 173–8.
MEYER, K. H. 1980. Von Kunstgeschichtlicher Werkinterpretazion zu Rezeptionsforschung, *Hephaistos*, 2 (1980), 7–51.
MONBRUN, P. 1989. Artémis et le palmier datier, *Pallas*, 35 (1989), 69–93.
MORET, J.-M. 1975. *L'Ilioupersis dans la céramique italiote* (Rome, 1975).
MOSSÉ, C. 1969. *La tyrannie dans la Grèce antique* (Paris, 1969).
MURRAY, O. 1980. *Early Greece* (Fontana, 1980).
NEILS, J. 1980. The Group of the Negro Alabastra: A Study in Motif Transferral, *Ant.K.* 23 (1980), 13–23.
NEUSER, K. 1982. *Anemoi: Studien zur Darstellung der Winde und Windgottheiten in der Antike* (Rome, 1982).
NEUTSCH, B. 1953/4. *Makaron Nesoi*: Zu einem lokrischen Relief in Heidelberg, *Röm. Mit.* 60/1 (1953/4), 62–74.
NILSSON, M. P. 1906. *Griechische Feste von religiöser Bedeutung* (Leipzig, 1906).
—— 1967. *Geschichte der griechischen Religion³*, i (1967).
OAKLEY, J. H. 1982. The Anakalypteria, *AA* 1982, 113–18.
OLDFATHER, A. W. 1910. Funde aus Lokroi, *Philologus*, 69 (1910), 114–25.
—— 1912. Die Ausgrabungen zu Lokroi, *Philologus*, 71 (1912), 321–31.
OLIVIER, J.-P., GODART, L., SEYDEL, C., and SOURVINOU, C. 1973. *Index Généraux du linéaire B* (Rome, 1973).
OOST, S. I. 1972. Cypselus the Bacchiad, *C.Phil.* 67 (1972), 10–30.
ORLANDOS, A. K. 1958. Τὰ Ὑλικὰ δομῆς τῶν ἀρχαίων Ἑλλήνων, ii (Athens, 1958).
ORSI, P. 1909. Locri Epizefiri: Resoconto sulla terza campagna di scavi locresi, *BdA* 3 (1909), 1–42 (of the separately printed extract, pp. 406 ff. and 463 ff. of the full journal).
OSBORNE, R. G. 1985. *Demos: The Discovery of Classical Attika* (Cambridge, 1985).
PADEL, R. 1983. Women: Model for Possession by Greek Daemons in: A. Cameron and A. Kuhrt, eds., *Images of Women in Antiquity* (London and Canberra, 1983), 3–19.
PAGE, D. L. 1955. *Sappho and Alcaeus* (Oxford, 1955).

PALAIOKRASSA, L. Τὸ ἱερὸ τῆς Ἀρτέμιδος Μουνιχίας (Ph.D., Univ. of Salonica, 1983).
PAPASPYRIDI-KAROUZOU, S. 1945–7. Αἱ ἑπτὰ θυγατέρες τοῦ Ἄτλαντος, *Arch. Eph.* 1945–7 (1949), 22–36.
PAPOUTSAKI-SERBETI, E. 1983. Ὁ ζωγράφος τῆς *Providence* (Athens, 1983).
PARKE, H. W. 1977. *Festivals of the Athenians* (London, 1977).
—— 1985. *The Oracles of Apollo in Asia Minor* (London, Sydney, and Dover, 1985).
—— and WORMELL, D. E. W. 1956. *The Delphic Oracle*, i. *The History* (Oxford, 1956).
PARKER, R. 1983. *Miasma: Pollution and Purification in Early Greek Religion* (Oxford, 1983).
—— 1987. Myths of Early Athens, in Bremmer 1987: 187–214.
—— 1989. Spartan Religion, in A. Powell, ed., *Classical Sparta: Techniques behind Her Success* (London, 1989), 142–72.
PAYNE, H. 1931. *Necrocorinthia* (Oxford, 1931).
PEARSON, A. C., ed. 1917. *The Fragments of Sophocles*, ii (Cambridge, 1917).
PÉLÉKIDIS, C. 1962. *Histoire de l'ephébie attique* (Paris, 1962).
PEMBROKE, S. 1965. The Last of the Matriarchs: A Study in the Inscriptions of Lycia, *JESHO* 8 (1965), 217–47.
—— 1967. Women in Charge: The Function of Alternatives in Early Greek Tradition and the Ancient Ideas of Matriarchy, *JWI* 30 (1967), 1–35.
—— 1970. Locres et Tarente: Le rôle des femmes dans la fondation de deux colonies grecques, *Annales: Économies, Sociétés, Civilisations*, 5 (1970), 1240–70.
PETSAS, F. 1972. Symperipoloi Artemiti, *AAA* 5 (1972), 252–4.
PFUHL, E. 1923. *Malerei und Zeichnung der Griechen* (Munich, 1923).
PHILIPPAKI, B. 1967. *The Attic Stamnos* (Oxford, 1967).
PICARD, C. 1922. *Ephèse et Claros* (Paris, 1922).
—— 1940. L'Ephésia, les Amazones et les abeilles, *REA* 42 (1940), 270–84.
PINI, I. 1981. Echt oder falsch?—Einige Fälle, in W.-D. Niemeier, ed., *Studien minoischen und helladischen Glyptik: CMS Beiheft 1.* (Berlin, 1981), 135–57.
POLIGNAC, F. DE. 1984. *La Naissance de la cité grecque* (Paris, 1984).
PÖTSCHER, W. 1960. Moira, Themis und Timè in homerischen Denken, *WS* 73 (1960), 5–39.
POUILLOUX, J. 1960. *Fouilles de Delphes II: Topographie et architecture: La Région nord du sanctuaire* (Paris, 1960).
PRELLER-ROBERT. 1921. L. Preller, *Griechische Mythologie*, ii, 4th edn. by C. Robert (Berlin, 1921).
PRICE, S. 1985. Delphi and Divination, in P. E. Easterling and J. V. Muir, eds., *Greek Religion and Society* (Cambridge, 1985), 128–54.

PRÜCKNER, H. 1968. *Die lokrischen Tonreliefs* (Mainz, 1968).
PUGLIESE-CARRATELLI, G. 1964. Atti e dottrine religiose in Magna Grecia, *Convegno Magna Grecia*, 4 (1964), 19–46.
QUAGLIATI, Q. 1908. Rilievi votivi arcaici in terracotta di Lokroi Epizephyrioi, *Ausonia*, 3 (1908), 136–234.
RADERMACHER, L., ed. 1931. *Der homerische Hermeshymnus*, Sitzungsberichte der Akademie der Wissenschaft in Wien, 213/1. (Vienna, 1931).
RAECK, W. 1981. *Zum Barbarenbild in der Kunst Athens im 6. und 5. Jahrhundert v.Chr.* (Bonn, 1981).
RANSOME, H. M. 1937. *The Sacred Bee in Ancient Times and Folklore* (London, 1937).
REDFIELD, J. 1982. Notes on the Greek Wedding, *Arethusa*, 15 (1982), 181–201.
REISNER, G. A., FISHER, C. S., and LYON, D. G. 1924. *Harvard Excavations at Samaria* (Cambridge, Mass., 1924).
RENFREW, C. 1985. *The Archaeology of Cult: The Sanctuary at Phylakopi* (London, 1985).
RICHARDS-MANTZOULINOU, E. 1979. Melissa Potnia, *AAA* 12 (1979), 72–92.
RICHARDSON, N. J., ed. 1974. *The Homeric Hymn to Demeter* (Oxford, 1974).
RICHTER, G. M. A. 1950. *The Sculpture and Sculptors of the Greeks*[3] (New Haven, Conn., and London, 1950).
—— 1958. Erotes, *Arch. Cl.* 10 (1958), 255–7.
—— and MILNE, M. J. 1935. *Shapes and Names of Athenian Vases* (New York, 1935).
RIDGWAY, B. S. 1974. A Story of Five Amazons, *AJA* 78 (1974), 1–17.
ROBERTS, C. R. 1978. *The Attic Pyxis* (Chicago, 1978).
ROBERTSON, D. S. 1941. The Delphian Succession in the Opening of the Eumenides, *CR* 1941, 69–70.
ROBERTSON, M. 1975. *A History of Greek Art* (Cambridge, 1975).
—— 1983. The Berlin Painter at the Getty Museum and Some Others, *Occasional Papers on Antiquities, 1. Greek Vases in the J. Paul Getty Museum*, 1 (1983), 55–72.
ROLLEY, C. 1977. *Fouilles de Delphes V.3: Les Trépieds à cuve clouée* (Paris, 1977).
—— 1983. Les grands sanctuaires panhelléniques, in Hägg 1983: 109–14.
ROSE, H. J. 1941. Greek Rites of Stealing, *H Th. R* 34 (1941), 1–5.
ROUX, G. 1976. *Delphes: Son oracle et ses dieux* (Paris, 1976).
RÜHFEL, H. 1974. Göttin auf einem Grabrelief? *Ant.K.* 17 (1974), 42–9.
RUMPF, A. 1927. *Chalkidische Vasen* (Berlin and Leipzig, 1927).
—— 1964. Bienen als Baumeister, *Jahrbuch der Berliner Museen*, 6 (1964), 5–8.

SALMON, J. B. 1984. *Wealthy Corinth: A History of the City to 338 BC* (Oxford, 1984).
SAYCE, A. H. 1883. *The Ancient Empires of the East: Herodotos I-III*, with notes, introductions, and appendices (London, 1883).
SCHAEFER, H. 1939. *Die Laubhütte: Ein Beitrag zur Kultur- und Religionsgeschichte Griechenlands und Italiens* (Leipzig, 1939).
SCHAUENBURG, K. 1961. Göttergeliebte auf unteritalischen Vasen, *AuA* (1961), 77–101
SCHEFOLD, K. 1975. *Wort und Bild* (Mainz, 1975).
—— 1981. *Die Göttersage in der klassischen und hellenistischen Kunst* (Munich, 1981).
—— 1982. Oineus, Pandions Sohn, *RA* 1982, 233–6.
SCHELP, J. 1975. *Das Kanoun: Der griechische Opferkorb* (Würzburg, 1975).
SCHMIDT, M. 1980. Review of J.-M. Moret, *L'Ilioupersis dans la céramique italiote* (Rome, 1975) in *Gnomon* 52 (1980), 751–60.
—— ed. 1988. *Kanon: Festschrift Ernst Berger zum 60. Geburtstag am 26. Februar 1988 gewidmet* (Basle, 1988).
SCHMITT, P., and SCHNAPP, A. 1982. Image et société en Grèce ancienne: les représentations de la chasse et du banquet, *RA* 1982, 57–74.
SCHMITT-PANTEL, P., and THELAMON, F. 1983. Image et histoire: Illustration ou document, in F. Lissarague and F. Thelamon, eds., *Image et céramique grecque: Actes du Colloque de Rouen* (Rouen, 1983), 9–20.
SCHNAPP, A. 1979. Pratiche e immagini di caccia nella Grecia antica, *Dial. di arch.* 1 (1979), 36–59.
—— 1984. Eros en chasse, *Cité*, pp. 67–83.
SCHNEIDER, L., FEHR, B., and MEYER, K.-H. 1979. Zeichen-Kommunikation-Interaktion, *Hephaistos*, 1 (1979), 7–41.
SCHNEIDER-HERRMANN, G. 1970. Spuren eines Eroskultes in der italischen Vasenmalerei, *BABesch.* 45 (1970), 86–117.
—— 1971. Der Ball bei den Westgriechen, *BABesch.* 46 (1971), 123–33.
SHAPIRO, H. A., ed. 1981. *Art, Myth and Culture: Greek Vases from Southern Collections* (New Orleans, 1981).
SIEWERT, P. 1977. The Ephebic Oath in Fifth Century Athens, *JHS* 97 (1977), 102–11.
SIMON, E. 1953. *Opfernde Götter* (Berlin, 1953).
—— 1959. *Die Geburt der Aphrodite* (Berlin, 1959).
—— 1967. Boreas und Oreithyia auf dem silbernen Rhyton in Triest, *AuA* 13 (1967), 101–26.
—— 1977. Criteri per l'esegesi dei pinakes locresi, *Prospettiva*, 10 (1977), 15–21.
—— 1983. *Festivals of Attica: An Archaeological Commentary* (Madison, Wis. 1983).
—— 1985. Zeus und Io auf einer Kalpis des Eucharidesmalers, *AA* 1985, 265–80.

SMITH, P. 1980. History and the Individual in Hesiod's Myth of Five Races, *CW* 74 (1980), 145–63.
SNELL, B. 1938. Identifikationen von Pindarbruchstücken, *Hermes*, 73 (1938), 434–9.
—— ed. 1964. *Pindarus: Pars altera. Fragmenta*³ (Leipzig, 1964).
SNOWDEN, F. M., jun. 1983. *Before Color Prejudice: The Ancient View of Blacks* (Cambridge, Mass., and London, 1983).
SOURVINOU-INWOOD, C. 1971*a*. On the Authenticity of the Ashmolean Ring 1919.56, *Kadmos*, 10 (1971), 60–9.
—— 1971*b*. Theseus Lifting the Rock and a Cup Near the Pithos Painter, *JHS* 91 (1971), 94–109.
—— 1971*c*. Aristophanes, *Lysistrata*, 641–7, *CQ* 21 (1971), 339–42.
—— 1973. The Young Abductor of the Locrian Pinakes, *BICS* 20 (1973), 12–21.
—— 1974*a*. The Boston Relief and the Religion of Locri Epizephyrii, *JHS* 94 (1974), 126–37.
—— 1974*b*. The Votum of 477/6 B.C. and the Foundation Legend of Locri Epizephyrii, *CQ* 24 (1974), 186–98.
—— 1979. *Theseus as Son and Stepson: A Tentative Illustration of the Greek Mythological Mentality* (London, 1979).
—— 1983. A Trauma in Flux: Death in the 8th Century and After, in Hägg 1983: 33–48.
1988*a*. *Studies in Girls' Transitions: Aspects of the Arkteia and Age Representation in Attic Iconography* (Athens, 1988).
—— 1988*b*. Le mythe dans la tragédie, la tragédie à travers le mythe: Sophocle, Antigone vv. 944–87, in C. Calame, ed., *Métamorphoses du mythe en Grèce antique* (Geneva, 1988), 167–83.
1989*a*. Assumptions and the Creation of Meaning: Reading Sophocles' Antigone, *JHS* 109 (1989), 134–48.
—— 1989*b*. The Fourth Stasimon of Sophocles' Antigone, *BICS* 36 (1989), 141–65.
STÄHLER, K. 1980. Archäologisches Museum der Universität Münster, *Heroen und Götter der Griechen* (Münster, 1980).
STEIN, H., ed. 1883. *Herodotos*⁵ i (Berlin, 1883).
STELLA, L. A. 1956. *Mitologia greca* (Turin, 1956).
STENGEL, P. 1910. *Opferbräuche der Griechen* (Leipzig and Berlin, 1910).
STINTON, T. C. W. 1976. Iphigeneia and the Bears of Brauron, *CQ* 26 (1976), 11–13.
—— 1985. Heracles' Homecoming and Related Topics: The Second Stasimon of Sophocles' *Trachiniae*, *PLLA* 5 (1985), 403–32.
STUDNICZKA, F. 1911. Das Gegenstück der Ludovisischen 'Thronlehne', *JdI* 26 (1911), 50–192.
STURROCK, J., ed. 1979. *Structuralism and Since: From Levi-Strauss to Derrida* (Oxford, 1979).

THALMANN, W. G. 1984. *Conventions of Form and Thought in Early Greek Epic Poetry* (Baltimore and London, 1984).
THEMELIS, P. G. 1983. Δελφοὶ καὶ περιοχὴ τὸν 80 καὶ 70 π.Χ.αἰῶνα (Φωκίδα-Δυτικὴ Λοκρίδα), *Annuario*, NS 45 (1983), iii. 213–55.
THOMPSON, D'A. W. 1936. *A Glossary of Greek Birds* (Oxford, 1936).
TOHER, M. 1989. On the Use of Nicolaus' Historical Fragments, *Classical Antiquity*, 8 (1989), 159–72.
TÖLLE-KASTENBEIN, R. 1976. *Herodot und Samos* (Bochum, 1976).
TORELLI, M. 1977. I culti di Locri, in *Locri Epizefirii: Atti del sedicesimo convegno di studi sulla Magna Grecia*, Taranto 3–8 ottobre 1976 (Naples, 1977), 147–84.
TOULOUPA, E. 1973. The Sanctuaries of Mount Ptoion in Boeotia, in E. Melas, ed., *Temples and Sanctuaries of Ancient Greece* (London, 1973), 117–23.
TRUMPF, J. 1958. Stadtgründung und Drachenkampf, *Hermes*, 86 (1958), 129–57.
TYRRELL, W. B. 1984. *Amazons: A Study in Athenian Mythmaking* (Baltimore and London, 1984).
VALK, M. VAN DER, ed. *Eustathii archiepiscopi Thessalonicensis, Commentarii ad Homeri Iliadem pertinentes*, i (Leiden, 1971).
VALLET, G. 1958. *Rhégion et Zancle* (Paris, 1958).
VALLOIS, R. 1944. *L'Architecture hellénique et hellénistique à Délos jusqu'à l'éviction des Déliens*, i (Paris, 1944).
—— 1966. *L'Architecture hellénique et hellénistique à Délos jusqu'à l'éviction des Déliens*, ii (Paris, 1966).
VERMEULE, E. T. 1977. Herakles Brings a Tribute, in V. Höckmann and A. Krug, eds., *Festschrift für Frank Brommer* (Mainz, 1977), 295–301.
—— 1979. *Aspects of Death in Early Greek Art and Poetry* (Berkeley, Los Angeles, and London, 1979).
VERNANT, J.-P. 1965. *Mythe et Pensée chez les Grecs* (Paris, 1965).
—— 1974. *Mythe et société en Grèce ancienne* (Paris, 1974).
—— 1976. *Religion grecque, religions antiques* (Paris, 1976).
—— 1982. From Oedipus to Periander: Lameness, Tyranny, Incest in Legend and History, *Arethusa*, 15 (1982), 19–38.
—— 1985. *La Mort dans les yeux* (Paris, 1985).
—— 1987. Entre la honte et la gloire, *Mètis*, 2 (1987), 269–99.
—— and VIDAL-NAQUET, P. 1972. *Mythe et Tragédie en Grèce ancienne* (Paris, 1972).
VIAN, F. 1963. *Les Origines de Thèbes: Cadmos et les Spartes* (Paris, 1963).
VICKERS, M. 1981. Recent Acquisitions of Greek Antiquities by the Ashmolean Museum, *AA* 1981, 541–61.
VIDAL-NAQUET, P. 1983. *Le Chasseur noir*² (Paris, 1983).
—— 1986. The Black Hunter Revisited, *PCPS* 32 (1986), 126–44.
VOS, H. 1956. *Themis* (Assen, 1956).

WALCOT, P. 1978. Herodotus on Rape, *Arethusa*, 11 (1978), 137–47.
WALKER, S. 1983. Women and Housing in Classical Greece: The Archaeological Evidence, in Cameron and Kuhrt 1983: 81–91.
WALTER, O. 1923. *Beschreibung der Reliefs im kleinen Akropolismuseum in Athen* (Vienna, 1923).
—— 1937a. Die Reliefs aus dem Heiligtum der Echeliden in neu-Phaleron, *Arch. Eph.* 1937, 1/97–119.
—— 1937b. Die heilige Familie von Eleusis, *ÖJh.* 30 (1937), 50–70.
WEHGARTNER, I. 1983. *Attisch Weissgrundige Keramik* (Mainz, 1983).
WEST, M. L., ed. 1966. Hesiod, *Theogony* (Oxford, 1966).
—— 1978. Hesiod. *Works and Days* (Oxford, 1978).
—— 1985. Hesiod's Titans, *JHS* 105 (1985), 174–5.
WHITE, M. E. 1979. Die griechische Tyrannis, in Kinzl 1979: 184–210.
WILL, E. 1955. *Korinthiaka: Recherches sur l'histoire et la civilisation de Corinthe des origines aux guerres mediques* (Paris, 1955).
WILLEMSEN, F. 1955. Der delphische Dreifuss, *JdI* 70 (1955), 85–104.
WILLETTS, R. F. 1962. *Cretan Cults and Festivals* (London, 1962).
WILLIAMS, D. 1983. Women on Athenian Vases: Problems of Interpretation, in: Cameron and Kuhrt 1983: 92–106.
YALOURIS, N. 1986. Le mythe d'Io: les transformations d'Io dans l'iconographie et la littérature grecque, in L. Kahil, C. Augé, and P. Linant de Bellefonds, eds., *Iconographie classique et identités régionales*, *BCH* supplement xiv (Paris, 1986), 3–23.
ZANCANI MONTUORO, P. 1940. Tabella fittile locrese con scena di culto, *RIA* 7 (1950), 205–24.
—— 1954a. Note sui soggetti e sulla tecnica delle tabelle di Locri, *Atti Soc. MGrec.* 1954, 71–106.
—— 1954b. Il rapitore di Kore nel mito locrese, *RAAN* 29 (1954), 79–86.
—— 1955. *Archivio storico per la Calabria e la Lucania*, 24 (1955), 283–308.
—— 1959. Il tempio di Persefone a Locri, *Rend. Acc. Lincei*, 1959, 225–32.
—— 1960. Il corredo della sposa, *Arch. Class.* 12 (1960), 37–50.
—— 1964. Persefone e Afrodite sul mare, *Marsyas: Essays in Memory of K. Lehmann* (New York, 1964), 386–95.
—— and ZANOTTI BIANCO, U. 1951. *Heraion alla Foce del Sele*, i (Rome, 1951).
ZEITLIN, F. I. 1978. The Dynamics of Misogyny: Myth and Myth-Making in the Oresteia, *Arethusa*, 11 (1978), 149–84.

INDEX OF VASES

alabastra
    Athens NM 15002: 107, 131 n.
      69, 134n. 100
    Athens Kerameikos HS 107: 115,
      117, 136 nn. 125, 130
    in Athens, National Museum,
      Stathatos Collection 110
    in Athens, in the Vlastos Collection
      108–9, 115, 132 n. 84
    in Barcelona, by the Syriskos
      Painter 110
    Basle Kä 403: 116–17, 136 n. 124
    Berlin 2289: 115
    Berlin 4037: 131 n. 70
    Berlin 4285: 174
    in Brummer Sale 1979: 131 n. 63
    Brussels R 397: 109–10, 138
      n. 147
    Cambridge 145: 132 n. 76
    Cracow 1292: 113–14, 136
      nn. 121, 122
    Dunedin F 54.78: 107, 109
    Eleusis 2404: 134 n. 100
    Kassel 551: 111, 112, 134 nn.
      103–4, 135 n. 112, Pl. 13
    London B 668: 134 n. 100
    London D 17: 136 n. 131
    Geneva 20851: 134 n. 100
    New York 21.131: 114–15,
      116, 136 nn. 124, 130
    in a Swiss private collection, by
      the Syriskos Painter 109
    Virginia Museum of Fine Arts,
      Richmond, The Williams
      Fund (78.145) 136 n. 123
amphorae
    London 1948.10–15.2: 90 n. 33
    Louvre G 197: 97 n. 131
    Munich 2309: 92 n. 55
    Oxford 1965.118: 140 n. 172
    Ionic amphora: Munich, Staatliche
      Museen 585: 142

neck-amphorae
    Berne 12214: 54 n. 38
    Leningrad 709: 53 n. 29, 93
      n. 69, 95 n. 108, Pls. 9–10
    London 285: 53 n. 35
    London 1928.1–17.58: 54
      n. 45, 88 n. 16
    Madrid 11097: 94 n. 85
    at Mykonos, by the Oinokles
      Painter 95 n. 107
    Mykonos 1424: 52 n. 21
    New York 41.162.155: 44–5,
      52 n. 7
    Oxford 1914.733: 97 n. 127
    Vienna 741: 90 n. 33, 127 n. 32
    Würzburg H 4533: 125 n. 11
Nolan amphorae
    Bonn 77: 52 nn. 19–20, 93
      n. 69
    Leningrad 697: 90 n. 33
    London E 310: 97 n. 127
    London E 313: 55 n. 62
    Munich 2334 (J.257): 55 n. 52
    Syracuse 20537: 53 n. 22, 88
      n. 16
    Warsaw 142334: 53 n. 25, 55
      n. 52
    Yale 134: 52 nn. 19–20

bobbin (white-ground)
    Athens NM 2350: 127 n. 36
bottle
    Béziers 22: 174
cups
    formerly in the Basle market 54
      n. 41
    Berlin 2289: 115, 133 n. 99
    Berlin 2294: 53 n. 35
    Carlsruhe 59.72: 53 n. 25, 80,
      95 n. 111
    Ferrara 2501: 141 n. 172

cups (cont.):
>Ferrara 44886: 80, 134 n. 101
>Ferrara T. 264: 31, 52, n. 10, 55 n. 47
>Florence 3 B 3: 125 n. 19
>Frankfort, Museum VF, X 14628: 61, 81, 96 n. 120
>Harvard 1917.149: 38
>Leningrad Hermitage N 658: 31
>London E 76: 137 n. 134
>London, Victoria and Albert Museum 4807.1901: 96 n. 102, 123 n. 1, 126 n. 27, Pls. 11–12
>Louvre C 10932: 54 n. 39
>Louvre G 22: 53 n. 27
>Louvre G 115: 53 n. 35
>Louvre G 265: 93 n. 70
>Munich 2645: 137 n. 134
>Oxford 1913.311: 89 n. 24
>Oxford 1925.73: 133 n. 90
>Oxford 1927.71: 87 n. 3
>Oxford 1929.464: 133 n. 87
>in Philadelphia, American Philosophical Society 53 n. 25, 69, 71, 77, 81
>Toledo, Ohio 72.55: 135 n. 112

Chalcidian cup
>Würzburg 354: 128 n. 41

hydriai
>Athens NM 17469: 120–1
>Bowdoin 08.3: 88 n. 21
>Chicago University fragment 54 n. 39, 61
>Florence 4014: 55 n. 49, 94 n. 79
>Genoa 1155: 118–20
>Hague 634: 87 n. 4
>Leningrad 4309: 133 n. 89
>London E 170: 56 n. 63
>London E 198: 70, 72–3, 77, 96 n. 114
>London E 224: 128 n. 44, 130 n. 62
>Louvre G 427: 84
>Malibu S. 80. AE. 185: 55 n. 53
>Naples 2422: 100, 132 n. 79
>Oxford 294: 133 n. 99
>Oxford 531: 133 n. 99
>in the Peiraeus Archaeological Museum 55 n. 62
>Syracuse 36330: 82, 87 n. 10, 134 n. 100
>at Taranto, from Ceglie 84
>Vatican 16553: 128 n. 43
>Vatican 16554: 133 n. 99
>Vatican 17882: 126 n. 30
>Worcester (Mass.) 1903.38: 60

kantharoi
>Athens NM 1236: 38
>Boston 95.36: 126 n. 31

kraters
>Corinth C 33.129 and 138 (fragmentary): 89 n. 27
>in Depositi della Soprintendenza alla antichità della Basilicata (fragmentary, probably a volute-krater) 89 n. 28

bell-kraters
>frr. in Athens, in manner of Dinos Painter 82
>Harvard 1960.344: 124 n. 4
>Istanbul 2914: 96 n. 113
>Leningrad 777: 29, 33, 42–6, 53 n. 30, 55 nn. 56–7, 68–9, 88 n. 14, 95 n. 107, Pls. 5–6
>Louvre A 488: 55 n. 53
>Louvre G 423: 53 n. 28, 60, 95 n. 101, 96 n. 118
>Tübingen E 104: 129 n. 52

calyx-kraters
>at Aachen, Ludwig 96 n. 114
>Bologna Pell. 300: 120
>Boston 95.23: 126 n. 29
>Geneva MF 238: 53 n. 31, 70–1, 81, 95 n. 109
>London E 466: 54 n. 41
>in the Lucerne market, in manner of Niobid Painter 54 n. 45
>New York 06.1021.173: 96 nn. 117, 120
>New York 08.258.21: 53 n. 35
>New York 41.83: 41–2, 55 n. 52

Oxford 1924.929: 55 n. 53
column-kraters
  Chiusi, Museo archeologico
    nazionale, già Coll. Civica
    n. 1822: 97 n. 125
  Ferrara T. 375: 97 n. 125
  Fogg Museum Harvard
    1925.30.126: 129 n. 52
  in Leningrad from Kerch 97
    n. 127
  Louvre G 362: 88 n. 13
  New York 96.19.1: 55 n. 62
  Oxford 527: 52 n. 6
  Göteborg 171–62: 87 n. 1, 94
    n. 79
  Tübingen 67.5806: 126 n. 32
volute-kraters
  Bologna 269: 42, 70–3, 78–9,
    95 n. 99
  Bologna 275: 54 n. 46, 94 n. 82
  Bologna PU 283: 37
  Boston 33.56: 71–3, 75–9, 95
    n. 99, 104, 108, 126 n. 26
  Halle inv. 211: 127 n. 36
  Izmir Inv. 3361: 90 n. 35
  London E 468: 53 n. 35
  Louvre G 343: 53 n. 35
  Naples 2421: 70, 72–3, 77–8,
    96 n. 120
  Oxford 525: 29, 54 n. 45, Pl. 2
  Palermo Museo nazionale G
    1283: 53 n. 35
kyathos
  Brussels, Musée du cinq
    A2333: 66

Lebetes gamikoi
  Basle BS 410: 131 n. 67, 133
    n. 99, 139 n. 162
  in the Robinson collection 52
    n. 16, 62, 89 n. 22, 91 n. 38,
    92 n. 53
lekane
  Leningrad fragment 54 n. 39,
    60, 62, Pl. 8
lekanis lid
  Tübingen s./10 1665: 134
    n. 100

lekythoi
  Athens Agora P 24.067: 140
    n. 168
  Athens NM 1818: 38
  Athens NM 12890: 133 n. 90
  formerly in Basle market 132
    n. 76
  Cleveland 66.114: 98 n. 139
  in Gela 140 n. 172
  Laon 37.951: 53 n. 22
  Münster Arch. Mus. 24: 137
    n. 134
  New York 17.230.35: 55 n. 62
  New York 41.162.117: 132
    n. 76
  Oxford 1890.23: 132 n. 76
  Oxford 1917.58: 138 n. 147
  Oxford 1920.104: 55 n. 53, 89
    n. 24
  Oxford 1938.909: 55 n. 53, 89
    n. 24
  Palermo NI 1886: 100, 101
loutrophoroi
  once in Berlin (ex Sabouroff)
    93 n. 70, 95 nn. 103, 106
  Boston 10.223: 92 n. 67, 95
    nn. 103, 106
loutrophoros-hydria
  Copenhagen 9080: 80, 93
    n. 70, 95 nn. 106, 111, 133
    n. 86, 139 n. 159

oinochoai
  Athens 2186: 134 n. 100
  in the Basle market 140 n. 168
  Bologna 346: 52 n. 21
  Ferrara sequestro Venezia
    2505: 36, 52 n. 21, 54
    n. 39, 88 n. 11, 95 n. 105, 96
    n. 113
  Florence 21B 308frr. 97 n. 127
  Heidelberg Univ. 71/2: 129
    n. 52
  Louvre C 10729: 53 n. 35
  fragment Louvre 11029: 55 n. 52
  Marseilles Mus. no. 2093, inv.
    Roberty 3593: 55 n. 53
  Münster 586: 132 n. 76

oinochoai (*cont.*):
    Paris, Petit Palais 315: 140
      n. 168
    Vatican H 525: 127 n. 32

pelikai
    once Coghill near the Phile
      Painter 54 n. 39
    Leningrad 728: 29, 53
      nn. 24–5, 55 nn. 50, 61, 87
      n. 4, Pl. 7
    Manchester iii I.41: 29, 55
      n. 61, Pl. 3
    Naples ex Spinelli 2041: 141
      n. 173
    Naples RC 155: 55 n. 52
    formerly in New York
      market 124 n. 10
    Oxford 285: 132 n. 77
phiale
    Berlin F 2310: 126 n. 32
plate (black-figure)
    Bologna 149: 142
pointed amphoriskos
    Oxford 537: 133 n. 90
psykter-neck-amphora (Chalcidian)
    Rome, Villa Giulia, Castellani
      47: 128 n. 41
pyxides
    Ancona 3130: 139 n. 153
    Athens 1288: 128 n. 41
    Athens 1585: 133 n. 90
    Athens Acr. 569: 93 n. 70, 95
      n. 111
    Athens Ceramicus 1008: 131
      n. 63
    Athens NM 1630: 133 n. 91,
      139 n. 162
    Ferrara 20298 (T 27 C VP) (or
      12451) 141 n. 172
    London D 11: 79, 131 n. 75
    London E 1920.12–21.1: 77
    Louvre CA 1857: 71
    Louvre G 605: 133 n. 90
    Munich 2720: 90 n. 33
    Munich 7741: 174

    Munich, formerly Lugano, Schon
      126 n. 27
    Paris Bibliothèque Nationale,
      Cabinet des médailles 94: 174
    in Toledo Ohio 63.21: 133 n. 90

rhyton
    Ruvo, Jatta 1116: 141 n. 173

skyphoi
    Boston 13.186: 93 n. 70
    London E 140: 160
    New York x. 22.25: 52 n. 6
    Providence 25.072: 34–5, 83–4,
      87 n. 1, 90 n. 34
    Reggio 3877: 35, 71, 84
    Reggio 4134: 35, 84
    in Vatican, by Lewis Painter 83,
      84
    Vienna 3710: 137 n. 134
stamnoi
    Athens NM 18063: 96–7
    Brooklyn 09.3: 71, 102
    Brussels R 311: 97 n. 127
    Krefeld Inv. 1034/1515: 33–4,
      44, 53 nn. 23, 33, 54 n. 45,
      88 nn. 12–13, 94 n. 84
    London E 446: 29, 44–5, 54
      n. 37, Pl. 4
    Munich, Museum Antiker
      Kleinkunst 2415: 38
    Oxford 1911.619: 29, 37, 53,
      nn. 32, 34, 88 n. 15, 95
      n. 101, 96 n. 121, Pl. 1
    Paris G 370: 137 n. 134
    in private collection in
      Switzerland 137 n. 134
    Vienna, Kunsthistorisches Museum
      3729: 141 n. 173
    Villa Giulia 5241: 52 n. 16, 62,
      89 n. 22
    Warsaw 142353: 33, 44, 95
      n. 108
stemmed dish
    Copenhagen inv. 6: 111–12,
      135 nn. 109–12

GENERAL INDEX

abduction 40, 54 n. 41, 56 n. 66, 65–70, 73, 75–8, 80, 81, 84–5, 88 n. 21, 89–90, 92 nn. 62, 66, 93 n. 71, 94 nn. 85, 88, 93, 97 n. 124, 102–5, 108, 110, 113, 120, 124 n. 6, 126 nn. 25, 27, 127 nn. 32, 36, 128 nn. 41, 48, 130–1 nn. 61–3, 131 n. 70, 136 n. 120, 141 n. 173, 153, 161
Adonis 174, 186 n. 113
Aeschylus 226, 227, 232–3
Agamedes 192–4, 204–6, 210 n. 11, 215–16
Agrigento Painter 42, 54 n. 39
Aiakos 205
Aigeus 60, 276 n. 80
Ajax 31, 64
Akraia 249, 271 n. 40, 275 n. 68
alabastron 106–18, 120, 131 nn. 64–7, 133 nn. 90, 99, 134 n. 100, 137 nn. 144–5, 138 n. 147, 154, 155, 158, 170–1, 175, 177
Alcaeus, *Hymn to Apollo* 218, 231
Alcmaeon 273 n. 55
altar 71, 77, 93 n. 77, 95 n. 93, 100, 103, 105, 107–8, 111–13, 118, 124 n. 6, 126 n. 25, 129 n. 49, 130 n. 63, 132 n. 76, 135 n. 111, 138 n. 149, 140–1, 164, 166, 176, 178 *and see* Artemis
 combined with laurel-tree 99
 combined with olive-tree 99, 103, 128 n. 43
 combined with palm-tree 71–2, 75, 76–7, 95 n. 93, 99–143
Althaimenes 272 n. 51
Alyattes 247
Amazons 38, 81, 83–4, 97 n. 129, 107, 112–18, 131 nn. 63, 65, 134 n. 100, 135 nn. 114–16, 136 nn. 118–21, 126, 137 nn. 137–8, 143, 146, 138 n. 147, 142

Amphiktyones, the 194, 216 n. 58
Androgeos 247, 255, 275 n. 68, 276 n. 80
animals, animality 49, 56 n. 67, 66, 68, 75–6, 86, 91 nn. 43, 46, 48, 141–2, 188 n. 146, 237 n. 9, 239 n. 38
Anthesphoria 161
Antiope 64, 84, 97 n. 131, 116, 136 n. 120
Apollo 56 n. 63, 99, 100–1, 125 nn. 11, 19, 127 nn. 32, 33, 140–1 n. 172, 149, 155, 192–8, 202, 204–5, 207, 210 n. 11, 211–12 nn. 14, 17, 21, 22, 24, 213–14, 215 n. 57, 217–35, 237 n. 9, 239 nn. 35, 38, 45, 240 nn. 47–8, 241 nn. 51–2, 59, 242 nn. 73, 75, 246, 255, 259, 274 n. 64, 277 n. 84
Aphrodite 80, 100, 103, 105, 118–19, 127 n. 32, 130 n. 61, 139 n. 154, 140 n. 167, 150, 151–2, 154, 155, 158, 164, 165, 171–2, 175–80, 183 n. 40, 185 n. 84, 186 n. 113, 187 n. 143, 188 nn. 144, 146, 151
 cult objects and symbols 175–7
 cultic association with Eros 176
 cultic association with Hermes 177–8, 180
 at Locri 175–8
 and Locrian *votum* 178–80, 188 n. 147
 and Persephone at Locri 178–80
Ares 182 n. 31
Argos 52 n. 6, 141–2
Ariadne 56 n. 66, 155
Aristogeiton 30, 43, 44–6, 54 n. 35
arkteia 75–6, 101, 103, 120, 121, 125 n. 20, 141 n. 172, 249, 269 n. 17
Artemis 75–7, 78, 86, 100–14, 117–18, 125 nn. 11, 16, 127 nn. 32, 33, 129 n. 52, 130 nn. 57, 61, 131 n. 71, 135 nn. 107, 112, 138–9

307

Artemis (cont.):
　　nn. 149–50, 140–1 nn. 167–9, 172–3, 153, 182 n. 31, 196, 214, 247, 249
　　altar of 75, 77, 99, 101, 103, 106, 107–8, 112, 113–14, 118, 119, 121, 122, 124 nn. 9–10, 125 n. 19, 128 n. 41, 129 n. 52, 130–1, 135 n. 112, 136 n. 118, 249, 270 n. 34
Artemis Agrotera 124 n. 10, 129 n. 52
Artemis Brauronia 75, 94 nn. 90, 92–3, 101, 102–3, 105, 108–9, 120, 121, 128 n. 41, 129 nn. 49–50, 130–1, 132 n. 81, 133 n. 86, 139 n. 159, 204–5, 212 n. 30, 249, 270 n. 26
Artemis Mounichia 94 n. 92, 101–3, 105, 108, 121, 125 n. 17, 129 n. 52, 132 n. 81, 140 n. 169
Artemis Orthia 94 n. 88, 128 n. 48, 249
Artemis Tauropolos 231
Asklepios 257, 276 n. 80
Atalanta 85–7, 97 nn. 134–5, 98 n. 139, 120–1, 140 n. 167
a-ta-na po-ti-ni-ja 221, 238 n. 17
Athena 89 n. 27, 99, 100, 103, 105–6, 124 n. 1, 127 n. 32, 130 nn. 59–60, 137 n. 134, 155, 172, 174, 193, 201, 203, 205–6, 209, 210 n. 3, 221–2
Athena Pronaia, sanctuary of (Marmaria) 220–2, 237 n. 10
athletes 81
attack 30–1, 32, 35–9, 41, 42–6, 48–51, 53–4, 56 n. 62, 90 n. 33, 91 n. 45
axe 83–4, 114

ball 154, 155, 158, 159–60, 163, 165–7, 179–80, 183 nn. 53, 55
bear 97 n. 134, 121, 128 n. 41, 212 n. 30, 270 n. 26
beautification 106–7, 109–10, 120, 162, 165
bees 196–200, 205, 207–8, 213–15
　　divination by 197–200, 213 n. 41
bee-women 197–8, 200, 213 n. 41

birds 170–2, 196, 200, 205, 207–8, 212 nn. 22, 24, 242 n. 75
Boreas 56 n. 66, 83, 88 n. 21, 96 n. 119, 97 n. 126, 102, 128 n. 38
bow 114, 155, 226
bowl of fruit 154
branch 80, 96 n. 114, 111
bricolage 218, 224, 245, 268 n. 8
bride 69, 73, 79, 80, 93 n. 70, 95 n. 106, 98 n. 139, 107, 120, 139 nn. 157, 162, 156–61, 164, 166, 168–70, 175, 178, 183 n. 40, 184 n. 69
bronze
　　as building material 192, 203–4, 209
　　temple 192–3, 194, 201, 202–4, 206, 208–10
bull 140 n. 168, 276 n. 80

castration 246–7, 248, 250, 264–5
chariot 97 n. 124, 153, 175
　　winged 152, 155
chasm, prophetic 225–6, 239 nn. 38, 43
cheir' epi karpo 68
chest 82, 154, 155, 157–8, 162, 167, 169–70, 171–3
child in basket 169–74
Chiron 96 n. 114
citharode 125 n. 19
Circe 30, 31, 41
Claros 224, 239 n. 35
cock 153–5, 157–9, 161, 163, 164–7, 171, 179–80, 183 n. 51
cognitive studies 5–6
colonnade 71
column 70–1, 72, 140 n. 172
　　combined with altar 70, 72
　　combined with altar and chair 72
　　combined with chair 72
　　combined with door 70–3, 78–9
　　combined with door and altar 70, 72–3
Corcyra 246–50, 255, 258, 262–3, 264, 270 nn. 20, 29, 34, 274 n. 61, 278–9 nn. 98, 100, 105, 280 nn. 107–8, 281 n. 111
Corinth 246, 249, 262–3, 266, 274 n. 61, cf. 244–84
crane, *see* heron

Crete 211 n. 17, 227, 235, 237 n. 11, 248, 269 n. 17
crow 196
crown 166–7
    laurel 194–5

Daphne 56 nn. 63, 68
daphnephorein 195
daphnephoreion 211 n. 21, 214–15
deities, divine personalities 5, 99, 147–88
    in context of development of religion 148–9
    definition of 147–51, 181 nn. 2–3
    local 148–50
    at Locri 150–88
    methodology for study 147–51, 152
    Panhellenic 148–50
    presentation of children to 186–7 nn. 113–14, 117–18
Delos 101, 125 n. 19, 127 n. 33, 141 n. 172, 213–14, 248
Delphi 100, 217–24, 227–31, 233–6, 241 n. 59
    myth of first temples at 192–216
Delphic oracle 17, 192–243
    previous owners of 217–43
Demeter 158, 173–5, 180, 184 n. 80, 185 n. 86, 187 n. 124, 196–7, 212 n. 30
diadem 53 n. 23, 79, 80, 95 n. 106, 118, 119, 125 n. 10, 132 n. 76
Didyma 211 n. 14, 224, 226, 239 n. 35
Dinos Painter 82
Dionysos 100, 125 n. 11, 137 n. 134, 155, 158, 160, 165, 171–2, 186 nn. 113–14, 218, 236, 241 n. 62, 242 n. 73, 243 nn. 77, 81
Dionysos Zagreus 243 n. 77
Dioskouroi 155
Dodona 200, 212 n. 30
dolphin 62, 88 n. 21, 131 n. 63
door 70, 72–3, 77–8, 90 n. 33, 170
'Dorian Invasion' 149
dove 154, 160, 172, 175, 176–7, 183 n. 57, 187 n. 142, 196, 200, 212 n. 22

eagle 226–7, 235–6
Eetion 275 n. 65, 282 n. 122
Eleusis 159
endromides 60, 91 n. 46
Eos 37, 54 n. 41, 56 n. 66, 94 n. 85, 96 n. 120
Epaulia 92 n. 53, 95 n. 111
ephebe, ephebic concerns 40–1, 50, 55 n. 53, 63–4, 65–7, 76, 79, 80, 82–6, 89 nn. 25, 28, 91 nn. 41, 42, 46, 97 n. 135, 98 n. 136, 129 n. 52, 139 n. 150, 255–8, 266, 269 n. 16, 270–1 nn. 32, 34, 274 n. 63, 275 n. 68, 276 n. 80, 283 n. 123, 284
Ephesos 136 n. 118, 196, 212 n. 30, 214
Ephoros 233, 280–1
Epidaurus 278–9
epidikasia 93 n. 67
epikleros 93 n. 67
epistemology 5, 8–9
Eretria 211–12, 214–15
Erginos 193
Eriboia (Periboia, Phereboia) 64, 90 n. 29
Erichthonios 174, 186 n. 113
Eros 59, 82, 92–3, 98 n. 139, 118–19, 132 n. 77, 133 n. 89, 139 nn. 154, 158, 141 n. 173, 172, 175–6, 180, 187 nn. 137, 139
erotic pursuit 11, 14–15, 18, 29–51, 55 nn. 51, 55, 57, 58–98, 102–5, 110, 113, 120–1, 126 nn. 25, 29, 32, 127 n. 36, 128 n. 41, 129 n. 52, 130–1, 133 n. 99, 136 n. 120, 137 n. 138, 139 n. 159, 141–2
    and capture 65–6
    combined with other subjects 80–2
    and consensual erotic intimacy 68–70
    and father figure 73–4, 80, 82
    by a god 22 nn. 22–3, 47–50, 55–6 nn. 62, 66, 67, 96 n. 123, 184 n. 80
    by a hero 22 n. 23, 47–8, 94 n. 85
    locations: courtyard 71–9, 95 n. 111; house 71–3; outdoor 70, 72–3, 77, 78; sanctuary 71, 73, 75–8; street 72

erotic pursuit (*cont.*):
  spatial indicators 70–9, 93 nn. 74, 77, 95 n. 111, 130 n. 54
  and spears 30, 32–43, 55 n. 55, 97 n. 125
  by Theseus 48, 58–98, 102
  and wedding/marriage 64, 66–70, 73–6, 77–80, 82, 85–7
eroticism, consensual 11, 43–4, 68–70, 86, 116–17
Essenes 196
Eumolpos 160
Euripides, *Iphigenia in Tauris* 227, 230–2

father 52 n. 21, 68, 73–4, 76–8, 80, 82, 94 nn. 79–80, 84–5, 89 n. 22, 96 n. 114, 250–3, 256–8, 263–4, 272–3, 273–4, 275 nn. 67, 71–2, 276 n. 80, 277 nn. 87–8, 278 n. 95, 281 n. 115
father–son hostility 245, 246–65, 272–3 nn. 42, 51, 58, 268 n. 9, 274 n. 61, 275 nn. 71–2, 277 n. 87, 278–9, 281 nn. 110–11, 283 n. 61
fawn/female deer 121, 122, 140–2 nn. 168–9, 172–3
fig-tree 240 n. 47
flower 65, 83, 98 n. 139, 108–10, 120, 152, 154, 158, 161–3, 166–7, 170, 179, 184 nn. 69, 75, 187 n. 128
  gathering, picking 65, 90 n. 36, 152, 161–3, 166–7, 184 n. 69
  'rose-like' 154, 175
flute 108–9, 133 n. 86, 176, 178
fruit 160–2, 166–7, 170, 176, 178, 184 n. 76
funerary realm 153, 159, 163, 166, 175, 176, 236

Gaia 217–34, 237 n. 9, 238 nn. 12–13, 16, 20, 28–9, 239 nn. 35, 38, 240–1, 242 nn. 65, 67, 73
Geneva Painter 42, 70, 72
Glaukos 276 n. 80
gnorismata 60, 88 n. 17
goat 237 n. 9, 240 n. 47
gods:
  and heroes, relationship of 15, 30, 47–8
  and men, relationship between 48–9
goose 157–8, 160, 165
gyne 79–80, 86, 103, 104, 111–12, 116, 122, 124 n. 2, 128 n. 41, 188 n. 152

Hades 67, 150, 152–3, 154, 156–61, 163, 171, 178, 182 n. 20, 183 n. 40, 184 n. 61, 239 n. 43, 241 n. 62
hare 142
Harpies 201
Hasselmann Painter 42, 46
Helen 30, 31, 42, 64, 89–90, 93 n. 70, 126–7, 128 n. 48, 155, 273 n. 51
Hephaistos 193, 203, 205–6, 209, 210 n. 3
Hera 137 n. 134, 153, 175, 183 n. 60, 187 n. 125, 249
Heracles 56 n. 70
Hermes 102, 105, 141–2, 154–5, 158, 160, 171, 175–8, 180, 187 n. 143, 198, 241 n. 56
Hermonax 42
Herodotos 244–84
heron/crane 80, 82, 110, 116–17, 133 n. 97, 134 n. 100
Hesiod 209
  Theogony 228, 231, 232, 233
hetaira 91 n. 39, 96 n. 122, 137 n. 143, 183 n. 53
hind (deer) 91 n. 43, 176, 178
Hippolytos 252, 254, 256–7, 258, 259, 272–3
Hippomenes 86, 97 n. 135, 120
Hipponion 161
Homeric Hymn to Apollo 200, 204–7, 209, 210 n.11, 218–19
  dragon-killing in 228–30
Homeric Hymn to Hermes 197–9, 213 n. 41
homosexuality 91 n. 46, 137 n. 135, 250
hoplite 79, 82, 125 n. 10, 129 n. 52
hydria 167, 170–3

General Index

Hyperboreans 192–4, 196, 209, 237 n. 9

Iakchos 186 n. 113
Iamos 56 n. 70
Ikaria stele 173
Ilioupersis 42
initiation 16, 65–6, 75, 79, 129 n. 52, 247–50, 255–60, 262, 264–5, 269 nn. 17–18, 270–1, 274 nn. 64–5, 275 n. 71, 276 nn. 80–1, 277 n. 84, 284
Ino-Leukothea or Albani relief 173
Io 49, 141–2
Iphigeneia 101, 125 n. 24, 128 n. 41, 205, 232, 242 n. 67

kalathos 110, 115–17, 133 nn. 87, 90, 99, 134 n. 100, 153, 157–8, 161–2, 164, 168, 170–1, 184 n. 80
Kalaureia 233
Kallisto 49, 56 n. 67
kanephoros 111, 130 nn. 59–60, 134 n. 105, 135 n. 112
kanoun 111, 164, 167
kantharos 155, 157–8, 160, 163, 164, 168, 170–3
Kassandra 31, 100–1
Kastalia 223
Kato Syme Viannou, sanctuary of Hermes and Aphrodite at 187 n. 143
Katreus 272 n. 51, 274 n. 61, 275 n. 71
Keledones 193, 194, 201–3, 215 n. 47
Kephalos 37, 94 n. 85
kerykeion 175
killer, polluted 259–60, 262, 265, 277 nn. 82–7, 89
kirkos 196, 242 n. 75
Knidians 269 n. 12, 270 n. 20
Kore 212 n. 30
Koronis 56 n. 66
Korythos 273 n. 51
Kronos 235, 242 n. 64
krypteia 255, 274 n. 62
Kypselids 267, 275 n. 65, 277 n. 82, 282 nn. 116, 121, 283 n. 123

Kypselos I 266, 282–3 nn. 118, 121–3
Kypselos II, *see* Psammetichos

Labda 275 n. 65, 282 n. 122
Labyrinth, the 248, 270 n. 23
Laconian reliefs 159, 160, 184 nn. 67, 75
Laios 276–7
lainos oudos 204, 215 n. 56
Lamia 201
laurel 125 n. 11, 194, 200, 205–8, 210 n. 12, 211 n. 14, 225, 226, 239 n. 35, 240 n. 47
from Tempe valley 194–5, 211 nn. 15, 17, 225
temple 192–3, 194–6, 201, 205–8, 210 n. 3, 211–12
Lebadeia 160
Leophron 179
Leto 100–1, 125 n. 11, 141 n. 172, 213 n. 44
Leukippids 102, 103, 105, 127 n. 36, 129 n. 49, 130 nn. 61–2, 212 n. 30
Leukothoe 56 n. 66
Lewis Painter 34–5, 42, 71, 83–4
literary theory 5, 9
Locri Epizephyrii 150–88, 248–9, 250
Locrian maidens 248–50, 270 nn. 24, 26, 29
Locrian pinakes 151–78, 181 n. 8, 188 n. 152
louterion 140 n. 168
Louvre Centauromachy, Painter of 42
Lydia 248, 264–5, 281 n. 114
Lykophron 244, 246–7, 250–67, 268 nn. 5, 7, 269 n. 12, 272 nn. 44, 49, 50, 273 n. 55, 274 n. 65, 276 n. 77, 277 nn. 82, 87, 278 n. 95, 278–9 nn. 98, 100, 280–1 nn. 107–8, 110, 282 n. 117, 283–4
lyre 81, 82, 155, 175

Maenad 81, 87 n. 3, 107, 113, 131 n. 74, 134 n. 100, 135 n. 115, 136 n. 129
Makron 160

Marathon, battle of 124 n. 10
marriage/wedding 64, 66–70, 73–6, 77–80, 82, 85–7, 89–90, 91 n. 47, 92–3, 94 nn. 78, 80, 95 nn. 102–3, 111, 96 n. 114, 98 nn. 137, 139, 99, 102–5, 107–12, 120–2, 125 n. 24, 126 n. 26, 127 n. 33, 130 n. 60, 131 nn. 70–1, 132 nn. 76, 77, 133 nn. 86, 93, 134 n. 105, 135 n. 112, 136 nn. 120–1, 137 n. 137, 139 nn. 154, 157–9, 162, 141–2, 153, 156–8, 159, 161–3, 164–7, 168–9, 177–8, 184 n. 69, 188 n. 151
'maschera' 161, 184 n. 71
matricide 251, 254, 258, 260, 273 n. 52, 276 n. 81, 277 n. 89, 278 nn. 92, 95, 273 n. 55
Medea 51, 55 n. 57, 56–7, 276 n. 80
Melanion 86, 97 n. 135
Melissa (wife of Periander) 246, 272 n. 50, 274–5, 278–9
Melissa (priestess) 196
Melissas Delphidos 197
Melissonomoi 196
Menelaos 30, 31, 42, 126–7
Minoan oval stone 226–7, 235–6, 239–40
Minoan young god 226–7, 235, 242–3
Minos 235–6, 247, 250, 265, 276 n. 80
Minotaur 52 n. 6, 247, 258, 269 nn. 16–17, 276 n. 80
mirror 109, 114, 133 nn. 89, 90, 154, 164, 165, 168, 170–1, 176–7
Mnemosyne 201
mother 51, 57 n. 75, 74, 94 n. 84, 107–8, 246, 250–1, 253, 254, 258, 273 nn. 55, 58, 278 n. 95, 279 n. 98, 282 n. 122
mourning 277 n. 85
Muses, 96 n. 122, 201, 242 n. 73
Mycenaean figurines 220, 238 n. 12, 240 n. 51
Mycenaean religion 148–9, 219–23, 237–8, 240 n. 51

myth:
of first temples at Delphi 192–216
in Herodotos 244–84
methodology for study of 16–20, 219, 244–5
of previous owners of Delphic oracle 217–43

Nausika 127 n. 33
Negro 112, 114, 117–18, 131 n. 65, 134 n. 100, 135 n. 116, 137 nn. 144, 146, 138 nn. 147, 149–50
Nereids 62, 88 n. 21, 95 nn. 93, 102, 126 nn. 27, 29, 129 n. 49
Nereus 94 n. 79, 126 n. 29
Nicolaos 279 n. 100
Nicolaos of Damascus 263–4, 276 n. 77, 279 nn. 100, 102, 104–5, 280–1 nn. 108, 110–11
Niobid Painter 42, 70–1, 72, 90 n. 31, 96 n. 114, 100, 127 n. 36
Nyx 242 n. 73

oak tree 200
Odysseus 30, 31, 41–2, 127 n. 33
Oedipus 273 n. 51, 274 nn. 58, 61, 275 nn. 67, 70–1, 276–7
Oinone 273 n. 51
oionoskopeia 196
olive-tree 99, 103, 105, 128 n. 43, 142
omphalos 141 n. 173, 225–7, 235–6, 239 n. 46, 243 nn. 77–8
Oreithyia 56 n. 66, 88 n. 21, 96 n. 119, 97 n. 126, 102–3, 105–6, 128 n. 38
Orestes 84, 231–3, 273 n. 55, 276 n. 81
Oschophoria 247, 269 n. 16
Ouranos 242 n. 64

Palladion 31, 100
palm-tree 71, 93 n. 75, 94–5, 99, 100–1, 102–3, 105–14, 117–18, 120–2, 123–4, 128 n. 41, 129 nn. 49, 52, 130–1 nn. 61, 63, 71, 134 nn. 100, 105, 138 n. 147, 139 n. 150, 140–2 nn. 168–9, 172–3

General Index

pantheon 147–50, 157, 225
Paris 93 n. 70, 273 n. 51
Parnassus 198
parthenos, parthenoi 73, 74–7, 81, 85, 89 n. 28, 95 n. 94, 97 n. 134, 98 n. 137, 101, 102–14, 116–17, 118–22, 124 n. 2, 125 n. 10, 127 n. 33, 128 n. 41, 129 n. 52, 131 n. 71, 133 n. 94, 134 n. 100, 135 n. 112, 136 nn. 119, 121, 137 n. 143, 139 nn. 154, 164, 140 n. 167, 141–2 nn. 172–3, 188 n. 152
patricide 16, 20, 251–4, 256–60, 272 n. 42, 273 n. 52, 274 n. 61, 273 n. 55, 275 nn. 70–1, 276 n. 80, 277 n. 89, 278 nn. 92, 95, 282 n. 119
Peirithous 62, 84
Peitho 127 n. 32
Peleiai/Peleiades 200, 212 n. 30
Peleus 32, 58–9, 62, 66–7, 88 nn. 18, 21, 89 n. 22, 92 nn. 48, 50, 95 n. 102, 96 n. 120, 102, 105
Penthesilea Painter 42, 53 n. 25, 71
peplophoria 167–9
peplos 132 n. 76, 165–70, 182 n. 31, 185 n. 98
Periander 244, 246–7, 250–63, 265–7, 268 n. 5, 269 n. 12, 272 nn. 44, 49, 50, 274 n. 65, 275 n. 67, 277 nn. 82, 87, 278–9 nn. 98, 100, 105, 280 n. 108, 281 nn. 110–11, 113, 282 nn. 117, 120–1, 283 nn. 123–4
Periboia, *see* Eriboia
perioikoi 279–80, 281 n. 110
Persephone 62, 67, 90 n. 36, 150–80, 182–3 nn. 26, 31, 40, 46, 184 nn. 69, 83, 186 nn. 113–14, 187 n. 142
  and Aphrodite at Locri 178–80
  as bride 156–61, 164, 166, 168, 178, cf. 90 n. 36
  cult objects and symbols 153–5, 157–73
  kourotrophic function 153, 169–75, 180
  at Locri 150–75, 188 n. 152
  Manella sanctuary of 151, 159, 160, 162, 165, 168
  and offering girls 163–7, 168–9, 179, 185 n. 84
  Panhellenic personality 153, 163, 175, 180
  as protectress of marriage 153, 156, 159, 165–6, 169, 174–5, 178, 180
  rape by Hades 67, 153, 158, 161
Persephone Painter 41
Persian Wars 51, 56 n. 74, 88 n. 21, 117, 137 nn. 144, 146
petasos 32–3, 40, 42, 45, 55 n. 57, 59, 60–1, 63, 91 n. 46
Phaedra 254
Phereboia, *see* Eriboia
Pherekydes 64, 90 n. 32
phiale 111, 154, 157–8, 160–1, 163, 164–6, 167, 176, 179, 185 n. 52
Phiale Painter 52 nn. 19, 20, 53 n. 22, 54 n. 39
phialophoros 167–8, 185 n. 98
Philochoros 213 n. 41
Phineids 272–3
Phoebe 227, 232–3
Phoenix 252–3, 256, 257, 259–60
pilos 61
Pindar 197, 207, 209–10, 228
  eighth *Paean* 192–3, 196, 201–3
Pitys 56 n. 68
Pleiades 139 n. 164
poloi 212 n. 30
pomegranate 157–8, 160, 162, 163, 164, 167, 183–4 nn. 57, 60, 76, 78
pompe Lydon 270–1
poppy-flowers, -heads 164–5, 185 n. 86
Poseidon 153, 212 n. 30, 222–3, 225, 233–4
Po-si-da-e-ja 222–3
post-structuralism 3, 10–11
Priam 100
Prokles 246, 268–9, 275 n. 67, 277 n. 87, 278–9, 284
Prometheus 241 n. 62

General Index

Psammetichos (Kypselos) 258, 263–6, 276 n. 76, 279 nn. 104–5, 282 nn. 116, 120–1
psychology of perception 5, 9
Pteras 192
Ptoion 224, 239 n. 35
Pyanopsia 247, 269 n. 16
Pylades 84
Pylos tablets 222–3
Pythia 194, 197, 199–200, 202–3, 211 n. 14, 225–6, 234, 240–1
Pythian Games 195
Pythias 195
Python 236, 242 n. 73, 243 n. 81

quince 184 n. 76

ram 154–5
rape 31, 40, 44, 56 n. 70, 84, 92 n. 54, 94 n. 90, 153, 158, 161
   see also abduction, erotic pursuit, Persephone, Kassandra
raven 196
reading images/texts 3–23
   and archaeological evidence 7–8
   assumptions and expectations: images 11–15; texts 15–16, 18
   and 'common-sense' empiricism 3–4, 9
   dangers of culturally determined preconceptions 4, 6–7, 9–10, 12–14, 19
   external sources of interpretative aid 5–6
   methodology: general 4–20; for study of divine personalities 147–51, 152; for study of myths 16–22, 219
   perceptual filters 8–11, 16, 19, 264
   post-structuralist approach 3, 10–11
   process of signification 10–11, 15, 18, 21 n. 10
   schemata in texts 16, 19–20, 245, 251–67, 268 n. 7
   separate analyses and cross-checking 7, 12, 14, 19, 22 n. 20
rhyta 237 n. 11
ritual stealing 249–50, 270 nn. 32, 34

Sabouroff Painter 42, 53 n. 25
Samos 246–9, 255, 269 n. 12
sanctuary 73, 75–8, 94 n. 88, 98 n. 138, 100, 101–6, 113, 118, 121–2, 127 n. 32, 130 n. 61, 136 n. 118, 140 n. 168, 141 n. 172, 148, 151, 159, 160, 162, 165, 168, 184 n. 83, 185–6, 200, 204, 205, 214, 217, 223, 236, 247, 249
Saon 197, 199
Sappho 96 n. 122
satyr 81, 87 n. 3, 128 n. 41, 176, 178
scabbard 30, 35, 39, 45, 60
sceptre 70, 74, 79, 82, 94 n. 84
   incorporating figurine of sphinx 157, 162, 163
Schuvalov Painter 52 n. 21, 54 n. 39
Semele 49
semiology 5, 46, 47
shield 155
Sicily 161
Sirens 166, 171, 201–2
Skiron 61, 81
snake 131 n. 63
Sparta 150, 159, 249, 255
spear 11, 15, 29–30, 50, 53–4
   and erotic pursuit 30, 32–43, 45–6, 50, 55 n. 55, 58–61, 63, 83–5, 87 n. 1, 89 n. 25, 97 n. 125
Sphinx 162, 201
stalk of grain 154–5, 157–9, 163, 164–5, 167, 183 n. 46
stepmother 57 n. 75, 254, 272–3 nn. 51, 56, 278 n. 92
Stesichoros 155
stool 108–10, 112, 114, 117, 133 nn. 87, 90, 134 n. 100, 138 n. 147
   folding 115–16
swan 196
sword 15, 29–32, 33, 35, 36, 38, 39, 41–7, 51, 55 n. 57, 60, 61, 84–5
Syrinx 56 n. 68
Syriskos Painter 107, 112, 117, 138 n. 147

table 165–6, 169
taenia 80, 92 n. 67, 95 n. 111, 118, 120, 170–1, 173, 176, 186 n. 109

General Index

Tainaron 233
Tartaros 227
Tauric cult 231
Telegonos 275 nn. 70–1
Tempe valley 192–5, 211 nn. 15, 17
temple, temples 71, 72, 77, 176
  of Aphrodite and Hermes 176, 178
  at Delphi: bronze 192–3, 194, 201, 202–3, 208–10; laurel 192–3, 194–6, 201, 205–8, 210 n. 3, 211–12; stone 192–3, 204–8; of wax and feathers 192, 193, 194, 196–200, 205, 206, 208–10, 213–16
Tenes 252, 256, 257, 260, 272–3
Thasos 213 n. 44
Themis 217–18, 223, 226, 227, 230–4, 238 n. 13, 240–1, 242 nn. 65, 73
theogamia 168, 169
Theseus 38, 48, 52 n. 6, 63–5, 67, 77, 78, 80, 81, 83, 84–5, 88 nn. 19, 20, 89–90, 92 n. 55, 93 n. 70, 94 n. 88, 96 n. 124, 97 n. 131, 104, 105, 116, 127 n. 32, 128 n. 48, 247–8, 252, 254, 255–6, 258, 269 nn. 16–17, 272 n. 49, 276 n. 80, 283 n. 123
  with drawn sword 41–2
  as ephebe 40–1, 50, 58, 63–4, 67
  erotic pursuit by 48, 51, 58–98, 102
  with sword 14–15, 29–32, 33, 36, 38, 41, 43–7, 51, 55 n. 57, 56–7, 91 n. 45
Thetis 32, 58–9, 60, 62, 64, 65, 66–7, 88 n. 21, 89 n. 22, 91–2, 95 n. 102, 96 n. 114, 100, 102, 105, 121–2, 124 n. 6, 126 nn. 27, 29, 130–1 nn. 63, 70
throne incorporating goose's head 154, 158, 160, 165, 167–8, 169–71

thymiaterion 175
  surmounted by cock 154, 158, 164, 167
Titans 241 n. 62
torches 96 n. 114, 108, 132 nn. 76, 77, 81
transvestism 250
tree, miraculous 140 n. 167
tripod, Delphic 194, 226, 227, 234, 240–1, 242 n. 73
Triptolemos 155
Trojans 117, 137 n. 144
Trophonios 192–4, 204–6, 210 n. 11
  oracle of 197, 199, 212 n. 31, 215–16
twig, twigs 120, 154, 158–9, 163
Typhoeus 229
Tyrannicides statues 30, 52 n. 8
tyranny 263, 265, 266–7, 280–1 nn. 107–8, 110–11, 282–3 nn. 116, 121–3

vine 155

wand 165–6, 167, 185 n. 92
whip 98 n. 139
wolf 239 n. 38, 255, 258, 274 n. 64
wreath 60, 80, 83, 95 nn. 101–3, 96 n. 114, 98 n. 139
wryneck 202, 215 n. 47

Xenokrateia relief 174

Yale oinochoe, Painter of 33–4, 44–5, 52 n. 18

Zeus 102, 126 n. 29, 141–2, 153, 157, 183 n. 44, 229–33, 235–6, 241 nn. 55, 62, 242 nn. 64, 67, 76
Zeus Hellanios 205
Zeus Herkeios 72, 75, 100, 132 n. 79